Adjudication Practice and Procedure in Ireland

This adjudication textbook uniquely brings together a comprehensive analysis of, and commentary on, the Construction Contracts Act 2013 with a real-world perspective of adjudication, considering the knowledge, process and skills parties and adjudicators require in order to successfully participate in the adjudication process. Drawing on combined experience of 40 years in construction law, the authors provide invaluable guidance for all stakeholders in the adjudication process. The authors analyse and comment on the adjudication provisions of the Construction Contracts Act and describe prudent practice and procedure required to comply with Irish adjudication law, including case studies, case law and sample documentation for those to be involved as the parties, or those who want to act as adjudicators.

Aimed at contractors, sub-contractors, developers, employers, construction, engineering and legal professionals and students, all of whom are either involved, or have an interest, in dispute resolution and adjudication.

Damien Keogh is ranked as one of Ireland's leading construction lawyers. He is a Fellow of the Chartered Institute of Arbitrators, a lecturer in Arbitration Law at Trinity College Dublin and the Law Society of Ireland and frequently acts as an arbitrator in both domestic and international commercial arbitrations. Damien is also an Accredited Mediator with the Centre for Effective Dispute Resolution (CEDR) and the Chartered Institute of Arbitrators (CIArb).

Niall Lawless is a Chartered Arbitrator and Engineer, Adjudicator and Mediator, and he has been involved with resolving international commercial disputes valued in excess of US$50 million. Niall is a Fellow of the Chartered Institute of Building, the Chartered Institution of Building Services Engineers, the Chartered Institution of Mechanical Engineers, the Chartered Institute for Information Technology (British Computer Society) and the Chartered Institute of Arbitrators. Niall is a Chartered Institute of Arbitrators Construction Adjudication and Mediation Course Director, Tutor and Assessor. Moreover, Niall has coached gold medal international teams in a wide spectrum of dispute resolution competitions. For six years, Niall chaired the Construction Industry Council's ADR Management Board and was responsible for leading the production of CIC ADR publications including the CIC Users' Guide to Adjudication: Ireland. He is a strong voice in the industry with articles and commentaries published on various international platforms.

Adjudication Practice and Procedure in Ireland

Construction Contracts Act 2013

Damien Keogh and Niall Lawless

Routledge
Taylor & Francis Group
LONDON AND NEW YORK

First published 2020
by Routledge
2 Park Square, Milton Park, Abingdon, Oxon OX14 4RN

and by Routledge
605 Third Avenue, New York, NY 10017

First issued in paperback 2022

Routledge is an imprint of the Taylor & Francis Group, an informa business

Publisher's Note
The publisher has gone to great lengths to ensure the quality of this reprint but points out that some imperfections in the original copies may be apparent.

British Library Cataloguing-in-Publication Data
A catalogue record for this book is available from the British Library

Library of Congress Cataloging-in-Publication Data
A catalog record has been requested for this book

ISBN: 978-0-367-49114-7 (pbk)
ISBN: 978-1-138-02030-6 (hbk)
ISBN: 978-1-315-67531-2 (ebk)

DOI: 10.1201/9781315675312

Typeset in Bembo
by Swales & Willis, Exeter, Devon, UK

Contents

First published 2020
by Routledge
2 Park Square, Milton Park, Abingdon, Oxon OX14 4RN

and by Routledge
52 Vanderbilt Avenue, New York, NY 10017

Routledge is an imprint of the Taylor & Francis Group, an informa business

British Library Cataloguing-in-Publication Data
A catalogue record for this book is available from the British Library

Library of Congress Cataloging-in-Publication Data
A catalog record has been requested for this book

ISBN: 978-1-138-02030-6 (hbk)
ISBN: 978-0-36-749114-7 (pbk)
ISBN: 978-1-315-67531-2 (ebk)

Typeset in Bembo
by Swales & Willis, Exeter, Devon, UK

Acknowledgements

Damien Keogh wishes to thank his wife Kristin for all her support throughout and, in particular, assistance with the early drafts, and also Stephen Hanaphy BL for the independent review he provided for the final edit.

Damien Keogh and Niall Lawless thank Sanja Tseveenjav for her editorial guidance and support.

Foreword

On 13 April 2016 the Minister for Business and Employment signed the necessary commencement order to enable the Construction Contracts Acts, 2013 ("the Act") to come into force and effect. As from 25 July, 2016, the Act applies to all construction contracts within the statutory definition of that term. The Act brings in new rules relating to payments and in relation to the adjudication of payment disputes. It is with regard to the latter aspect of the Act that this new work by Damien Keogh and Niall Lawless is directed. As a consequence of the provisions of the Act dealing with adjudication, there is now a new legal framework in place which parties to construction contracts (*and those who advise them*) will have to address.

As the Act begins to be utilised, there will undoubtedly be a learning process for participants in the construction industry and their advisors. In due course, as parties seek to enforce adjudication awards, there will also be a learning process for the courts. That process has now been made significantly less onerous for all concerned thanks to the generosity of the authors in sharing their extensive knowledge and expertise in this book.

The book falls into two quite distinct – yet interrelated - parts. The first part of the book (*written by Damien Keogh*) takes a microscope to the provisions of the Act dealing with adjudication and provides a detailed commentary on them with extremely helpful and extensive reference to decisions of the courts of the United Kingdom in relation to similar provisions in the equivalent legislation in the U.K. The work undertaken by Damien Keogh will be of enormous assistance to anyone seeking to understand how the provisions of the Act should be interpreted and applied. In fact, it will be an essential point of reference for parties, practitioners, adjudicators and the courts.

The second part of the book (*written by Niall Lawless*) offers a unique insight into the thought processes of an expert adjudicator. Niall Lawless brings his very considerable practical experience and expertise to bear in offering masterful views as to how parties, advocates and adjudicators might best grapple with the adjudication process and use it to full advantage. It offers a thought provoking and illuminating perspective on the adjudication process.

Each part of the book complements the other. The reader gets the benefit of two books in one. Crucially, the book offers the reader invaluable guidance and assistance in understanding and applying the adjudication provisions of the Act. New legislation will often give rise to teething problems for those having to apply it for the first time. However, with the benefit of this book, those teething problems will be minimised. The treatment of the UK case law is extremely helpful in this context. It will save the reader hours of difficult and time-consuming research.

I have no doubt that the book will become the definitive guide to the adjudication provisions of the Act and I wish the authors every success with it.

Denis McDonald
The High Court
Four Courts
Dublin 7
16 April, 2020

Table of cases

Case Name	Paragraph Number Part 1	Section Number Part 2
A		
ABB Limited v BAM Nuttall Limited [2013] EWHC 1983 (TCC)		15.16
ABB Power Construction Limited v Norwest Holt Engineering Limited [2000] EWHC 68 (TCC)	4.20	
Absolute Rentals Limited v Glencor Enterprises Limited [2000] CILL 1637	10.25	
AC Yule & Son Limited v Speedwell Roofing and Cladding Limited [2007] EWHC 1360 (TCC)	8.23	
Adyard Abu Dhabi v Sd Marine Services [2011] EWHC 848 (Comm)		17.3
Aedifice Partnership Limited v Mr Ashwin Shah [2010] EWHC 2106 (TCC)		16.15
Air Design (Kent) Limited v Deerglen (Jersey) Limited [2008] EWHC 3047 (TCC)	5.58, 6.52	
A J Brenton v Jack Palmer [2001] Unreported, 19 January 2001 (TCC)	5.6	
Allied P&L Limited v Paradigm Housing Group Limited [2009] EWHC 2890 (TCC)		16.15
Alstom Limited v Yokogawa Australia Pty Limited and Anor (No 7) [2012] SASC		17.3
Amber Construction Services Limited v London Interspace HG Limited [2007] EWHC 3042 (TCC)		16.19
Amec Civil Engineering Limited v Secretary of Transport [2004] EWHC 2339 (TCC)		16.2
All-In-One Building & Refurbishments v Makers UK [2005] EWHC 2943 (TCC)	5.31	
Amec Capital Projects Limited v Whitefriars City Estates Limited [2004] EWCA Civ 1418		15.15, 16.14, 16.15

(*Continued*)

(Cont.)

Case Name	*Paragraph Number* *Part 1*	*Section Number* *Part 2*
Amec Group v Thames Water Utilities Limited [2010] EWHC 419 (TCC)	5.58	16.14
Amec Civil Engineering v Secretary of State for Transport [2004] EWHC 2339 (TCC)	5.12, 5.13, 5.19, 5.27, 5.31	16.2
Amey LG Limited v Cumbria County Council [2016] EWHC 2856 (TCC)		17.4
Anglian Water Services v Laing O'Rourke Utilities Limited [2010] EWHC 1529 (TCC)	13.11	
Anglo Group plc v Winther Brown & Co. Limited and BML (Office Computers) Limited [2000] EWHC Technology 127		17.16
Arcadis UK Limited v May and Baker Limited (t/a Sanofi) [2013] EWHC 87 (TCC)		16.7
Aspect Contracts (Asbestos) Limited v Higgins Construction Plc [2013] EWHC 1322 (TCC)		16.20
Atlas Ceiling & Partition Co Limited v Crowngate Estates (Cheltenham) [2000] CILL 1639 (TCC)	3.7, 3.8	
Atlas Ceiling & Partition Co Limited v Crowngate Estates (Cheltenham) Limited [2002] 18 Const LJ 49	11.19	
Attorney General for the Falkland Islands v Gordon Forbes Construction (Falklands) Limited (2003) 6 BLR 280		17.3
Aveat Heating Limited v Jerram Falkus Construction Limited [2007] EWHC 131 (TCC)	8.4	
AWG Construction Services Limited v Rockingham Motor Speedway Limited [2004] EWHC 888	5.28	
Axis M&E UK Limited v Multiplex Construction Europe Limited [2019] EWHC 169 (TCC)		17.13
B		
Balfour Beatty Construction Northern Ireland v London Borough of Lambeth [2002] EWHC 597 (TCC)	9.12, 9.13, 9.21, 9.34	17.9
Balfour Beatty Construction Limited v The Mayor and Burgesses of the London Borough of Lambeth [2002] BLR 288		16.18
Balfour Beatty Construction v Serco Limited [2004] EWHC 3336 (TCC)	10.12, 10.13	
Banham Marshalls Services Unlimited v Lincolnshire County Council, and Others [2007] EWHC 402 (QB)		15.18
The Bank of Ireland UK Plc v Watts Group Plc [2017] EWHC 1667 (TCC)		17.16
Barr Limited v Law Mining Limited [2001] Scot CS 152	5.59	

(*Continued*)

(Cont.)

Case Name	*Paragraph Number* *Part 1*	*Section Number* *Part 2*
Barrie Green v (1) GW Integrated Building Services Limited and (2) G&M Floorlayers (Derby) Limited [2001] LE014261 Leicester CC		17.19
Barrs v British Wool Marketing Board, 1957 SC 72		15.16
Barnes & Elliott v Taylor Woodrow Holdings Limited [2003] EWHC 3100 (TCC)	8.11	
Beck Interiors Limited v UK Flooring Contractors Limited [2012] EWHC 1808 (TCC)	5.17, 5.35	16.2
Beck Peppiatt Limited v Norwest Holst Construction Limited [2003] EWHC 822 (TCC)		16.2
Billingford Holdings Limited and BFL Trade Limited v SMC Building Solutions Limited and Another [2019] EWHC 711 (TCC)		16.15
Bloor Construction (UK) Limited v Bowmer & Kirkland (London) Limited [2000] EWHC 183 (TCC)	8.10, 10.30, 10.31, 10.34	
BM Services Inc. Limited v Greyline Builders Limited [2018] EWHC 3884 (TCC)		17.19
Bolam v Friern Hospital Management Committee [1957] 1 WLR 582; [1957] 2 All ER 118		15.19
The Board of Trustees of National Museums and Galleries on Merseyside v AEW Architects and Designers Limited & PIHL UK Limited & Another [2013] EWHC 2403 (TCC)		17.16
Bouygues (UK) Limited v Dahl-Jensen (UK) Limited [2000] EWCA CIV 507 Bouygues UK Limited v Dahl-Jenson UK Limited [1999] EWHC Technology 182	10.8, 10.9, 10.21, 10.23, 10.25	17.13
Bovis Lend Lease v The Trustees of The London Clinic [2009] EWHC 64 (TCC)	5.30	
Bremer Handelsgesellschaft v Westzucker (No. 2) [1981] 2 Lloyd's Report 130		17.12
Bridgewater Construction Limited v Tolent Construction Limited (2000) CILL 1662		15.17
Browne & Son v Crosby North West Homes [2009] EWHC 3503 (TCC)	5.58	
C		
Calderbank v Calderbank [1975] 3 All ER 333		17.14
Camillin Denny Architects Limited v Adelaide Jones & Company Limited [2009] EWHC 2110 (TCC)	5.53	
Cantillon Limited v Urvasco Limited [2008] EWHC 282 (TCC)	4.24, 4.25	16.2, 16.13
Cantillon Limited v Urvasco Limited [2008] BLR 250		15.16

(Continued)

(Cont.)

Case Name	*Paragraph Number* *Part 1*	*Section Number* *Part 2*
Capital Structures v Time & Tide Construction Limited [2006] EWHC 591 (TCC)	4.16	
Carillion Construction Limited v Devonport Royal Dockyard Limited [2003] HT-02-395	5.21	
Carillion Construction Limited v Devonport Royal Dockyard Limited [2005] EWHC 778 (TCC); [2005] EWCA CIV 1358	6.39, 6.40, 10.4, 10.5	15.18, 17.12
Carillion Construction Limited v Devonport Royal Dockyard Limited [2006] 1 BLR 15		16.19
Carrilion Utility Services v Scottish Power Systems Limited 2012 SLT 119		15.16
Cassell & Co Limited v Broome and Another [1972] UKHL 3; [1972] 2 WLR 645; [1972] AC 1027		15.10
Chan Shun Kei, trading as Chan Shun Kei Construction Works v Hong Kong Construction (Hong Kong) Limited [2014] HKEC, 7 April 2014		17.16
Christiani & Neilsen Limited v The Lowry Centre Development Company Limited [2000] HT 001/59 (TCC), 26 June 2000	3.5, 3.8, 6.45, 6.46	16.15
CIB Properties Limited v Birse Construction Limited [2004] EWHC 2365 (TCC)	8.22, 10.32	
CIB Properties Limited v Birse Construction [2005] BLR 173		16.1
City Inn Limited v Shepherd Construction Limited 17 July 2001 Scotland Outer House, Court of Session CA101/00		16.20
CJP Builders Limited v William Verry Limited [2008] EWHC 2025 (TCC)	9.27	16.13
Cleveland Bridge (UK) Limited v Whessoe-Volker Stevin Joint Venture [2010] EWHC 1076 (TCC)	3.67, 3.68	
Connex South Eastern Limited v MJ Building Services Group Plc [2005] EWCA Civ 193	5.71	15.6, 15.16
Collins (Contractors) Limited v Baltic Quay Management Limited [2004] EWCA Civ 1757	5.27, 5.31	
Costain Limited v Strathclyde Builders Limited [2003] Scot CS 316	9.7, 9.8	
Costain Limited v Strathclyde Builders Limited [2004] SLT 102		15.16
Cowlin Construction Limited v CFW Architects [2003] EWHC 60	6.34	
CSK Electrical Contractors Limited v Kingwood Electrical Services Limited [2015] EWHC 667 (TCC)		16.2, 16.6

(*Continued*)

(Cont.)

Case Name	*Paragraph Number* *Part 1*	*Section Number* *Part 2*
Cubitt Building & Interiors Limited v Fleetglade Limited [2006] EWHC 3413 (TCC)	7.15, 7.16, 7.27, 8.10, 8.17, 9.41	17.19
Cubitt Building & Interiors Limited v Richardson Roofing (Industry) Limited [2008] EWHC 1020 (TCC)	5.77	
Cynthia Jacques and Elsie Jacques Grombach (t/a C&E Jacques Partnership) v Ensign Contractors Limited [2009] EWHC 3383 (TCC)		15.13
D		
David McLean Housing Contractors Limited v Swansea Housing Association Limited [2001] EWHC 830 (TCC)	5.38	
Deko Scotland Limited v Edinburgh Royal Joint Venture and Others [2003] ScotCS 113		17.9
Devon County Council v Celtic Composting Systems Limited [2014] EWHC 552 (TCC)		16.2
DGT Steel & Cladding v Cubitt Building & Interiors Limited [2007] EWHC 1584	5.75, 5.76, 5.77	
Digi-Tech (Australia) Limited v Brand [2004] NSWCA 58 (61 IPR 184)		17.8
Director of Public Prosecutions v Kilbourne [1973] AC 729		15.13
Discain Project Services v Opecprime Developments Limited [2001] EWHC 450 (TCC)	9.16	
Dorchester Hotel Limited v Vivid Interiors Limited [2009] EWHC 70 (TCC)		16.1
E		
Earls Terrace Properties v Waterloo Investments [2002] HT 02/237 (TCC)	3.9	
Ellis Building Contractors Limited v Vincent Goldstein [2011] EWHC 269		15.13
Ellis Mechanical Services Lid v Wates Construction Limited (1976) 2 BLR 57		15.4
Edenbooth Limited v Cre8 Developments Limited [2008] EWHC 570 (TCC)	4.6	
Edmund Nuttall Limited v RG Carter Limited [2002] EWHC 400 (TCC)	5.24, 5.25	16.2, 17.5, 17.16
English v Emery Reimbold & Strick Limited [2002] EWCA Civ 605		17.12
Epping Electrical Company v Briggs and Forrester [2007] EWHC 4 (TCC)	8.15, 8.17	16.9
Eurocom Limited v Siemens PLC [2014] EWHC 3710 (TCC)		16.6, 16.7

(*Continued*)

(Cont.)

Case Name	*Paragraph Number* *Part 1*	*Section Number* *Part 2*
F		
Fastrack Contractors Limited v Morrison Construction Limited [2000] EWHC 177 (TCC)	5.14, 5.39	
Fence Gate Limited v J.R. Knowles Limited [2001] 25/01 SF102200 (TCC)	3.50, 3.66, 3.69, 4.20	
Fenice Investments Inc v Jerram Falkus Construction Limited [2011] EWHC 1678 Fenice Investments Inc v Jerram Falkus Construction Limited [2011] EWHC 1678 (TCC)	9.43, 11.18	17.19
Fileturn Limited v Royal Garden Hotel Limited [2010] EWHC 1736 (TCC)		15.15
Fiona Trust and Holding Company and Others v Yuri Privalov [2007] 4 All ER 951 (HL); [2007] 4 RERS ALLERS 951	3.63, 5.52, 5.53, 5.54, 5.60	15.6
Floyd v Barker, 77 Eng. Rep. 1305 (Star Chamber 1607)		15.19
G		
Gabrielle and Christopher Shaw v Massey Foundation & Pilings Limited [2009] EWHC 493 (TCC)	4.5	
Geoffrey Osborne Limited v Atkins Rail Limited [2009] EWHC 2425 (TCC)		17.13
Gibson Lea Retail Interiors v Makro Self Service Wholesalers Limited [2001] HT 01 226 (TCC)	3.28	
Gibson (Banbridge) Limited v Fermanagh District Council [2013] NIQB 16		16.13
Gillespie Bros & Co Limited v Roy Bowles Transport Limited CA [1973] 1 QB 400		15.19
Gillies Ramsay Diamond v PJW Enterprises [2002] Scots 340 Gillies Ramsay Diamond v PJW Enterprises Limited [2003] P629/01	3.17, 3.21	17.12
Glencot Development and Design Co v Ben Barrett & Son (Contractors) Limited [2001] EWHC 15 (TCC)	9.6	
GPS Marine Contractors Limited v Ringway Infrastructure Services Limited [2010] EWHC 283 (TCC)		16.14, 16.15
Griffin v Midas Homes Limited [2000] EWHC 182 (TCC)	5.101	
Grindrod Shipping Pte Limited v Hyundai Merchant Marine Co. Limited [2018] EWHC 1284 (Comm)		17.9

(*Continued*)

(Cont.)

Case Name	*Paragraph Number* *Part 1*	*Section Number* *Part 2*
H		
Halki Shipping Corporation v Sopex Oils Limited [1997] EWCA Civ 3062		16.2
Halsey v Milton Keynes General NHS Trust [2004] EWCA Civ 576		17.14
Hart Investments v Fiddler & Another [2006] EWHC 2857 (TCC)	7.5, 7.6, 7.9, 7.23	16.6, 16.12
Herbosh–Kiere Marine Contractors Limited v Dover Harbour Board [2012] EWHC 84 (TCC)	9.35	15.16
Herschel Engineering Limited v Breen Property Limited [2000] EWHC Technology 178	5.69, 10.25	15.6, 16.18, 16.20
Herschel Engineering Limited v Breen Properties Limited (No2) [2000] Adj.L.R. 07/28		
Hillcrest Homes v Beresford & Curbishley [2014] EWHC 280 (TCC)	5.60	15.6
HG Construction Limited v Ashwell Homes (East Anglia) Limited [2007] EWHC 144 (TCC)	5.89	
Holt Insulation Limited v Colt International Limited [2001] EWHC 451	5.84, 5.85	
I		
Ian McGlinn v Waltham Contractors Limited & Others (No. 5) [2007] EWHC 698 (TCC)		17.14
IDE Contracting Limited v RG Carter Cambridge Limited [2004] EWHC 36 (TCC)	6.4	
The Ikarian Reefer – National Justice Compania Naviera SA v Prudential Assurance Company Limited [1993] 2 Lloyd's Rep 68 (Commercial Court); [1995] 1 Lloyd's Rep 455 (CA)		17.16
J		
J Murphy & Sons Limited v W Maher and Sons Limited [2016] EWHC 1148 (TCC)		15.6
John Mowlem v Hydra-Tight & Co [2002] 17 Const LJ 358	5.66	
John Stirling v Westminster Properties Scotland [2007] Scot EF CSOH 117	5.18	
K		
A&S Enterprises Limited v Kema Holdings Limited [2005] BLR 76		15.15
A&S Enterprises Limited v Kema Holdings Limited [2004] EWHC 3365 (QB)		16.16

(*Continued*)

(Cont.)

Case Name	*Paragraph Number* *Part 1*	*Section Number* *Part 2*
KNS Industrial Services (Birmingham) Limited v Sindall Limited [2001] 75 Con LR	7.8, 7.9	
KNN Coburn LLP v GD City Holdings Limited [2013] EWHC 2879 (TCC) KNN Colburn LLP v GD City Holdings Limited [2013] EWHC 2879 (TCC)	8.27	16.12, 16.12
L		
Laker Vent Engineering Limited v Jacobs E & C Limited [2014] EWHC 1058 (TCC)	10.38	
Lanes Group Plc v Galliford Try Infrastructure Limited [2011] EWCA Civ 1617; [2011] EWHC 1679 (TCC) Lanes Group Plc v Galliford Try Infrastructure Limited [2011] EWHC 1035 (TCC)	5.79, 7.18	15.15, 15.16, 16.6
Lead Technical Services Limited v CMS Medical Limited [2007] EWCA Civ 316		16.19
Letchworth Roofing Company v Sterling Building Company [2009] EWHC 1119 (TCC)	5.95, 5.96, 8.26	
Christopher Michael Linnett v Halliwells LLP [2009] EWHC 319 (TCC)	7.26, 9.47, 11.13, 11.16	16.12, 16.15
Locabail (UK) Limited v Bayfield Properties Limited [2000] QB 451		15.15
London & Amsterdam Properties v Waterman Partnership Limited [2003] EWHC 3059 (TCC)	9.14	17.4, 17.19
London Borough of Camden v Makers UK Limited [2009] BWHC 605 (TCC)	5.70	
London Borough of Merton v Stanley Hugh Leach Limited [1986] 32 BLR 51		17.10
London, Chatham and Dover Railway Co v South Eastern Railway Co [1893] AC 429		15.18
Lovell Projects v Legg and Carver [2003] C.I.L.L. 2019	4.15	
Lulu Construction v Mulalley & Co Limited [2016] EWHC 1852 (TCC)	11.3, 11.4, 11.5	
M		
Macob Civil Engineering v Morrison Construction Limited [1999] EWHC Technology 254	6.48, 10.3, 10.5	
Makers UK Limited v London Borough of Camden [2008] EWHC 1836 (TCC) Makers UK Limited v The Mayor and Burgesses of the London Borough of Camden [2008] EWHC 1836 (TCC)		15.15, 16.7

(*Continued*)

(Cont.)

Case Name	*Paragraph Number* *Part 1*	*Section Number* *Part 2*
Maymac Environmental Services Limited v Farraday Building Services Limited [2001] 75 Com LRHT 00/222	6.36, 6.37, 6.38, 12.8	
Mecright v TA Morris Developments [2001] HT01 84	5.92	
Michael John Construction v Richard Henry Golledge & Others [2006] EWHC 71 (TCC)	5.59	17.7
Michael J. Lonsdale (Electrical) Limited v Bresco Electrical Services Limited (in Liquidation) [2018] EWHC 2043 (TCC)	5.72, 5.73	15.6
MI Electrical Solutions Limited v Elements (Europe) Limited [2018] EWHC 1472 (TCC)	5.99	16.2
Miller v Minister of Pensions [1947] 2 All ER 372		15.12
Mott MacDonald Limited v London & Regional Properties Limited [2007] EWHC 1055 (TCC)	8.18	
Murphy & Sons Limited v W Maher and Sons Limited [2016] EWHC 1148 (TCC)	5.61	
N		
Nageh v Giddings [2006] EWHC 3240 (TCC)	13.9	
National Museums and Galleries on Merseyside Board of Trustees v AEW Architects and Designers Limited [2013] EWHC 3025 (TCC)	11.6, 11.7	
Niru Battery Manufacturing Co v Milestone Trading Limited (No 1) [2002] EWHC 1425 (Comm)		
Nordot Engineering Services Limited v Siemens [2001] CILL, 1778 (TCC)	4.12	
North Midland Construction Plc v AE & E Lentjes UK Limited [2009] EWHC 1371 (TCC)	4.20	
O		
O'Brien v Chief Constable of South Wales Police [2005] UKHL 26		15.13
Oceanbulk Shipping & Trading SA (Respondents) v TMT Asia Limited and Others (Appellants) [2010] UKSC 44		17.14
O'Donnell Developments v Build Ability [2009] EWHC 3388 (TCC)	10.36	
OFC Building Services Limited v Interior Dimensions Contracts Limited [2009] EWHC 248 (TCC)		16.11
O'Gorman & Co Limited v JES Holding Limited [2005] 11 ICLMD 124		17.16
Orange EBS Limited v ABB Limited [2003] EWHC 1187 (TCC)	5.15, 5.16, 5.17	
OSC Building Services Limited v Interior Dimensions Contracts Limited [2009] EWHC 248 (TCC)	5.103	
Ove Arup and Partners International Limited v Coleman Bennett International Consultancy plc [2019] EWHC 413 (TCC)		16.15

(*Continued*)

(Cont.)

Case Name	Paragraph Number Part 1	Section Number Part 2
P		
Paice & Anr v MJ Harding (t/a MJ Harding Contractors) [2015] EWHC 661 (TCC)		16.7
Palmac Contracting Limited v Park Lane Estate Limited [2005] EWHC 919; [2005] ELR 301	6.9, 6.10	
Palmers Limited v ABB Power Construction Limited [1999] HT 99 000090 (TCC)	3.39, 3.40, 4.20	
Paul Jensen v Staveley Industries Plc [2001] WN 101245	11.19, 11.20	
Parkwood Leisure v Laing O'Rourke Wales & West [2013] EWHC 2665 (2CC)	3.52, 3.53, 3.55	
PC Harrington Contractors Limited v Systech International Limited [2012] EWCA Civ 1371 Systech International Limited v PC Harrington Contractors Limited [2012] EWCA Civ 1371	11.14, 11.15, 11.16	16.15, 17.9, 17.19
Petition of Mr and Mrs Jack Patton [2011] CSOH 40		15.16
Picardi v Cuniberti [2002] EWHC 2923 (TCC)	4.13, 4.14, 4.15	
Pihl UK Limited v Ramboll UK Limited [2012] CSOH 139		17.12
Pilon Limited v Breyer Group PLC [2010] EWHC 837 (TCC)	5.93	16.11
Pioneer Cladding v John Graham Construction Limited [2013] EWHC 2954 (TCC)	10.27	
Porter v Magill [2002] 2 AC 357		15.15
Post Systech International Limited v PC Harrington Contractors Limited [2012]		16.15
Potton Developments Limited v Thompson and Another [1998] NPC 49, ChD	3.29	
Primus Build Limited v Pompey Centre Limited (& Slidesilver Limited) [2009] EWHC 1487 (TCC)		15.17, 16.14, 17.14
Project Consultancy Group v The Trustees of The Gray Trust [1999] HT/99/29 (TCC)	6.48, 6.49, 10.21	
PT Building Services Limited v ROK Build Limited [2008] EWHC 3434 (TCC)	7.23	16.12
Q		
Quartzelec Limited v Honeywell Control Systems Limited [2008] EWHC 3315 (TCC)		16.13
Quietfield Limited v Vascroft Contractors Limited [2006] EWHC 174; [2006] EWCA Civ 1737	5.86, 5.87, 5.88	16.4

(Continued)

(Cont.)

Case Name	*Paragraph Number* *Part 1*	*Section Number* *Part 2*
R		
Rainford House (in Receivership) v Cadogan [2001] HT 01/014	10.25	
Re D (Fact-Finding Hearing: Medical Treatment) [2014] EWHC 121 (Fam) (Mostyn J)		15.12
A.E.I. Rediffusion Music Limited v Photographic Performance Limited [1999] 1 WLR 1507		17.14
Regina v Dennis John Adams [1996] 2 Cr. App. R. 467		15.13
RG Carter Limited v E Nuttal Limited [2000] HT-00-230 (TCC), 21 June 2000	5.59, 5.67	
Rohde Construction v Markham-David [2006] EWHC 814 (TCC)	13.7, 13.8	
ROK Building v Celtic Composting Systems Limited [2009] EWHC 2664 (TCC)	10.19, 10.33	
RSL (South West) v Stansell [2003] EWHC 1390 (TCC)	9.9, 9.10, 9.21	
S		
Samuel Thomas Construction v J&B Developments [2000] High Court at Exeter (Case No: Exeter ZN 900750)	4.3, 4.4	
SG South Limited v King's Head Cirencester LLP & Another [2009] EWHC 2645 (TCC)	4.18	15.4
Shepherd Construction v Mecright [2000] HT-00-28 (TCC)	3.62, 3.63	
Sherwood & Casson Limited v MacKenzie Engineering Limited [1999] EWHC 274 (TCC)	5.82	
Sherwood & Casson Limited v MacKenzie Engineering Limited [2000] 2 TCLR 418	10.21	
Sierra Fishing Company & Ors v Farran & Ors [2015] EWHC 140 (Comm)		15.14
Sindall Limited v Abner Solland [2002] Con LRHT 01/129 (TCC); [2001] 2 TCLR 30	5.26, 5.27, 5.40	
Squibb Group Limited v Vertase F.L.I Limited [2012] EWHC 1958 (TCC)	10.15, 10.16	
Sprunt Limited v Camden LBC [2011] EWHC 3199		16.7
Staveley Industries v Odebrecht Oil and Gas Services [2001] All ER (d) 359	3.31	
Stephens v Cannon [2005] EWCA Civ 222		15.12, 15.13, 17.16

(*Continued*)

(Cont.)

Case Name	*Paragraph Number* *Part 1*	*Section Number* *Part 2*
Strathmore Building Services v Colin Scott Greig (t/a Hestia Fireside Design) [2000] Scot CS 133	13.5	
Stubbs Rich Architects v WH Tolley & Sons Limited [2001] Gloucester County Court, Case No. BP001105	11.17	
Sutcliffe v Chippendale & Edmondson [1971] 18 BLR 149		17.10
T		
Thermal Energy Construction Limited v AE & E Lentjes UK Limited [2009] EWHC 408 (TCC)		17.12
Ticket2Final OU v Wigan Athletic AFC Limited [2015] EWHC 61b (CH)		16.11
Total M&E Services Limited v ABB Building Technologies Limited [2002] EWHC 248	5.8	
Tramountana Armadora SA v Atlantic Shipping Co. SA [1978] 2 All ER 870		15.17
Twintec Limited v Volkerfitzpatrick Limited [2014] EWHC 10 (TCC)		16.7
U		
V		
Van Gend en Loos [1963] ECR 1 Costa v ENEL [1964] ECR 585; Internationale Handelsgesellschaft [1970] ECR 1125		15.10
Van Oord UK Limited and SICIM Roadbridge Limited v Allseas UK Limited [2015] EWHC 3074 (TCC)		17.16
Vertase F.L.I. Limited v Squibb Group Limited [2012] EWHC 3194 (TCC)		17.13
Vision Homes Limited v Lancsville Construction Limited [2009] EWHC 2042 (TCC)	6.11	
Volker Stevin Limited v Holystone Contracts Limited [2010] EWHC 2344 (TCC)		15.13, 17.16
W		
Wales and West Utilities Limited v PPS Pipeline System GMBH [2014] EWHC 54 (TCC)		16.11
Walker Construction (UK) Limited v Quayside Homes Limited & Anor [2014] EWCA Civ 93		16.20
Whiteways Contractors (Sussex) Limited v Impresa Castelli Construction UK Limited [2000] EWHC Technology 67 (TCC)	6.50, 6.51	

(Continued)

(Cont.)

Case Name	*Paragraph Number* *Part 1*	*Section Number* *Part 2*
William Verry (Glazing Systems) Limited v Furlong Homes Limited [2005] EWHC 138 (TCC)	5.28	16.4, 16.11
William Verry Limited v The Mayor & Burgesses of the London Borough of Camden [2006] EWHC 761 (TCC)	10.10	
Wimbledon Construction Company 2000 Limited v Derek Vago [2005] EWHC 186 (TCC)	10.25, 10.26, 10.27	
Witney Town Council v Beam Construction (Cheltenham) Limited [2011] EWHC 2332 (TCC)	5.42, 5.44	
Working Environment Limited v Greencoat Construction Limited [2012] EWHC 1039 (TCC)	5.33	
X		
Y		
Yarm Road Limited v Costain Limited [2001] HT 01228 (TCC)	3.10, 3.11	
YCMS Limited v Grabiner [2009] EWHC 127 (TCC)	10.34	17.13
Yuanda (UK) Co Limited v WW Gear Construction Limited [2010] EWHC 720 (TCC)		15.17, 15.18
Z		

Table of statutes

Statute	*Paragraph Number Part 1*	*Section Number Part 2*
Arbitration Act 2010	14.10	
S 22(1)	9.42	
S 30	4.9	
UK Arbitration Act 1996		17.9
Building Control Act 2007, Section 2	6.30	
UK Civil Procedure Rules		17.14, 17.16
Practice Direction 32 – Evidence		15.13
Practice Direction PD 1 of 2015, Northern Ireland		17.16
Irish Constitution		17.19
Construction Contracts Act 2013	1.1	15.2, 15.6, 15.9, 15.11, 15.17, 15.18, 15.20, 16.1, 16.2, 16.3, 16.4, 16.6, 16.8, 16.9, 16.10, 16.11, 16.19, 16.20, 17.9, 17.12, 17.13, 17.14, 17.15, 17.17, 17.18, 17.19
S 1	2.3	15.6
S 1(1)	3.2, 3.12, 3.13, 3.14, 3.16, 3.18, 3.19, 3.21, 3.24, 3.28, 3.33, 3.35, 3.36, 3.38, 3.39, 3.56, 5.3	
S 1(1)(a)	3.24, 3.28, 3.46	
S 1(1)(b)	3.21, 3.24, 3.33	
S 1(1)(c)	3.24	
S 1(1)(d)	3.24, 3.36	
S 1(1)(e)	3.24, 3.38, 3.39, 3.41, 3.43, 3.44	

(Continued)

(Cont.)

Statute	*Paragraph Number* *Part 1*	*Section Number* *Part 2*
S 1(1)(f)	3.24, 3.45, 3.47	
S 1(1)(g)	3.24, 3.46	
S 1(2)	3.21, 3.47, 3.48, 3.49, 3.50, 3.51, 3.57, 3.60, 3.61, 4.19	
S 1(2)(a)	3.48, 3.50, 3.61	
S 1(2)(b)	3.21	
S 1(2)(c)	3.50, 3.60, 3.61	
S 1(3)	3.39, 3.41, 3.42, 3.64, 3.69, 4.11	
S 1(4)	3.65	
S 2	2.3, 4.1, 4.2, 4.4, 4.9, 4.10, 4.11, 4.19, 4.20, 4.26, 4.28, 5.64, 7.4	
S 2(1)	4.1, 4.2, 4.4	
S 2(1)(b)	4.2, 4.4, 4.11	
S 2(2)	4.9	
S 2(3)	4.10	
S 2(4)	3.50, 4.19, 4.20	
S 2(5)	4.26	
S 2(5)(b)	4.28, 5.64, 7.4	
S 4	5.32, 5.47, 5.94	
S 4(3)	5.32, 5.94	
S 5(1)	10.39	
S 6	2.1, 2.3, 5.1, 12.1	
S 6(1)	2.3, 5.2, 5.3, 5.11, 5.37, 5.48, 9.38, 10.17	
S 6(2)	2.3, 5.62, 5.63, 5.90, 6.5, 6.6	16.11
S 6(3)	2.3, 6.1, 6.3, 6.5, 6.6, 6.12, 6.13	16.7
S 6(4)	2.3, 6.2, 6.5, 6.13, 6.19, 6.20	
S 6(5)	7.1, 7.4	
S 6(5)(a)	2.3, 6.17, 6.25, 7.1, 7.2, 7.4, 7.5, 7.7, 7.11, 7.12, 7.13, 7.14, 7.25, 7.29, 8.1, 8.2	
S 6(5)(b)	2.3, 7.19, 7.21, 7.22, 7.24	
S 6(6)	2.3, 8.1, 8.2, 8.3, 8.5, 8.8, 8.14, 8.16, 8.19, 9.22	
S 6(7)	2.3, 8.5, 8.20	
S 6(8)	2.3, 6.18, 7.12, 8.3, 9.1, 9.2, 9.3, 9.5, 9.17, 9.18, 12.4, 12.6	

(*Continued*)

(Cont.)

Statute	*Paragraph Number* *Part 1*	*Section Number* *Part 2*
S 6(9)	2.3, 9.1, 9.23, 9.33, 9.38, 9.39	
S 6(10)	2.3, 10.1	
S 6(11)	2.3, 10.20, 14.1, 14.3	
S 6(12)	2.3, 10.1	
S 6(13)	2.3, 10.9, 10.28, 10.29, 10.30	
S 6(14) Paragraph 6.(14)	2.3, 9.1, 9.40, 9.41	15.19
S 6(15) Paragraph 6.(15)	2.3, 11.1	15.17
S 6(16)	2.3, 11.8, 11.9, 11.10, 11.11	
S 6(17)	2.3, 6.27, 8.23, 9.1, 9.44, 9.45	
S 6(18)	2.3, 6.27, 6.28, 6.29, 11.10	
S 7	2.3	
S 7(1)	10.18, 10.39, 10.40	
S 7(2)	10.39, 10.40	
S 7(3)	10.40	
S 7(3)(b)	10.41	
S 7(4)	10.42	
S 7(5)	10.42	
S 8	6.1, 6.19, 6.30, 9.2	
S 8(5)		16.7
S 8(6)		16.7
S 9	2.3, 9.2, 12.1	
S 10	2.3	
S 10(1)	13.1, 13.3	
S 10(3)	13.12	
S 12(2)	3.4	
Code of Practice Governing the Conduct of Adjudications (25 July 2016)		15.6, 15.11, 15.14, 15.15, 15.16, 15.17, 15.20, 16.6, 16.7, 16.8, 16.11, 16.12, 16.14, 16.16, 17.4, 17.9, 17.14, 17.15, 17.19
Paragraph 5	12.6, 12.7	16.11
Paragraph 9	6.15	
Paragraph 10	6.16	
Paragraph 11	6.16	
Paragraph 12	6.17, 7.12, 7.16	
Paragraph 13	6.20	
Paragraph 14	6.20	

(*Continued*)

(Cont.)

Statute	*Paragraph Number* *Part 1*	*Section Number* *Part 2*
Paragraph 15	6.20	
Paragraph 16	6.21, 6.22	
Paragraph 18	6.22, 7.13, 7.16	
Paragraph 19	6.26	16.8
Paragraph 20	6.26	
Paragraph 22	7.21, 7.24, 7.25, 7.30	
Paragraph 22(iv)	7.21	
Paragraph 22(vi)	7.21, 7.22, 7.25	
Paragraph 23	9.5, 9.18	
Paragraph 24	9.1, 9.19	
Paragraph 24(i)	9.20	
Paragraph 24(ii)	9.21	
Paragraph 25	8.3, 8.4, 9.22, 9.23	
Paragraph 26	8.7, 9.23	
Paragraph 27	8.7, 9.28	
Paragraph 28	8.7, 9.29	
Paragraph 32	9.17, 9.20	16.8
Paragraph 38	11.1	
SI No. XXX of 2014 – Construction Contracts Act 2013 (Code of Practice) (Adjudication) Order 2014		15.20
Construction Contracts (Northern Ireland) Order 1997		15.9
Construction Contracts Act 2013 (Code of Practice) (Adjudication) Order 2014		15.15
Construction Contracts Regulations in England and Wales (1998) ("UK Scheme")[1]		15.18, 15.20, 16.6, 16.7, 16.9, 16.11, 16.12, 16.14

(Continued)

1 While there are the separate, but similar, Scheme for Construction Regulations in England & Wales (1998), Scheme for Construction Contracts Regulations in Scotland (1988) and the Scheme for Construction Contracts Regulations in Northern Ireland (1999) Scotland and Northern Ireland, for convenience we refer to these collectively (for the purposes of this Part 1) as the "UK Scheme".

(Cont.)

Statute	*Paragraph Number Part 1*	*Section Number Part 2*
Paragraph 1(3)	12.7	
Paragraph 2(1)	6.4	
Paragraph 7(1)	7.2, 7.3, 7.5, 7.6, 7.11, 7.12, 7.23	
Paragraph 7(2)	7.19, 7.20, 7.23	
Paragraph 7(3)	8.4	
Paragraph 11	6.29, 11.20	
Paragraph 11(2)	6.29	
Paragraph 13	7.6	
Paragraph 19(1)	8.10, 8.27	
Paragraph 19(1)(b)	8.20, 8.23	
Paragraph 19(3)	8.10, 8.16	
Paragraph 25	9.43, 11.8, 11.9	
Paragraph 26	9.41	
Malaysia Construction Industry Payment and Adjudication Act 2012		15.20, 16.15, 17.18, 17.19
European Communities (Late Payment in Commercial Transactions) Regulations 2012 SI No. 580/2012		15.18
General Data Protection Regulation (GDPR) (EU) 2016/679		16.6, 16.19
Housing Grants, Construction and Regeneration Act 1996		15.2, 15.6, 15.9, 16.1
S 104(1)	3.14	
S 104(2)	3.48, 3.50	
S 104(2)(a)	3.50	
S 104(2)(b)	3.50	
S 104(5)	3.50, 4.20	
S 105(1)	3.40, 3.46, 4.21	
S 105(1)(a)	3.28, 3.46	
S 105(1)(c)	3.31	
S 105(1)(e)	3.39, 3.40	
S 105(2)	3.40, 3.69, 4.12, 4.20	
S 105(2)(c)	3.39	
S 105(2)(c)(i)	4.20	

(*Continued*)

(Cont.)

Statute	*Paragraph Number Part 1*	*Section Number Part 2*
S 105(2)(d)	4.6	
S 105(2)(c)(ii)	3.67	
S 106	4.2, 4.5, 4.13	
S 106(1)(a)	4.6	
S 108A	11.5	
S 108A(2)	10.30	
S 108	5.77, 8.22	
S 108(1)	5.11	
S 108(2)	11.8	
S 108(2)(a)	5.63, 7.2	
S 108(2)(b)	6.3, 7.3, 7.4, 7.5	
S 108(2)(c)	8.8, 8.10, 8.13	
S 108(2)(d)	8.20	
S 108(2)(e)	9.3	
S 108(4)	9.41, 11.17	
S 115(6)	13.3	
Insolvency (Northern Ireland) Order 1989		16.17
Insolvency Rules		15.6
UK Late Payment of Commercial Debts (Interest) Act 1998		15.18
Mediation Act 2017		15.3, 17.15
S14		15.3
Prompt Payment Code 2008 (the UK)		15.4
Public Works Contracts published by the Office of Government Procurement		15.3
SI No. 450/2016 – Rules of the Superior Courts (Construction Contracts Act 2013) 2016		15.2, 16.19
SI No. 12/2008 – Rules of the Superior Courts (Costs) 2008		15.17
Rules of the Superior Courts (Order 23)		16.14

(*Continued*)

(Cont.)

Statute	*Paragraph Number Part 1*	*Section Number Part 2*
Rules of the Court of Judicature (NI) 1980, Order 14		16.17
State Authorities (Public Private Partnership Arrangements) Act 2002	4.8	
SI No. 165 of 2016 – Construction Contracts Act 2013 (Appointed Day) Order 2016	3.3	
UK Local Democracy, Economic Development and Construction Act 2009	1.3, 10.3, 11.8	15.9
Unfair Terms in Consumer Contract Regulations 1999	4.13, 4.15	
United Nations Convention on International Settlement Agreements Resulting from Mediation (the "Singapore Convention")		15.3
United Nations Convention on the Recognition and Enforcement of Foreign Arbitral Awards (the "New York Convention")		15.3
UK Value Added Tax Regulations 1995		17.10
European Communities (Unfair Terms in Consumer Contracts) Regulations 1995 (SI 27-1995)	4.14	

Part 1

Construction Contracts Act 2013

Damien Keogh

1 Introduction

1.1. Since the Construction Contracts Act 2013 ("the Act") was given legal effect in Irish law on 25 July 2016 parties to a construction contract, governed by Irish law, are entitled by statute to refer to adjudication payment disputes arising under their contract.

1.2. The Act also prohibits "pay-when-paid" clauses and sets out a claims process for interim and final payment applications under construction contracts. This book focuses on and examines adjudication under the Act and does not deal with the payment provisions in detail other than in the context of adjudication where applicable.

1.3. The rationale for introducing statutory adjudication in Ireland is to create a fast-track (albeit perhaps interim) process for resolving construction contract payment disputes more economically. This underlying objective should be reflected in Irish Court judgments on cases requiring the interpretation of the Act. Strict interpretation of the Act is likely, as seen from English (and Scottish) courts' interpretation of their adjudication legislation under the Housing Grants, Construction and Regeneration Act 1996 (the "UK Act").[1] In broad terms, the adjudication process under the Act is similar to the process under the UK Act which incorporates the Scheme for Construction Contracts Regulations in England and Wales (1998) ("UK Scheme").[2] The UK Scheme applies to adjudications where a construction contract does not contain the UK Act's mandatory adjudication provisions. While the Act governs only construction payment disputes, the UK legislation (including the UK Act and Scheme) applies to all construction contract disputes and not only payment disputes.

1 As amended by the UK Local Democracy, Economic Development and Construction Act 2009.

2 While the Scheme for Construction Regulations in England & Wales (1998), Scheme for Construction Contracts Regulations in Scotland (1988) and the Scheme for Construction Contracts Regulations in Northern Ireland (1999) Scotland and Northern Ireland are similar but separate, for convenience we refer to these collectively (for the purposes of this Part 1) as the "UK Scheme".

1.4. This book has two parts. The first part examines and discusses the adjudication procedure under the Act. It incorporates references to the UK Act and the vast body of English and Scottish case law from the last 20 years which, whilst not binding on Irish Courts, will undoubtedly provide guidance and invaluable assistance to Irish construction companies, practitioners and adjudicators on interpreting provisions of the Act which are similar to those in the UK legislation. The second part of this book provides a real-world perspective of adjudication, considering the knowledge, process and skills required to act as adjudicator and to assist adjudicating parties.

1.5. The objective of this book is to provide an understanding of the practice and procedure of adjudication under the Act. We hope we have achieved that objective.

2 Overview of the Construction Contracts Act 2013

2.1. The Act has 12 sections. Sections 1, 2 and 6 to 10 relate to adjudication. Section 6 is the most substantive on adjudication procedure. Its provisions relate to, *inter alia*, initiating the process, appointing the adjudicator, the adjudicator's role and powers, adjudication procedure, the adjudicator's decision and adjudication costs.

2.2. Sections 3, 4 and 5, not addressed in this book, govern construction payments. Sections 11 and 12 respectively relate to the Minister's expenses in administering the Act and its short title and commencement.

2.3. Part 1 of this book deals with the adjudication procedure under the Act, with Chapters 3 to 13 addressing the following issues:

- What constitutes a construction contract under the Act? (Section 1).
- What construction contracts are exempted? (Section 2).
- The right to refer payment-related disputes to adjudication (Sections 6(1) and (2)).
- The adjudicator's appointment (Sections 6(3), 6(4), 6(18) and Section 8).
- Referral to adjudicator (Section 6(5)).
- Timing of adjudicator decision (Sections 6(6) and (7)).
- The adjudicator's role, responsibilities and powers (Sections 6(8), (9), 6(14) and 6(17)).
- The adjudicator's decision (Sections 6(10) to 6(13) and Section 7).
- Adjudication costs (Sections 6(15) and 6(16)).
- Code of Practice Governing the Conduct of Adjudications (Section 9).
- Service of notices under the Act (Section 10).

2.4. The final chapter in Part 1 examines Order 56B of the Rules of the Superior Courts on court enforcement of adjudicator decisions.

2.5. Our views on how the Irish Courts may interpret the Act are informed from the experience of other statutory adjudication jurisdictions, and in particular, case law in those jurisdictions. In this regard, any case law cited herein relates to English cases unless otherwise indicated.

2.6. As at the date of publication of this book there have been no cases before the Irish Courts in respect of adjudication. The Irish Courts' approach to adjudication and interpretation of the Act remains to be seen. It is anticipated, however, that the Irish Courts will fully support statutory adjudication as a dispute resolution mechanism.

3 What constitutes a construction contract?

Section 1 of the Act

3.1. The statutory right of a party to a construction contract to refer disputes relating to payments ("payment dispute")[1] to adjudication only arises where there exists an enforceable construction contract as defined under the Act. In such circumstances either party may refer the payment dispute to adjudication regardless whether the other party agrees to adjudication. If there's no enforceable construction contract, adjudication cannot be compelled unless otherwise agreed.

3.2. Section 1(1) defines a "construction contract" as

> an agreement (whether or not in writing) between an executing party and another party, where the executing party is engaged for any one or more of the following activities:
>
> (a) carrying out construction operations by the executing party;
> (b) arranging for the carrying out of construction operations by one or more other persons, whether under subcontract to the executing party or otherwise;
> (c) providing the executing party's own labour, or the labour of others, for the carrying out of construction operations.

Agreement entered into and commencement of the Act

3.3. While the Act was passed into law on 29 July 2013 its commencement was subject to a Ministerial Order which was not issued until 15 April 2016.[2] Pursuant to this Ministerial Order, under Statutory

1 For convenience the phrase "payment dispute" is used in this book to paraphrase the statutory phrase "dispute relating to payment". However, the two phrases do not necessarily have the same meaning and this distinction is addressed in Chapter 5.

2 SI No. 165 of 2016 – Construction Contracts Act 2013 (Appointed Day) Order 2016.

Instrument No. 165 of 2016 ("SI No. 165"), the Act only applies to construction contracts entered into after 25 July 2016.

3.4. SI No. 165 provides that 25 July 2016 is the date appointed pursuant to Section 12(2) of the Act and that the Act "shall apply in relation to construction contracts entered into after that date". Section 12(2) similarly provides that "This Act applies in relation to construction contracts entered into after such day as the Minister may by Order appoint". The critical phrase, in the Ministerial Order, SI No. 165 and Section 12(2), is "entered into". The interpretation as to when a contract is "entered into", as opposed to when it is executed or purports to take effect from, has been considered extensively by the UK Courts.

3.5. In *Christiani v Lowry*,[3] His Honour Judge (HHJ) Thornton QC held that a contract entered into after, but taking effect before, the commencement of the UK Act was nevertheless governed by the UK Act. Work under the contract began on foot of an 11 August 1997 letter of intent, which incorporated the UK standard-form engineering contract, ICE Conditions. A formal contract was signed on, and dated, 1 December 1998. The formal contract superseded the letter of intent and specified that "Notwithstanding the date of execution of this Agreement, the Agreement shall take effect from 11 August 1997". The UK Act provides that disputes under construction contracts entered into after 1 May 1998 may be referred to adjudication. A dispute arose when Lowry deducted liquidated damages from contractual sums due to Christiani. Lowry argued against the right to adjudicate the dispute because the contract was to take effect from 11 August 1997, thus predating the UK Act. The adjudicator disagreed, deciding that the dispute could be referred to adjudication, and he therefore had jurisdiction to decide the dispute. Judge Thornton agreed with the adjudicator, holding that the letter of intent was clearly intended to have a limited life and the contract aimed to ensure it covered work undertaken before its execution. The judge held that the "Notwithstanding" provision was an attempt to avoid the UK Act's operation and was ineffective as the contract was actually entered into on 1 December 1998, after the operative date of the UK Act. Consequently, the judge held that either party was entitled to refer a dispute to adjudication. Lowry also argued that even if the UK Act applied, Christiani was estopped from referring the dispute to adjudication as it waived its entitlement to rely upon the Act by agreeing to the clause that the contract would take effect from 11 August 1997 which predated the UK Act's operative

3 Christiani & Nielsen Limited v Lowry Centre Development Company Limited [2000] HT 001/59 (TCC), 26 June 2000.

date. However, this argument hadn't been advanced before the adjudicator. Consequently, Judge Thornton held that it would be unfair to allow Lowry to rely upon the ground, who, by failing to raise this argument previously, waived the right to argue it before the court. Judge Thornton emphasised that, in any event, an estoppel was not raised because such a clause depriving Christiani of its statutory right to adjudicate would be invalid as parties cannot contract out of the UK Act.

3.6. The key date, therefore, is the date on which the contract is entered into and not the date on which it is actually executed. In the construction industry it is common for contracts to be executed on a date following the date on which the parties actually entered into agreement. Under the Act, the date on which the contract is entered into is the one which determines whether any payment dispute thereunder may be referred to adjudication. The date on which the contract was actually entered into may be evidenced by an exchange of emails or correspondence or by the contractor's commencement of the works on site.

3.7. In *Atlas v Crowngate*,[4] Crowngate issued a letter of intent to Atlas on 18 December 1997 authorising expenditure up to £1.3m. The letter stated that if a contract was signed it would have retrospective effect and include works executed under the letter. On 3 April 1998 the parties signed a UK standard-form subcontract. However, there were material matters unfinished between the parties when signing the contract, including the extent of the work scope and a finalised accurate bill of quantities. Thus, Crowngate issued a further letter of intent on 15 May 1998 increasing the authorised expenditure but dating it the same as the first letter despite its actual date. One year later, in final account discussions, the parties agreed to execute the subcontract without delay and a signed version was sent to Crowngate on 12 April 1999. Crowngate argued that the retrospective nature of the contract meant that the right to adjudicate could not be implied. However, Judge Thornton held that while a contract may include a provision so as to operate with retrospective effect, that provision does not affect the date on which the contract was entered into. The judge concluded that the activities, actions, expressions and statements in documentation emanating from Crowngate clearly pointed away from there being in the minds of the parties a concluded contract on or after 3 April 1998. The judge held that the second letter was inconsistent with Crowngate's position for a contract being formed before 15 May 1998 and that the contract

4 Atlas Ceiling & Partition v Crowngate Estates (Cheltenham) [2000] CILL 1639 (TCC).

came into existence on 12 April 1999 and was therefore subject to adjudication.

3.8. Consequently, a distinction must be drawn between the date on which the parties enter into a contract and a provision within that contract which states that the contract is to have retrospective effect in respect of any work carried out prior to that date. Under the Act where a contract is entered into after the operative date of 25 July 2016, even if the contract purports to have retrospective effect, then (based on the *Christiani* and *Atlas* principles) either party may refer a payment dispute to adjudication.

3.9. In *Earls v Waterloo*,[5] Judge Seymour QC held that where parties entered into a construction contract before the operative date, but varied the contract thereafter, the variation wouldn't usually bring the contract within the scope of the UK Act unless the variation by itself can be construed as a construction contract. The parties entered into an agreement dated 4 December 1996 but agreed (solely for the purpose of the court application) that the agreement came within the definition of a construction contract under the UK Act. Both parties also agreed that the UK Act did not apply to the agreement as it pre-dated the operative date. The agreement was subsequently amended by a deed of variation on 20 July 1998, after the operative date of the UK Act. Earls sought a court declaration that the agreement, amended by the subsequent deed, was not a construction contract within that definition under the UK Act but merely amended the fee payable to Waterloo and deleted a sub-clause in the principal agreement. Earls accepted that while the deed was signed after the operative date, the variation itself did not amount to a construction contract. The main issue before the court was whether the 20 July 1998 deed (not in itself a construction contract) which varied the terms of the 4 December 1996 contract had the effect to bring the entirety of the two agreements within the scope of the UK Act. Judge Seymour held that the deed simply modified the fee provisions and so did not bring the principal agreement within the UK Act. He commented, however, that it is possible that a variation (which post-dated the operative date) to a construction contract (which pre-dated the operative date) could constitute a construction contract where the entirety of the varied agreement came within the scope of the UK Act. As the deed in itself in this case did not amount to a construction contract, the judge held that the adjudicator's decision related to a construction contract entered into before the operative date. Consequently, the adjudicator had no jurisdiction to act and accordingly his subsequent decision was a nullity.

5 Earls Terrace Properties v Waterloo Investments [2002] HT 02/237 (TCC).

3.10. In *Yarm v Costain*,[6] Judge Havery held that a novated contract constitutes a new agreement. Consequently, where the original contract is entered into before the operative date, if the novated agreement is entered into after the operative date, it comes within the UK Act and disputes arising may be referred to adjudication. Costain, the main contractor, entered into a subcontract with a company called CSE in August 1995. CSE subsequently changed its name and thereafter entered into a novation agreement on 14 August 1998 novating the subcontract to Yarm. In contemplation of the novation agreement, Costain and the CSE newly named entity concluded a supplemental agreement on 10 August 1998 which was designed to vary the subcontract provisions relating to the calculation and reimbursement of the subcontract price. Costain contended that if the novation agreement was found to constitute a construction contract the UK Act would have retrospective effect on the 1995 agreement and this could not have been the legislature's intention. Costain also argued that the novation agreement did not make a new provision for undertaking construction operations since the same operations were already provided by the original subcontract, and that if the novation agreement was deemed a construction contract Yarm would have the benefit of referring any dispute to adjudication going as far back as 10 August 1995. Despite this argument, Judge Havery held that the UK Act's wording was clear, and the novation agreement fell within the construction contract definition. The judge emphasised, however, that the court was not to be taken to have decided that the UK Act had retrospective effect.

3.11. The *Yarm* decision is authority in the UK for the principle that an agreement novating rights and obligations under a construction contract constitutes, in its own right, a construction contract under the UK Act and that the contracting parties may therefore refer disputes to adjudication, provided that the novation agreement took effect after the UK Act's operative date. Irish courts will likely conclude that while the Act itself has no retrospective effect, a novation agreement (in itself being a construction contract) has a retrospective effect on the parties' rights and therefore any payment dispute under a construction contract entered into before the operative date of 25 July 2016 but novated thereafter can be referred to adjudication.

The agreement can be oral or written

3.12. Pursuant to Section 1(1) construction contracts do not have to be in writing in order to come within the scope of the Act. They may also be

6 Yarm Road Limited v Costain Limited [2001] HT 01228 (TCC).

oral agreements and be encompassed by the Act. It is largely irrelevant whether the parties actually formally sign or execute the contract. While Section 1(1) refers to the "executing party" and "another party", the former term should not be construed literally as an executed contract need not exist for the parties to have entered into a construction contract for the purposes of the Act. Provided the existence of the agreement can be demonstrated, and the agreement constitutes a "construction contract" under Section 1(1), either party under that agreement may refer any payment dispute to adjudication.

Executing party

3.13. The term "executing party" is defined in Section 1(1) as meaning "the contractor" where the parties to the construction contract are a contractor and the person for whom the contractor is doing work under the contract, or "the subcontractor" where the parties to the construction contract are a contractor and a subcontractor, or two subcontractors. The term "work" is defined as any act done in furtherance of a construction contract under the terms of the construction contract.

3.14. Section 1(1) of the Act is very similar to Section 104(1) of the UK Act, save in one significant respect. The UK Act defines a "construction contract" as "an agreement with a person" whereas the Irish Act defines a "construction contract" as an agreement between "an executing party and another party, where the executing party is engaged for any one or more of the following activities", and then goes on to list similar activities as are listed in the UK Act.

3.15. The term "executing party" is particularly pertinent to the payment provisions under Sections 3 to 5 of the Act and this is the likely reason why the Act diverges from the UK Act in this aspect of the definition of a construction contract. However, the term's use may have inadvertently created confusion in the context of consultants' collateral warranties and professional appointments.

3.16. For example, in the case of a collateral warranty between an architect and a funder, the funder would not be an "executing party" as that term is defined in Section 1(1) of the Act. Similarly, in a traditional build arrangement where the architect is engaged by the employer the architect is neither a contractor nor a subcontractor and as such would not be an "executing party". However, in a design-build arrangement, where an architect is engaged by a contractor, it is arguable the architect would come within the scope of the definition of a "subcontractor" and therefore within the definition of an "executing party". This is because a "subcontractor" is defined as "a person to whom the execution of work under a construction contract is subcontracted by the contractor or another subcontractor" and in a design-build arrangement the contractor would be subcontracting the design work to its architect. In the traditional build arrangement, where neither the architect nor the funder is an executing party, each of them could only therefore be construed as "another party" under the definition of a "construction contract", and consequently, on this interpretation, the architect-funder collateral warranty would not fall within the definition of a construction contract. This anomaly, where an architect-funder collateral warranty would constitute a construction contract in a design-build arrangement but not under a traditional build arrangement, may be an unintended consequence of the use of the term "executing party" for the purposes of operation of the payment provisions in Sections 3, 4 and 5 of the Act. This type of anomaly is unlikely to have been anticipated by the legislature and the legislative intention is more likely to be that the definition of "executing party" is only relevant for the purposes of the operation of the payment provisions under the Act.

3.17. It will be interesting to observe whether the Irish Courts adopt a wide or narrow interpretation as regards the parties to a construction contract. The Scottish Courts have adopted a wide approach and in the *Gillies Ramsay*[7] case held that "adjudication ... was intended to cover the whole range of persons involved ... in order that disputes could be resolved on an interim basis". If the Irish courts adopt a similar approach then they may not feel restricted by the statutory terms "executing party and another party" in construing the parties to a construction contract. However, if the Irish Courts were to adopt a strict interpretation as to what constitutes a construction contract in terms of an "executing party" one could certainly envisage difficulties arising in relation to some of the ancillary contractual arrangements that are often entered into, including collateral warranties, performance bonds and parent company guarantees.

Arranging for carrying out construction operations

3.18. Section 1(1) relates to agreements, by an executing party and another, for carrying out, arranging for and providing labour for, construction operations.

3.19. Section 1(1)(b) provides that the definition of a construction contract includes arranging for the carrying out of construction operations, whether such arrangements are made under a subcontract with the executing party or otherwise. The Act does not explain as to what constitutes "arranging for the carrying out of construction operations". However, such wording would include management

7 Gillies Ramsay Diamond v PJW Enterprises [2002] Scots 340.

contracts or construction manager contracts, where the managing contractor or construction manager does not carry out the work itself but arranges for other contractors to carry out the work.

3.20. As construction procurement frameworks and contracts continue to evolve, the broadly worded Section 1(1)(b) will serve to ensure that irrespective of the construction procurement approach the majority of contracts within the varying contractual frameworks will nonetheless come within the scope of the meaning of a construction contract. The fragmented nature of contracting in the construction industry is addressed in Part 2 and concludes that multiple forms of contracting are likely to come within the scope of a construction contract as defined under the Act.

3.21. While Section 1(2)(b) provides that project management contracts are encompassed by the Act, it is arguable that Section 1(1)(b), relating to construction arrangements, would also extend to include project management contracts. Indeed, in *Gillies Ramsay v PJW Enterprises*,[8] the court held that a surveyor's agreement "administering the contract" was part of the function of "arranging for others to do work".[9] PJW, who had signed a building refurbishment contract with R&R Construction, employed Gillies Ramsay as the contract administrator. During the works there were five adjudications, resulting in PJW's liability to the contractor for additional payments. PJW alleged that its liability had been caused by Gillies Ramsay's negligence and breach of contract. PJW subsequently employed others in Gillies Ramsay's place, referring a claim to adjudication for professional negligence against Gillies Ramsay. The adjudicator found in PJW's favour and awarded damages. Gillies Ramsay refused to pay contending that the adjudicator had no power to award damages on the basis that a contract administrator appointment is not a construction contract under the UK Act. In delivering her decision Lady Paton, in the Outer House of the Scottish Court of Sessions, made some informative comments relating to what is meant by the phrase "arranging for the carrying out of construction operations by others" as provided for under Section 104(1)(b) of the UK Act. Lady Paton commented that:

> arranging covered a broad range of activities, including securing the proper administration of the contract … and ensuring that payments were made so that work would continue.

3.22. As regards the contract administrator role, Lady Paton stated:

> the purpose of the adjudication scheme was to maintain progress under contracts, and to keep cash flowing, a contract administrator's

8 Gillies Ramsay Diamond v PJW Enterprises [2002] Scots 340.

9 Paragraph 24 of the judgment.

remit encompassed both elements. It was irrelevant that the … administrator was the man in the middle who made decisions about parties' rights under the contract … the … scheme was intended to cover the whole range of persons involved, from sub-subcontractor to employer to the professional, in order that disputes could be resolved on an interim basis to allow the contract to proceed.[10]

3.23. The Irish courts will likely adopt a similar approach to Lady Paton's wide interpretation as to what services fall within the term "arranging for the carrying out of construction operations" and should they do so it would represent a recognition of one of the Act's primary purposes to facilitate a smooth and continuous payment operation under construction contracts.

Construction operations

3.24. The common phrase in the three categories (a) to (c) within the definition of construction contract is "construction operations", which is defined under Section 1(1) (subject to Sections 1(3) and (4)) as "any activity associated with construction, including operations of any one or more" of the descriptions set out in Sections 1(1)(a) to (g).

3.25. There are several issues to consider concerning the definition of construction operations, including: What is meant by forming part of the land? What operations come within the definition of cleaning and restoration? What operations would be considered integral, preparatory or rendering complete other construction operations?

Forming part of the land

3.26. Sections 1(1)(a) to (c) of the definition of construction operations relate to works to buildings and structures "forming, or to form, part of the land". Consequently, if a building or structure is not, or will not become, part of the land then it will not come within the definition of construction operations at Sections 1(1)(a) to (c) nor within the scope of the Act.

3.27. The phrase "forming, or to form, part of the land" has been extensively considered and adjudicated on in the context of land law and, in particular, as to whether a structure represents a fixture or a fitting. Land includes things attached to the land such as buildings and other permanent structures.[11] Consequently, in seeking to ascertain whether a construction operation relates to a structure, building

10 At paragraph 26 of the judgment.

11 Garner "Meaning of Land in Administrative Statutes" (1957) 21 Conv. 141.

or works forming, or to form, part of the land, the key words are "attached" and "permanent". In simple terms, if the structure, building or works are, or will be, "attached" to the land and are "permanent", then the activity will, ostensibly, come within the definition of construction operations.

3.28. In *Gibson v Makro*,[12] Gibson was engaged under four contracts to install shop fittings for Makro, including moveable furniture, business counter islands and column cladding fixed to the walls by screw fittings, primarily for the purposes of stability. Makro argued that the fittings that Gibson had supplied and installed were not fixtures and consequently did not form part of the land, and the works were therefore construction operations. HHJ Seymour QC agreed and held that the shop fittings installation did not amount to "construction operations" unless they were structures. The judge held that the reference to "forming or to form part of the land" imported the real property law concepts and tests relating to fixtures and, as such, a factor relevant to the determination of whether an item formed part of the land is whether the attachment is to be permanent or not. Significantly, Judge Seymour held that the reference in Section 105(1)(a) (the UK Act equivalent of Section 1(1)(a) of the Act) to "whether permanent or not" related to the buildings or structures themselves as opposed to the works being done to those buildings or structures.

3.29. In *Potton v Thompson*,[13] the claimant was a distributor for a company which produced individual pods containing a bedroom with an en-suite bathroom. The pods were transported intact and placed on large concrete bases, to which they were affixed using a relatively small fixing. The purchaser then simply had to connect electricity and plumbing to the pod making them ready for occupation. The respondent purchaser failed to keep up lease payments on the pods. On becoming aware that the respondent's pub (which contained the pods) was up for sale, the claimant sought to assert that the pods were chattels and could not form part of the property sale. The issue before the court was whether the pods, not being firmly affixed to the land, but in all other respects apparently permanent, were fixtures or chattels. The court held that the "pods" were not part of the land as the pod units were "clearly designed for both delivery and removal entire … can be removed without the units in any way being destroyed".

3.30. Consequently, contracts involving the installation of pod units are likely, prima facie, not to fall within the definition of a construction contract unless the underlying works on those contracts relate to

12 Gibson Lea Retail Interiors v Makro Self Service Wholesalers Limited [2001] HT 01 226 (TCC).

13 Potton Developments Limited v Thompson and Another [1998] NPC 49, ChD.

structures "forming or to form part of the land". Most commercial fit-out contracts will involve substantial works relating to floors, ceilings, partitions and other permanent-type works. Such contracts will unquestionably involve works forming or to form part of the land, constituting construction operations. However, if a fit-out contract merely involves works to install shop fittings which can be removed without causing any damage to the building in which they are fitted, and are therefore not permanent, the fit-out contract will not come within the constructions operations definition and not constitute a construction contract. Similarly, if pod units can be removed entirely without the pod, or the building, being damaged, then the pods would not be construed as fixtures of a permanent nature. In those circumstances, the pods would not form part of the land and the works in respect of the installation of those pod units would not constitute construction operations.

3.31. In *Staveley v Odebrecht*,[14] the phrase "forming, or to form, part of the land" was considered in the context of an offshore structure. Odebrecht subcontracted Staveley to supply and install specific equipment into steel modules which were being constructed in England. The modules were for use as living quarters for operatives on an oil platform in the Gulf of Mexico. The modules were to be towed to the location and welded onto platforms which were to be supported by legs founded in the seabed. A dispute arose and Staveley, in seeking to refer the dispute to adjudication, applied for a declaration from the court that the four subcontracts were construction contracts within the meaning of the UK Act. There was no material difference between the subcontracts which were for design, engineering, procurement, supply, delivery to site, installation, testing and commission of instrumentation, fire and gas, electrical and telecommunications equipment. The equipment Staveley designed and supplied was for installation in the steel modules. Staveley argued that as the modules were to be attached to oil platforms they would form part of the land, and thus the modules came within the construction operations definition (under Section 105(1)(c) of the UK Act) as they were for "installation in any building or structure or fittings forming part of the land". Staveley argued that as the UK Interpretation Act 1978 defined land as including land covered by water, and the platform's feet were founded in the seabed, the platforms constituted land. Odebrecht argued that the platforms were not land and that the reference to "the land" in the UK Act referred to the land on which the relevant construction operation was carried out. They argued that as the modules were to be moved they could not form part of the land in England and as the UK Act only applied to construction operations in England, Wales or

14 Staveley Industries v Odebrecht Oil and Gas Services [2001] All ER (d) 359.

Scotland the modules could not form part of those lands to which the UK Act applied. Judge Havery QC held that the expression "forming part of the land" requires the fittings to form part of the land at least prospectively, if not immediately upon their installation. Judge Havery held that the absence of an express inclusion of offshore platforms in the UK Act suggested an intention to exclude them from the ambit of the UK Act. Furthermore, the judge held that on its true construction the expression "the land" referred only to the land where the building or structure forming part of the land was situated when built, and that land had to be in England, Wales or Scotland. Consequently, the judge held that the modules could not form part of the land in England and thus the UK Act did not apply to the subcontracts.

3.32. Accordingly, oil platforms and oil rigs are structures founded on the seabed below the watermark and not structures forming, or to form, part of land for the purposes of the definition of construction operations under Sections 1(1)(a), (b) or (c) of the Act. Similarly, an offshore wind farm would not come within the definition of construction operations as it clearly would not form part of the land.

3.33. The situation is different where a structure founded on the seabed below the watermark is also in some way attached to the land. Section 1(1)(b) refers to the construction, alteration, repair, maintenance, extension, demolition or dismantling of "docks and harbours" and "coast protection or defence" which could encompass works constructed on the seabed but nonetheless in some manner attached to the land. The determining issue is whether the works constructed on the seabed are also attached in any way to the land. In most cases docks and harbours are in some way attached to the land, thereby forming part of the land, despite the fact that they may be within the sea, and off the shoreline. In addition, it is undoubtedly the case that the same could be said of coastal protection works. However, it is unlikely that a pontoon bridge or floating harbour would be considered as forming part of the land, as they are floating structures not attached to the land, and works in respect of them would fall outside the scope of the Act.

3.34. Section 1(1)(b) of the definition of "construction operations" also includes "pipe-lines" as construction operations. Clearly, a pipe-line running under the land from one point to another would form part of the land, be construed as a construction operation, and a contract in respect of work on the pipe-line would come within the definition of a construction contract under the Act. Similarly, where a pipe-line is on the seabed running from one coastline to another, below the watermark, if the pipe-line is attached to the land at either end, work on that pipe-line would constitute a construction operation. It will be interesting to observe how the courts construe such operations if and when contracts of this nature come before them. Each case will no doubt be considered on its own merits. Where there is uncertainty as to whether a specific operation constitutes a construction operation or not, the answer to the question as to whether it forms part of the land will be significant.

External and internal cleaning of a building, restoration and maintenance

3.35. The definition of construction operations under Section 1(1) also includes, at Section (d), "external and internal cleaning of buildings and structures" operations provided the cleaning is "carried out during the course of the construction, alteration, repair, extension or restoration of the building or structure".

3.36. Site clean-up at the completion of a contract, before handover, would therefore come within Section 1(1)(d) as it is the external and internal cleaning of a building during construction, albeit at the end of the construction period. If the cleaning of a building's external façade is the sole operation being carried out on the building it would not be construed as an operation being carried out "during the course of construction, alteration, repair, extension" of the building. However, depending on the precise nature of the façade cleaning operation, and the building, the cleaning operation in its own right might constitute restoration of the building, thus falling within Section 1(1)(d).

3.37. The regular internal cleaning of an office building could be construed as its upkeep as opposed to restoration, and therefore not as a construction operation. However, one could certainly envisage a fine line of distinction being drawn as to whether the external (or internal) cleaning of a building constitutes in itself restoration and therefore construction operations or the upkeep of the building. To be construed as restoration, the cleaning operation would have to be of such nature so as to constitute returning the building to a former condition, as opposed to the upkeep of the building. The building would first need to lapse into disrepair before the cleaning of the building could constitute restoration. Depending on the nature of the operation, it may also be arguable that cleaning is building maintenance, thus coming within the construction operations definition under Section (1)(1)(a).

Operations integral or preparatory

3.38. Section 1(1)(e) contains broad language seeking to encompass a range of ancillary activities as construction operations, from enabling works to work rendering complete the operations previously described in Sections 1(1)(a) to (d). Section 1(1)(e) also provides that operations which form an integral part of, or are preparatory, to construction operations, are to be construed in their own right as construction operations. Section 1(1)(e) also includes, for example, the "erection, maintenance or dismantling of scaffolding", an issue which has given rise to interesting debate in the interpretation of a similar provision in the UK Act.

3.39. In *Palmers v ABB,*[15] HHJ Thornton QC analysed Section 105(1)(e) of the UK Act, which is worded similarly to Section 1(1)(e) of the Act. ABB were assembling and erecting a heat recovery steam generator boiler for the plant in connection with a co-generation project contract and subcontracted Palmers for scaffolding work. Palmers provided scaffolding for several parts of the overall project including for ABB's works on the boiler. In July 1999 ABB alleged that Palmers breached its contract by failing to supply appropriately qualified personnel in adequate numbers, thereby delaying ABB's works. ABB sought to exercise its set-off rights by issuing notices to withhold payment against monies contractually due to Palmers. Palmers issued a notice of adjudication claiming non-payment of interim invoices. ABB responded to the notice arguing that the contract with Palmers was not a construction contract or operation pursuant to the UK Act, as the boiler would not form part of the land, and hence scaffolding erected for the purposes of installing the boiler could not be a construction operation. Palmers issued court proceedings seeking declarations that the scaffolding operation was a construction operation and an adjudicator would have jurisdiction. ABB argued that its works were not "construction operations" as the boiler, and its supporting structures and pipework, were either not works forming part of the land or were excluded operations under Section 105(2)(c) of the UK Act, as they related to a power generation plant. The parties agreed that ABB's contract with the main contractor was not a "construction contract" since it fell within the exclusion in Section 105(2)(c) of the UK Act, which provides that certain operations are not construction operations if they constitute, *inter alia,* the installation of plant or machinery on a site where the primary activity is power generation. No equivalent provision exists under the Act. However, the *Palmers* principle is nonetheless relevant in the context of Section 1(3) of the Act which excludes certain construction operations relating to the manufacture or delivery to a construction site of, *inter alia*, materials, plant or machinery or components for the systems of power supply.

3.40. HHJ Thornton QC held that it was possible for some work on a site to fall within exclusions and for other work not to be excluded. Therefore, some subcontractors might be able to seek adjudication where the main contractor is not. In this case, Palmers' scaffolding works constituted "construction operations" and therefore their

15 Palmers Limited v ABB Power Construction Limited [1999] HT 99 000090 (TCC).

subcontract with ABB was a "construction contract". However, Judge Thornton stated that there may be some operations that fall within the definition of construction operations and yet are excluded pursuant to Section 105(2) of the UK Act. The judge held that the word "described", in Section 105(1)(e) of the UK Act (operations which are preparatory to operations as previously described in that section), rather than "defined" meant that the works related to Section 105(1) of the UK Act even if the related operations were excluded operations.

3.41. Therefore, under Section 1(1)(e) of the Act an operation that is preparatory to an excluded process would nonetheless be subject to the Act if the preparatory operation is itself a construction operation. It is therefore possible for a contract between an employer and a contractor to fall within the meaning of construction operations but be excluded by Section 1(3) of the Act while a contract between the contractor and a subcontractor providing, for example, scaffolding services to the contractor may come within the definition of a construction contract. Consequently, some subcontractors will be entitled to refer payment disputes to adjudication while their main contractor may not be able to seek adjudication against the employer because its contract relates to construction operations excluded by Section 1(3) of the Act.

3.42. Under Section 1(3), operations involving manufacture or delivery of specific items and not installation, do not constitute construction operations, including the "manufacture or delivery to … site of … components for … heating, lighting, air-conditioning, ventilation, power supply".

3.43. In considering whether an ancillary activity comes within Section 1(1)(e) the first step is to determine whether the primary activity, to which the ancillary activity relates, might be described as a construction operation, notwithstanding that it may be an excluded operation. If the primary activity is a construction operation (albeit excluded) then the next step is to determine whether the ancillary activity itself forms an integral part of, is preparatory to, or is for the purpose of rendering complete the primary activity. Whether an activity forms "an integral part of" the construction operation depends on the specific circumstances. Section 1(1)(e) seeks to provide examples of the type of works which form an integral part of, or are preparatory to, or are for rendering complete construction operations, including "site clearance, earth moving, excavation, tunnelling and boring, laying of foundations, erection, maintenance or dismantling of scaffolding, site restoration, landscaping and the provision of roadways and other access works and traffic management". Where any of these aforementioned ancillary activities are carried out in relation to a primary activity which comes within the definition of construction operations under the Act, then the contract under which the ancillary activity is being carried out will be a construction contract, and come within the scope of the Act, irrespective of whether the primary activity is an excluded construction operation or not.

Rendering complete

3.44. Section 1(1)(e) also encompasses operations for "rendering complete" construction operations as those previously described in Sections 1(1)(a) to (c) of the Act. Operations which are carried out "for rendering complete" qualifying construction operations might not be immediately apparent as constituting a construction contract. For example, a contract for commissioning services on a pharmaceutical plant would fall within Section 1(1)(e) as it is an operation required to render complete the construction of the plant. So too would a traffic management services contract relating to motorway repair or construction works. Consequently, whilst on first consideration a particular contract for ancillary works or services might not immediately appear to fall within the definition of construction operations, and therefore not constitute a construction contract under the Act, on closer scrutiny the particular works or services could be construed as construction operations within the scope of the Act.

Painting and decorating

3.45. Section 1(1)(f) provides that "painting or decorating the internal or external surfaces of any building or structure" constitutes construction operations and therefore any contract in respect of the same would be considered a construction contract. The fact that the painting or decorating may relate to any "structure" significantly broadens the potential scope of this provision where contracts may involve the painting of a monument or other structure which might not immediately be considered as a structure to which the Act applies. Each case will undoubtedly be considered on the specific nature of the building or structure in respect of which the painting or decorating works have been carried out. While it may be straightforward determining what constitutes a "building", one could envisage significant debate as to what constitutes a "structure".

Artistic works

3.46. Section 1(1)(g) of the Act provides that "making, installing or repairing sculptures, murals and other artistic works that are attached to real property" constitutes construction operations. No equivalent provision exists under Section 105(1) of the UK Act, which may be because these activities would in any event come within the scope of

Section 105(1)(a) of the UK Act, a provision equivalent to Section 1(1)(a) of the Act. This is because if the sculpture, mural or other artistic work is attached to the land it would in itself be the construction of a structure forming part of the land and as such come within the scope of Section 1(1)(a).

3.47. However, it is unusual that the Act encompasses "murals" as in the normal course of works one would certainly not generally view the painting of murals as construction operations. In addition, a mural would have to be painted onto a structure of some description, normally a wall, and consequently the painting of the mural would likely fall within the scope of Section 1(1)(f) as it would be the painting of a structure, albeit a mural painting on the structure.

Consultant appointments

The definition of a construction contract is further broadened to include contracts and appointments for professional services in the industry. Section 1(2) provides that:

> In this Act references to a construction contract include an agreement, in relation to construction operations, to do work or provide services ancillary to the construction contract such as –
>
> (a) architectural, design, archaeological or surveying work,
> (b) engineering or project management services, or
> (c) advice on building, engineering, interior or exterior decoration or on the laying-out of landscape.

3.48. Section 1(2) is similar to Section 104(2) of the UK Act, save that the latter does not include project management services. This may be because project management services likely come within the scope of "arranging for the carrying out of construction operations" and as such fall within the construction contract definition. In addition, Section 1(2)(a) provides for archaeological work as constituting a construction contract, a provision which is not included in the UK Act, and there is no explanation as to why the Irish legislature decided to include such work.

3.49. For a professional services appointment to come within the definition of a construction contract, pursuant to Section 1(2), it must (a) relate to construction operations, and (b) be an agreement to do work or provide services ancillary to the construction contract. Therefore, if a professional services agreement involves services ancillary to a construction contract, but those services do not relate to construction operations then the professional services agreement would not come within the scope of the definition of a construction contract pursuant to Section 1(2). Consequently, the Irish Courts are likely to only consider a consultant's appointment as a construction contract where the services being carried out are services associated with actual construction.

3.50. In *Fence Gate v Knowles,*[16] HHJ Gilliland QC held that acting as an expert and giving evidence at a hearing was not an activity "in relation to construction contracts", but rather was advice in relation to the arbitration proceedings arising out of a dispute under a construction contract. Fence Gate appointed Knowles under a contract to provide architectural and surveying services described as a "preliminary report on defective kitchen floor". Fence Gate also engaged Knowles to provide evidence as a witness of fact and assisting in an arbitration as an architect or engineer. Knowles claimed payment under four invoices in respect of their services under the contract. Following a dispute regarding the invoices the matter was referred to adjudication. The adjudication did not relate to any invoices for the preliminary report work on the defective kitchen floor but rather invoices for assisting in the arbitration. The adjudicator awarded Knowles payment under three invoices, but Fence Gate sought a declaration that the adjudicator had no jurisdiction to make any award in respect of those invoices under the contract, contending that it was not a construction contract under the UK Act. Judge Gilliland referred to the two limbs of Section 104(2) of the UK Act, stating that "the substantive questions are whether providing evidence as a witness of fact ... and assisting in an arbitration ... are matters which fall within the definition of a construction contract" within Section 104(2) of the UK Act. The professional services/work must fall within subsection (a) "architectural, design, or surveying work" or (b) "provide advice on building, engineering, interior or exterior decoration", and secondly, it must be work "in relation to construction operations". In an Irish context, Sections 104(2)(a) and (b) of the UK Act are the same as Sections 1(2)(a) and (c) of the Act. Judge Gilliland held that Knowles' services were not themselves "construction operations" but this did not mean that these activities were not carried out "in relation to construction operations". Knowles submitted that the giving of factual evidence as an architect/surveyor was included under Section 104(2)(a) as "doing architectural, design or surveying work" and the assisting at the arbitration was the provision of advice "on building" or "engineering" and the arbitration itself concerned construction operations. Judge Gilliland disagreed with Knowles' submissions and found that the giving of factual evidence was a different and distinct activity

16 Fence Gate Limited v J.R. Knowles Limited [2001] 25/01 SF102200 (TCC).

from the performance of the architectural, design or surveying work and not incidental to it. The judge also held that disputes relating to litigation support work were not disputes "in relation to construction operations". Judge Gilliland also found that Section 104(5) of the UK Act, which is similar to Section 2(4) of the Act, applied. Section 2(4) of the Act provides that where an agreement relates to construction operations and also relates to other matters which are not construction operations, then the Act can apply to activities carried out under the agreement between the parties only so far as those activities relate to a construction contract.[17] On this basis the Act (and thereby adjudication) only applies to that part of the contract relating to construction operations. Consequently, Judge Gilliland held that the adjudicator did not have jurisdiction to rule on Knowles' entitlements for their services rendered as a witness of fact and assisting in the arbitration and that *Fence Gate* was entitled to the declaration sought that the adjudicator had no jurisdiction. The *Fence Gate* decision supports the proposition that disputes relating to payment for appearing as a witness of fact and for assisting at an arbitration are not disputes relating to a construction contract and do not come within the Act.

Collateral warranties

3.51. Professional appointments are encompassed by the Act under Section 1(2). However, the Act does not specify as to whether it also encompasses collateral warranties entered into between the relevant professionals (architects, engineers and other consultants) and parties who are invariably not the party directly engaging the relevant construction professional. There is no definitive position under the Act in relation to the question as to whether a collateral warranty constitutes a construction contract within the meaning of that definition under the Act.

3.52. In *Parkwood v Laing*,[18] Parkwood provided facilities management services on several projects including the Cardiff International swimming pool, which Cardiff City Council owned but had leased to Orion Land and Leisure (Cardiff) Limited for 25 years commencing on 25 April 2006. Orion sub-let the facility to Parkwood for 10 years under a sub-lease dated 11 January 2008. Orion engaged Laing under a standard form design and build contract. Under Article 10 of

17 Section 2(4) of the Irish Act is similar to Section 104(5) of the UK Act insofar as it provides that where a contract relates to both construction operations and other activities the contract was to be treated as severable between those parts.

18 Parkwood Leisure v Laing O'Rourke Wales & West [2013] EWHC 2665 (2CC).

the contract Laing were required to enter into a "Deed of Warranty" with any financier of the project, first purchaser and mortgagee. Laing did this and executed a collateral warranty in favour of Parkwood, the tenant, who was named as the beneficiary. The works were completed in 2008 and the facility opened to the public. Defects arose in July 2013 and Parkwood issued court proceedings seeking a declaration that the warranty was a construction contract for the purposes of the UK Act "on the basis that it contains [Laing O'Rourke's] express agreement to carry out construction work". In considering the precise wording of the warranty, HHJ Akenhead had no doubt that it was to be treated as a construction contract "for ... the carrying out of construction operations". Judge Akenhead noted that the recital to the warranty set out that the underlying construction contract was "for the design, carrying out and completion of the construction of a pool development" and that Clause 1 of the warranty related expressly to carrying out and completing the works. Judge Akenhead concluded that the collateral warranty was a construction contract for the carrying out of construction operations on the basis that:

(a) The recitals and a clause in the warranty provided that the underlying construction contract was for the design, carrying out and completion of a development.
(b) A clause in the warranty contained the express wording that the contractor "warrants, acknowledges and undertakes", to carry out the works.
(c) A clause in the warranty set out Parkwood's remedies in the event the contractor breached its warranties and undertakings, and such remedies included the reasonable cost of repair, renewal and/or reinstatement. It also provided that where the contractor breached these warranties it would be liable for losses incurred by the beneficiary.
(d) While the warranty was agreed toward the end of the construction period it did not solely constitute a guarantee as to a past state of affairs but also in respect of the carrying out and completion of the works that remained to be carried out under the contract at the time the warranty was entered into.

3.53. Judge Akenhead held that the warranty related to both the past and future, stating that "the fact that the warranty was retrospective in effect was not a bar to it being a construction contract. It was common for construction contracts to be finalised after the works had started". In the judge's view, the warranty recognised that the contractual works remained to be completed albeit Laing had already carried out a significant part of the works and the design. The undertaking in Clause 1 primarily related to the execution and completion of the remaining works. The warranty related to the work and design that had already been carried out or provided and to that work and design yet to be carried out and provided. The judge found that whilst Laing's undertaking in Clause 1 was to carry out and complete the works in accordance with the underlying contract, the undertaking that was being given to Parkwood (that in the execution and completion of the works it would comply with the contract) related to the quality and completeness of the works.

3.54. While Judge Akenhead held that in this particular case the collateral warranty was a "construction contract" for the purposes of the UK Act he specifically stated that:[19]

> It does not follow from the above that all collateral warranties given in connection with all construction developments will be construction contracts under the Act. One needs primarily to determine in the light of the wording of the relevant factual background each such warranty to see whether, properly construed, it is such a construction contract for the carrying out of construction operations. A very strong pointer to that end will be whether or not the relevant Contractor is undertaking to the beneficiary of the warranty to carry out such operations. A pointer against may be that all the works are completed and that the contractor is simply warranting a past state of affairs as reaching a certain level, quality or standard.

3.55. Judge Akenhead's judgment in *Parkwood* was clearly decided on the specific facts of the case and the precise wording of the collateral warranty. Consequently, the question as to whether a collateral warranty comes within the scope of the definition of a construction contract under the Irish Act will depend on the precise wording of the warranty in question and, where applicable, the timing as to when the warranty is furnished to the beneficiary.

3.56. The use of the term "executing party" in the Act creates further uncertainty as to in what circumstances a collateral warranty might not constitute a construction contract, thus not coming within the scope of the Act. This is because collateral warranties are generally furnished by contractors, consultants and subcontractors (amongst others) carrying out work (or services) on a construction project, in favour of parties for whom they are not actually carrying out the

19 At paragraph 27 of the judgment.

work. The term "work" in the context of the "executing party" definition further accentuates the problem. The fact that the term "work" is defined in terms of "any act done in furtherance of the construction contract" further brings into question whether collateral warranties fall within the definition of a construction contract. A beneficiary of a collateral warranty, while unlikely to be an "executing party", could be categorised as "another party" (in the contest of the construction contract definition under Section 1(1) of the Act). The question then arises, however, as to whether the party furnishing the collateral warranty comes within the definition of an "executing party". In circumstances, for example, where the collateral warranty is being furnished by a contractor to a funder (as the beneficiary) the contractor would be the "executing party" as it is clearly doing work under a contract. As such, a collateral warranty between a contractor and a funder would constitute "an agreement … between an executing party and another party" bringing the warranty within the definition of a construction contract under the Act. The fact that "work" is defined as "any act done in furtherance of the construction contract" further complicates the question as to whether warranties fall within the construction contract definition.

3.57. The position as regards collateral warranties furnished by architects, engineers, project managers or other design consultants may be construed differently. While Section 1(2) provides that reference to a construction contract includes agreements to do work or provide services ancillary to the construction contract, such as those works carried out by architects, engineers, project managers and others as set out within Section 1(2) of the Act, it is debatable whether a collateral warranty furnished by a consultant (who is not "an executing party") to a funder would fall within the ambit of the Act. If the collateral warranty is being furnished to a main contractor (or a subcontractor) then clearly the contractor or subcontractor would be an "executing party", and the consultant providing the warranty would represent "another party". Consequently, any payment dispute under this warranty could be referred to adjudication. If, however, the beneficiary of the consultant's warranty is not an executing party, which would be the case in most circumstances, and is, for example, a funder, then any payment dispute under the warranty would arguably not fall within the ambit of the Act. If the warranty contains step-in provisions with the effect that the beneficiary be treated as the employer under the building contract following step-in, then the issue as to whether any payment dispute under the collateral warranty may be referred to adjudication becomes even more complex.

Other agreements relating to or arising out of a construction contract

Performance bonds

3.58. The position regarding performance bonds is similar to that of collateral warranties. Whether a performance bond constitutes a construction contract will depend on the precise wording of the performance bond. Generally, a performance bond is furnished by a contractor's guarantor at the outset of the works and will contain a term that provides that the contractor has agreed to carry out and complete the works and failing which the guarantor will discharge the losses suffered by the recipient of the bond. Many bonds also entitle the guarantor to propose the appointment of a replacement contractor to carry out and complete the remaining works which the original contractor failed to perform. Where a performance bond contains the aforementioned common provisions it is likely that such a bond will be construed as a construction contract and any payment dispute arising under that bond may be referred to adjudication.

Parent company guarantees

3.59. Parent company guarantees are commonly furnished where a contractor has an ultimate parent company, which is financially more secure than the contractor. The parent company guarantee is a guarantee given by the contractor's ultimate or intermediate parent company in favour of the other contracting party to secure the performance of the contractor's obligations under the contract. Similar to performance bonds, a parent company guarantee will include a term providing that if the contractor fails to carry out and complete the works in accordance with the contract, the parent company will step in and continue to carry out and complete the works that remain following the contractor's default. Provided the parent company guarantee includes the aforementioned provisions, it is likely it will be construed as a construction contract and any payment dispute arising under the parent company guarantee may be referred to adjudication.

Project Supervisor appointments (PSCS & PSDP)

3.60. Agreements appointing a Project Supervisor for the Construction Stage (PSCS) or Project Supervisor Design Process (PSDP) are agreements that are incidental to the actual performance of construction operations and also directly connected to such operations. Such appointments would come within the scope of Section 1(2)(c) as representing advice on building.

Assigned Certifier and Design Certifier appointments

3.61. Assigned Certifier appointments pursuant to the Building Control (Amendment) Regulations 2014 ("B(C)AR"), constitute an agreement relating to construction operations as they are both directly connected to, and incidental to the actual performance of, construction operations. Furthermore, it is a service ancillary to the construction contract and consequently would come within the scope of Section 1(2)(c) as advice on building. A Design Certifier appointment would come within the definition of Section 1(2)(a) of the Act as it would constitute a "design" service associated with, and ancillary to, construction operations.

Settlement agreements

3.62. The English Courts have held[20] that a dispute under a settlement agreement in relation to a disputed construction contract effectively constitutes a dispute under the relevant construction contract, and consequently the settlement agreement dispute may be referred to adjudication. However, earlier English case law cautions against the assumption that a settlement agreement, which is in full and final settlement of all claims under a construction contract, will automatically be a construction contract simply because it compromises a construction contract. In *Shepherd v - Mecright*,[21] HHJ Lloyd QC held that a dispute under a settlement agreement was not a dispute under the original construction contract. Shepherd had engaged Mecright as its subcontractor. A dispute arose but the parties reached a settlement agreement which was recorded in writing and signed by Mecright as "in full and final settlement of all claims". Mecright subsequently wrote to Shepherd expressing their dissatisfaction about the amount that had been agreed and served a notice of adjudication in July 2000, claiming a greater sum. Mecright's July 2000 letter did not mention the settlement agreement, nor did their subsequent referral notice to the adjudicator. However, Shepherd informed the adjudicator of the settlement agreement, following which Mecright then claimed that the settlement had been reached under duress alleging that Shepherd had taken advantage of Mecright's financial impecuniosity at the time. Shepherd applied to the court for a declaration that there was no dispute under the subcontract capable of referral to adjudication and the duress claim was similarly not a dispute referable to adjudication. Judge Humphrey Lloyd QC agreed with Shepherd and held that the effect of the settlement agreement was that there could thereafter be no dispute

20 Fiona Trust & Holding Corporation v Yuri Privalov [2007] 4 RERS ALLERS 951.

21 Shepherd Construction v Mecright [2000] HT-00-28 (TCC).

capable of being referred to adjudication. The judge held that any dispute under the settlement agreement could not be a dispute under the original contract even if it is argued that the settlement agreement is voidable for duress. The effect of the settlement agreement was to replace the original agreement and the adjudicator therefore had no jurisdiction as there was no dispute under the original agreement. Consequently, the court held that a dispute under a settlement agreement which settles a dispute under a construction contract is not a dispute under that contract.

3.63. However, whether a settlement agreement could be described as an agreement which "arises out of" a construction contract is a different argument. In the *Shepherd* case the subcontract had included (in the adjudication referral provision) the term "under" the subcontract, rather than "in connection with" or "arising out of". Consequently, Judge Humphrey Lloyd believed that a dispute about a settlement agreement of this kind could not be a dispute "under" the subcontract, since the effect of a settlement agreement is one which replaces the original agreement. In the *Fiona Trust* case, which was an English Court of Appeal case in relation to an arbitration, the court determined that the words "arising under", do not have a narrower meaning than the words "arising out of" and that the words "arising out of" should cover "every dispute except a dispute as to whether there was ever a contract at all". Consequently, the *Fiona Trust* case is authority for the view that the term "arising under" should be given a broad meaning and should cover "every dispute except a dispute as to whether there was ever a contract at all". The *Shepherd* case was clearly decided on its own particular facts. However, if the *Shepherd* case were to be revisited post *Fiona Trust* the court would likely maintain the view that a dispute under an original contract could not be referred to adjudication if the same dispute is the subject of a settlement agreement on the reasoning that the same dispute should not be heard twice. However, if a dispute arose in relation to the settlement agreement itself then such dispute could be referred to adjudication on the basis that the settlement agreement arose out of the original construction contract. Consequently, any dispute arising under a settlement agreement effectively constitutes a dispute arising under the construction contract, therefore coming within the scope of the Act and being referable to adjudication, save where the dispute amounts to a re-run of the settled dispute.

Excluded operations

3.64. Section 1(3) sets out operations which do not constitute construction operations and includes:

the manufacture or delivery to a construction site of –

(a) building or engineering components or equipment,
(b) materials, plant or machinery, or
(c) components for systems of heating, lighting, air-conditioning, ventilation, power supply, drainage, sanitation, water supply or fire protection, or for security or communications systems.

3.65. However, Section 1(4) provides that where a contract for these "excluded operations" also involves installation of the things referred to, then the supply and installation contract constitutes a construction operation and a contract within the scope of the Act.

3.66. In *Fence Gate v J.R. Knowles*[22] it was held that where an agreement relates to construction operations as well as other matters (in that case surveying, reporting back to the client and subsequently giving evidence in an arbitration) the UK Act applied to such an agreement "only so far as it relates to construction operations". The court was prepared to treat construction operations-related services as severable from the other matters and held that disputes in relation to those services could be referred to adjudication. Accordingly, an adjudicator will have jurisdiction over disputes arising out of a contract in relation to construction operations works or services but not over disputes involving elements of the works or services which do not relate to construction operations.

3.67. Parties entering into, or drafting, professional services appointments should consider what aspects of their works or services relate to construction operations and are within the scope of the Act, and those aspects of their works or services which do not relate to construction operations and are outside the scope of the Act. In this regard, it may be prudent to prepare two separate contracts or appointments so as to differentiate between the two types of works or services being carried out. Subsequently, if a payment dispute arises under the construction operations-related contract or appointment then the issue as to which works or services fall within and outside the scope of the Act would not need to be addressed, thus reducing the likelihood of challenges to the adjudicator's jurisdiction, saving the parties time and costs in having to litigate the matter before the courts. The worst-case scenario is where the contract or appointment for professional services is incapable of being severed into works or services which fall within the scope of the Act and those which do not. Such a situation could render it impossible to apply the adjudication provisions of the Act to only part of the payment dispute in respect of the

22 Fence Gate Limited v J.R. Knowles [2001] 25/01 SF102200 (TCC).

works or services provided. This is what occurred in *Cleveland Bridge v Whessoe-Volker*,[23] where HHJ Ramsey found that the adjudicator's decision concerned matters covered by the UK Act and matters excluded by the operation of Section 105(2)(c)(ii) thereof. Whessoe-Volker engaged Cleveland Bridge to carry out works at a natural gas terminal. A dispute arose in relation to an agreed final account on which certain monies remained outstanding to Cleveland Bridge. Cleveland Bridge referred the dispute to adjudication and were awarded the full amount outstanding. In the enforcement proceedings Judge Ramsey had to determine whether the work carried out constituted construction operations and whether any of the work was excluded. If the works did consist of both construction operations and excluded operations, the judge also had to determine whether the adjudicator's decision was severable and/or enforceable. Judge Ramsey held that the work carried out consisted of both construction operations and excluded operations, and as the adjudicator had decided the whole of the dispute the judge had to consider whether the adjudicator's decision was enforceable as a consequence. The judge held that where only one dispute was referred to an adjudicator (in respect of works which included both construction operations and excluded operations) and the adjudicator decided the whole of the dispute (without differentiating between the included and excluded operations) then the decision was neither severable nor enforceable. The judge stated that

> If the adjudicator had made a decision on the whole dispute but had also made a decision which dealt only with the part of the dispute which was within her jurisdiction then, in my judgment, the decision on the whole dispute would not be enforceable or valid but there would be a valid decision on the part of the dispute which was within her jurisdiction.

3.68. Consequently, if a referring party is seeking to refer a payment dispute to adjudication where the contract includes construction operations-related works or services and also includes works or services relating to excluded matters, the referring party should seek to refer to adjudication only those construction operations-related matters. Where a referring party considers that all matters in dispute may be referred to adjudication, but the responding party contests this view and submits that certain services relate to excluded operations, the adjudicator should address this in any decision. Based on the

23 Cleveland Bridge (UK) Limited v Whessoe-Volker Stevin Joint Venture [2010] EWHC 1076 (TCC).

Cleveland Bridge decision a prudent adjudicator should make a decision in relation to the matters referred which it is not contested relate to construction operations and make a separate decision on contested matters. In other words, the adjudicator would make two decisions within the overall decision. In such circumstances, at enforcement the court would be able to consider the adjudicator's decision in its entirety, and enforce the entire decision if the court found that the adjudicator had jurisdiction over all matters (thereby rejecting the responding party's argument on excluded operations). However, if the adjudicator is found not to have jurisdiction over the contested matters the court could sever the adjudicator's decision on those matters and enforce the remainder of the decision.

3.69. While the excluded operations under Section 1(3) of the Act are quite different to the excluded operations as provided for under Section 105(2) of the UK Act, the principle (as established by the *Fence Gate* decision) is nonetheless informative. Where a construction contract or a consultant's appointment involves works or services that fall within the scope of construction operations under the Act and also works or services which do not or are otherwise excluded, then, unless the adjudicator separately decided those matters, it may not be possible to sever the respective works or services and uphold an adjudicator's decision in relation to the entirety of the payment dispute.

4 Exempted contracts

Section 2 of the Act

4.1. Section 2 of the Act sets out details of those contracts which do not come within the term "construction contract" under the Act. Section 2(1) specifically provides that a contract is not a construction contract if the value of the contract is not more than €10,000, or

> (i) if the contract relates only to a dwelling, and (ii) the dwelling has a floor area not greater than 200 square metres, and (iii) one of the parties to the contract is a person who occupies, or intends to occupy, the dwelling as his or her residence.

The "private dwelling" exemption

4.2. Section 2(1)(b) of the Act is very similar to Section 106 of the UK Act which has been considered by the English Courts on numerous occasions and is commonly referred to as the "private dwelling" exemption.

4.3. In *Samuel v J&B*[1] the construction contract related to the renovation of two barns, one intended for occupation and the other intended for re-sale. In considering whether the contract (as a whole) was an exempted contract (on the basis that it partly related to a "private dwelling"), the court had to determine two issues: firstly, whether the construction work was being carried out at the time "on a dwelling"; and secondly, whether the contract constituted an agreement that principally related to a dwelling. HHJ Overend held that a residential occupier does not have to be in occupation for the contract to constitute a contract with the residential occupier. The judge held that it was irrelevant that the barn was not a dwelling at the time of the contract, since it was at some stage to become a dwelling. However, the works

1 Samuel Thomas Construction v J&B Developments [2000] High Court at Exeter (Case No: Exeter ZN 900750).

did not "principally" relate to a dwelling, they also concerned a barn to be sold, plus works on a garage block, a courtyard and drainage works. Judge Overend held that the work could not be said to be work that principally related to operations on a private dwelling when only one property was to be occupied as a private dwelling and the other was to be sold. Consequently, Judge Overend held that the contract was subject to the UK Act and the adjudicator's decision was enforceable.

4.4. The *Samuel* decision is informative on two principles concerning the private dwelling exemption –

(a) a residential occupier does not have to be in occupation at the time that the construction works are being carried out, nor does the building have to be a dwelling at that time, for the contract to be exempted under Section 2(1)(b) of the Act.

(b) if a contract is for works relating to a dwelling and works not related to a dwelling, the works may not "principally" relate to a dwelling, and not come within the Section 2(1)(b) exemption, thereby enabling referral of payment disputes to adjudication.

4.5. In *Shaw v Massey*,[2] the contract related to a gate lodge adjacent to a large residence. The gate lodge, known as the East Lodge, was a separate building located a quarter of a mile from the main residence. Disputes arose which Massey sought to refer to adjudication. The Shaws argued that as residential occupiers the adjudicator had no jurisdiction to make an award in favour of the contractor as the contract was caught by the "private dwelling" exemption. The adjudicator disagreed and decided in Massey's favour. The Shaws refused to pay and there followed a series of hearings culminating in proceedings before HHJ Coulson, who rejected the Shaws' position on a number of grounds. Firstly, the Shaws had argued that they were "consumers" when entering into the construction contract and for this reason the contract should not come within the scope of the UK Act.[3] The Shaws also argued that they were residential occupiers of the East Lodge and the contract was exempted under Section 106 of the UK Act. Judge Coulson held that the East Lodge was a separate building to the main residence despite the Land Registry entry referring to the East Lodge and the main residence as one building. Judge Coulson stated that what mattered was not how the Land Registry had registered the title but whether the Shaws were the residential occupiers of the East Lodge, which the judge determined they were not. Judge Coulson also commented that if a person could no longer afford to

2 Gabrielle and Christopher Shaw v Massey Foundation & Pilings Limited [2009] EWHC 493 (TCC).

3 See paragraphs 4.13 to 4.15 below in relation to "consumers" entering into a construction contract.

live in a terraced house and decided to convert the house into three small flats, one of which they intended to live in, with the other two being sold or rented, the "private dwelling" exemption would not apply because of the commercial element of the nature of the works that were to be carried out.

4.6. In *Edenbooth v Cre8*,[4] Judge Coulson had to consider whether a company could be considered a residential occupier. Cre8, a development company, engaged Edenbooth to carry out ground works at two adjoining properties, one of which was owned by a Cre8 director. Edenbooth commenced enforcement proceedings in relation to an adjudicator's decision in Edenbooth's favour. Cre8 had taken two jurisdiction points in the adjudication. First, that the work was not a construction operation pursuant to Section 105(2)(d) of the UK Act, and secondly, that Cre8 was a residential occupier and the contract was exempt under Section 106(1)(a) of the UK Act. Judge Coulson held that the use of the word "residential" in the legislation imported a meaning that a real person must be living and residing in the property. The judge found that while one of the properties was owned by Cre8's director, Cre8 were the contracting party, and as a company could not be a residential occupier it could not avail of the residential occupier exemption; thus, the adjudicator's decision was enforceable. In delivering his decision, Judge Coulson commented generally on the appropriate procedure the court should adopt in considering issues regarding an adjudicator's decision:

> First … look at any points that had been taken as to the adjudicator's jurisdiction. If the adjudicator did not have the necessary jurisdiction then the decision is a nullity. If, on the other hand, the adjudicator did have the necessary jurisdiction, then the court should go on to consider – if it is raised – any suggestion that the adjudicator acted unfairly. If the court concludes that the adjudicator did have jurisdiction and had not acted unfairly then the court is obliged to enforce the decision of the adjudicator.

4.7. While the Act exempts "private dwelling" contracts between a main contractor and a residential occupier, subcontracts between the main contractor (on a private dwelling construction) and subcontractors would nonetheless come within the definition of construction contract (provided not otherwise excluded) and the scope of the Act entitling either party to refer payment disputes to adjudication. Where a residential occupier engages a contractor on a private dwelling and a payment dispute arises under the contract, the contractor would not

4 Edenbooth Limited v Cre8 Developments Limited [2008] EWHC 570 (TCC).

be entitled to refer that payment dispute to adjudication due to the "private dwelling" exemption. However, any subcontractors of the contractor could refer payment disputes to adjudication under their subcontracts, despite the fact that the contractor has no entitlement to do so under its main contract. Although the main contract between the contractor and the residential occupier may be exempted under the Act, that does not mean any subcontracts are also exempted. One further example where this situation arises is on subcontracts in public private partnership arrangements, an issue which is dealt with below.

Other exempted contracts

4.8. Sections 2(2) and 2(3) of the Act provide for further categories of exempted contracts as follows:

> (2) A contract of employment[5] … is not a construction contract.
> (3) A contract between a State authority and its partner in a public private partnership arrangement, as those terms are defined in the State Authorities (Public Private Partnership Arrangements) Act 2002, is not a construction contract.

4.9. Section 2(2) is similar to Section 30 of the Arbitration Act 2010, which also excludes employment contracts from the scope of that legislation. This is an expected exemption as employment contract disputes are subject to a specific regime under Irish employment law.

4.10. Section 2(3) excludes a contract between a State authority and its contracting partner in a public private partnership (PPP) and there is no explanation for this exclusion. While PPP contracting entities generally consist of a number of partners, often including a funding partner, it is nonetheless equally important that a PPP contracting entity maintains its cash flow and seeks an adjudicator's interim decision to release cash flow from its State client when it is so entitled. The principle of pay now and argue later is as important to a PPP contracting entity as it is to any contracting entity, particularly where the PPP entity has subcontracts to which the Act applies. A PPP contracting entity may find itself in a situation where it is paying out monies to subcontractors (pursuant to payment provisions or as a consequence of an adjudicator's decision) when it has not been paid by the State authority and has no entitlement to refer a payment dispute with the State authority to adjudication. It might be argued that statutory adjudication is unnecessary in PPP contracts on the basis that such contracts generally incorporate sophisticated dispute

5 As defined in the State Authorities (Public Private Partnership Arrangements) Act 2002.

resolution provisions entitling parties to seek binding (albeit temporary) decisions in relation to disputes. However, in many circumstances (such as under the Public Works Contracts) such decisions are only binding when the party, to whom any payment is due on foot of the decision, furnishes a bond in respect of the amount due.

Contracting into the Act

4.11. While Sections 1(3) and 2(1) to (3) exclude certain contracts from the scope of the Act, the parties to any contract (including those contracts specifically excluded by the Act) may nonetheless choose to contract into the Act thereby providing for any payment disputes arising under their agreement to be referred to adjudication. Where a contractor enters into a construction contract with a residential occupier concerning a private dwelling, and such contract is likely to be exempted under Section 2(1)(b) of the Act, a prudent contractor (particularly where it is engaging subcontractors) should seek to include a provision in its contract that payment disputes will be referred to adjudication under the Act.

4.12. In *Nordot v Siemens*,[6] the English Courts enforced an adjudicator's decision (despite the contract being otherwise outside the scope of the Act) because the parties had contracted into adjudication. The case involved the provision of a gas turbine generating plant, which under Section 105(2) of the UK Act constitutes excluded works and is not a construction operation under the UK Act. Notwithstanding this, the parties agreed to engage in an ad hoc adjudication in respect of a dispute under the contract relating to payment. Siemens had indicated to the adjudicator that they would "abide by your decision in this matter and will comply with whatever direction you deem appropriate". The adjudicator agreed to act, conducted the adjudication and made a decision in Nordot's favour. Following the adjudicator's decision, Siemens argued that as the UK Act expressly excludes contracts such as the one between themselves and Nordot it was not open to the parties to confer an ad hoc jurisdiction on the adjudicator. HHJ Gilliland QC rejected that argument and held that the parties' agreement to abide by the adjudicator's decision, and Siemens' confirmation in respect of the same, was an agreement in itself and the parties were bound to comply with the adjudicator's decision. Judge Gilliland also emphasised the need for a clear jurisdictional objection to be lodged by the party who did not wish to enter into such an agreement or be bound by an adjudicator's decision. The judge held that whether this was apparent would depend on a fair

6 Nordot Engineering Services Limited v Siemens [2001] CILL, 1778 (TCC).

reading and interpretation of the correspondence between the parties. The judge held that the parties had agreed to confer ad hoc jurisdiction on the adjudicator as Siemens' letter to the adjudicator had stated that they would abide by the adjudicator's decision. Judge Gilliland therefore enforced the adjudicator's award. Consequently, where parties to a contract, which might otherwise be excluded or exempted from the Act, voluntarily choose to contract into the Act and be bound by an adjudicator's decision, in such circumstances it will not be open to those parties to later argue that because the contract came within the exclusions or exemptions under the Act the adjudicator's decision is not binding on them. Where there is evidence to support the contention that parties have voluntarily agreed to be bound by an adjudicator's decision, and there is no clear jurisdictional objection by either party, the Irish Courts would likely adopt an approach similar to the *Nordot* decision and find the adjudicator's decision binding on the parties and enforceable.

4.13. The situation may be less clear where it is not definitive as to whether the parties have agreed to refer their dispute to binding adjudication. In *Picardi v Cuniberti*,[7] HHJ Toulmin QC held that an adjudication clause in the English standard form RIBA Conditions of Engagement (CE/99) between the contractor and residential occupier had not been incorporated into the contract. The Cunibertis engaged Mr Picardi as architect on the refurbishment of their private dwelling house. A dispute arose concerning Mr Picardi's fees and the dispute was referred to adjudication. The adjudicator decided that the Cunibertis should pay Mr Picardi's outstanding fees, but they failed to comply and Picardi issued court proceedings seeking a declaration that the parties had concluded a contract incorporating CE/99, which included provisions allowing for disputes to be referred to adjudication. HHJ Toulmin QC found against Mr Picardi on several grounds. While the judge found that there was no contract between the parties incorporating CE/99 (because the letter of appointment and the accompanying copy of CE/99 required it to be signed and returned by the client to the architect, and they were not) the judge held that even if there was a contract, Picardi could not validly invoke any adjudication provision. The judge reached this decision on the basis that as the Cunibertis were consumers, the contract was subject to operation of the UK Unfair Contract Terms Regulations (UCTR).[8] The judge found that the RIBA guidance notes on the CE/99 contract state that the adjudication provisions are unusual terms (because adjudication would not otherwise be

7 Picardi v Cuniberti [2002] EWHC 2923 (TCC).

8 The Unfair Terms in Consumer Contract Regulations 1999.

available) and suggested that clauses such as the adjudication provisions should be individually negotiated with prospective clients in order to prevent exclusion of those clauses under the UCTR. Judge Toulmin found that even if the CE/99 contract had been incorporated, and that contract provided for adjudication in the event of a dispute, the Cunibertis would not have been bound to refer disputes to adjudication because the adjudication provisions in the contract had not been brought to the specific attention of the Cunibertis, separately negotiated and their inclusion in the contract explicitly agreed. The judge held that the Cunibertis did not have the benefit of any advice concerning the adjudication terms that Picardi had sought to incorporate in the contract. In the judge's opinion, Picardi should have brought these provisions to the specific attention of the Cunibertis and provided advice in respect of the consequences of incorporating adjudication provisions in the contract. The judge found, as fact, that Picardi had not, as he had alleged, offered to go through the RIBA terms with the clients and that none of the relevant terms had been drawn to their attention, let alone specifically negotiated. Judge Toulmin commented that a provision incorporating adjudication in a contract was an unusual provision which ought (in accordance with Regulation 5 of the UCTR) to be brought to the specific attention of a lay party if it is later to be validly invoked. The judge's opinion in this regard was informed by his view that because the UK Parliament, by virtue of Section 106 of the UK Act, had specifically excluded private dwelling houses from the adjudication legislation, that inferred that Parliament's intention must have been that if a private dwelling owner was requested to enter into a contract incorporating adjudication, such provision should be specifically brought to the dwelling owner's attention.

4.14. Similar unfair contract terms legislation exists in Irish law under the European Communities (Unfair Terms in Consumer Contracts) Regulations 1995.[9] Consequently, on the basis of the *Picardi* decision where a contractor or consultant enters into a contract with a residential occupier which incorporates adjudication provisions, those provisions may be unenforceable if they are not fairly and properly brought to the residential occupier's attention and individually negotiated and agreed. The client would need to be advised that they are not obliged to refer disputes to adjudication under the Act, as residential occupier contracts are excluded from the Act. The client should also be advised that should they agree to the incorporation of an adjudication provision they will be specifically agreeing

9 European Communities (Unfair Terms in Consumer Contracts) Regulations 1995 (SI 27-1995).

to refer payment disputes to adjudication should they arise. A prudent contractor or consultant would also be well advised to fully document the fact they have fairly and properly brought to the client's attention the adjudication provisions, the fact the client need not refer disputes to adjudication and that by agreeing to do so they will be bound by any subsequent adjudicator's decision.

4.15. In *Lovell v Legg and Carver*,[10] HHJ Mosley QC differentiated the *Picardi* decision on its own facts finding that it was not unfair to include an adjudication clause in an English standard form contract entered into between the contractor and a residential occupier. Mr Legg and Ms Carver ("Legg & Carver") engaged an architect to prepare drawings and specifications for proposed refurbishment works to their property. At the couple's insistence, following their architect's advice, the parties entered into the JCT Agreement for Minor Building Works, 1998 Edition incorporating amendments MW1–11, that included supplemental condition D which provided for adjudication in the event of a dispute. As a consequence, while adjudication under the UK Act did not apply (under the "private dwelling" exemption), the parties had nonetheless agreed to a contractual adjudication dispute resolution procedure by incorporating supplemental condition D. During the course of the work Mr Legg refused to pay Lovell in respect of informally certified valuations. Lovell suspended work and served a Notice of Dispute seeking an 11-week extension of time and payment of outstanding sums. An adjudicator was appointed who decided that Legg & Carver were liable to Lovell to the sum of £85,873.59. Legg & Carver, relying on the UCTR, refused to comply with the adjudicator's decision arguing that the adjudication clause was unfair because it caused a significant imbalance between the parties to the detriment of Legg & Carver. Legg & Carver argued that the adjudication clause in the JCT contract had not been brought to their specific attention. However, Judge Mosley believed the evidence suggested that Legg & Carver, through their architect and on his advice, had adopted the JCT contract. The judge explained that to be unfair under the UCTR, the adjudication clause must cause a significant imbalance of the parties' rights and obligations under the contract, to the detriment of the consumer. The judge held that this imbalance must be caused by the adjudication provisions and be contrary to the requirement for good faith. He held that there had been no breach of the requirement of openness, and that the adjudication terms were fully, clearly and legibly set out in the contract and contained no concealed pitfalls or traps. In respect of fair dealing, the contract was insisted upon by Legg &

10 Lovell Projects v Legg and Carver [2003] C.I.L.L. 2019.

Carver's architect; they were knowledgeable business people who had successfully engaged an architect and a contract administrator and had solicitors with whom they had the opportunity to consult and whom they may have consulted. Consequently, Judge Mosley found there was no departure from a "good standard of commercial morality and practice" concluding that the employers could not rely upon the UCTR as the adjudication clause in question did not produce a significant imbalance between the parties. The judge held that the adjudication clause was equally balanced between the parties as either party could, pursuant to the adjudication clause, refer a dispute to adjudication and that it was not to the detriment of the employers. Judge Mosley held that for these reasons Legg & Carver were bound by the adjudication clause and that the adjudicator's decision was enforceable.

Duress and fraud

4.16. Circumstances may arise where one party to the construction contract argues that they entered into the contract under duress and consequently the contract was voidable meaning any dispute arising under the contract could not be referred to adjudication. In *Capital Structures v Time & Tide*[11] HHJ Wilcox refused summary judgment, which sought to enforce an adjudicator's decision, on the basis that if a contract was reached by economic duress, and was therefore voidable, any dispute arising under the contract would constitute a dispute about the formation and the validity of the contract itself. Time & Tide engaged Capital as a subcontractor for the supply, delivery and installation of structural steel work and cladding on a project in respect of which Time & Tide were the main contractor. Following disputes in relation to valuations and interim payments Capital withdrew from the site. Time & Tide's employer threatened to take over the development if the matter of Capital's withdrawal was not resolved and the project not progressed. Time & Tide believed that such a threat could lead to their liquidation and consequently signed a settlement agreement with Capital in full and final settlement of all existing and future claims by Capital. The settlement agreement included an adjudication clause. Time & Tide made part payment of the sums due under the settlement agreement but refused to pay the balance. Capital referred the matter to adjudication. Time & Tide argued that it had entered into the settlement agreement under economic duress and elected to have the agreement set aside.

11 Capital Structures v Time & Tide Construction Limited [2006] EWHC 591 (TCC).

The adjudicator rejected Time & Tide's argument and made an award in Capital's favour, which Time & Tide refused to pay. Capital applied to the court for summary judgment to enforce the adjudicator's award. Capital submitted that the facts of the case did not support Time & Tide's plea of economic duress and even if the agreement had been voided for duress, the adjudication provisions of the agreement would survive. Judge Wilcox held that, on the facts, there was an arguable, albeit "shadowy", case for economic duress stating that:

> If there has never been a contract from which the adjudicator's jurisdiction derives because, on the grounds of duress, proper steps are taken to avoid an agreement, it logically follows that any arbitration or adjudication provision also becomes void.

4.17. Consequently, if there are allegations of fraud or deceit in respect of any claims referred to adjudication under the contract then such circumstances might give rise to an adjudicator's decision in respect of those claims being rendered unenforceable.

4.18. In *SG South v Kings Head*,[12] Judge Akenhead held that allegations of fraud relating to contract claims under the contract could be decided by an adjudicator and that fraudulent misrepresentation claims might also be within his jurisdiction, but claims in the tort of deceit probably could not be. The developer, King's Head, employed SG South as its contractor to carry out works. SG South had only been incorporated a few months prior to entering into the contract and had little or no trading record. King's Head reached an agreement with SG South that they would employ and directly pay some of the works contractors engaged on the project. Disputes arose between the parties and SG South referred two disputes to adjudication. The first related to the extent to which SG South had been paid a certified sum under an interim certificate number 8. The adjudicator decided that SG South had been paid the certified sum either directly, or by virtue of payments made by King's Head directly to SG South's works contractors on SG South's behalf. However, the adjudicator decided that some interest on late payments was due to SG South. The second adjudication related to the non-payment of interim certificate number 14. King's Head's response to the referral raised allegations of fraud against SG South, which SG South denied. The second adjudicator found that in the absence of payment and adequate withholding notices, monies were due to SG South, and that to decide on the issue of fraud went beyond his jurisdiction.

12 SG South Limited v Kings Head Cirencester LLP & Another [2009] EWHC 2645 (TCC).

King's Head did not pay the amounts and resisted enforcement on the basis that, *inter alia*, there was a *prima facie* case that SG South had behaved fraudulently on the project, and the court should not permit itself to be used by a party that had behaved fraudulently to enforce claims under contracts. Judge Akenhead found that there was a distinction between an adjudication taking place where an allegation of fraud is raised (or could be raised) as a defence, and where fraud only comes to light after an adjudication has taken place. He held that in the latter case, fraud (or an allegation of fraud) which only comes to light after the adjudication, could provide a basis for refusing to enforce an adjudicator's decision. However, the judge held that where King's Head had raised the allegation of fraud in the adjudication it could not later be raised as a defence in the enforcement proceedings. On the facts of the case, the judge held that a credible case for fraud had not been made out in evidence or argument such that it could be deployed to defeat enforcement, and in the circumstances the judge awarded summary judgment in favour of SG South.

Severability of activities

4.19. Section 2(4) of the Act provides that

> Where a contract contains provisions in relation to activities other than those referred to in the definition of a construction contract and section 1(2), it is a construction contract only so far as it relates to those activities.

4.20. Section 2(4) is similar to Section 104(5) of the UK Act and has the effect of allowing the qualifying construction operations under a contract to be treated as severable from any other matters or activities under that contract which may not come within the scope of construction operations, and would therefore not be subject to the application of the Act. The effect of this provision is that an adjudicator may have jurisdiction to determine payment disputes over part of a contract but not in respect of other aspects of the contract. In the *Fence Gate* case[13] Judge Gilliland held that Section 104(5) of the UK Act provided that a contract which relates both to construction operations and other activities was to be treated as severable between those parts that relate to construction operations and those parts which do not. Adjudication under the Act would only apply to those aspects of the contract relating to construction operations. This

13 [2001] CILL 1757–1759 at paragraph 7.

issue of severability of activities could, in practice, be quite challenging for an adjudicator. Indeed, in *North Midland v Lentjes*[14] HHJ Ramsey addressed this issue suggesting that situations could arise where it might be impossible to apply the adjudication provisions of the Act to only part of a dispute, and that in such circumstances the Act could potentially not be applied at all. Lentjes, the turnkey contractor for the installation of a flu gas desulphurisation plant at two coal-fired power stations, engaged North Midland under four agreements, one for enabling works and one for civil engineering works for each of the two power stations. The subcontracts were contained in four purchase orders in the same substantive form consisting of a purchase order agreement incorporating conditions of contract. The enabling works included erecting temporary fencing and gates, constructing temporary roads, offices and drainage, and demobilising concrete surfaces, buildings and workshops. The civil works involved excavation, piling and reinforced concrete foundations for the plant itself and for the steel structures that would support and provide access to the plant. None of the subcontracts contained clauses providing for adjudication. A final account dispute arose and North Midland, in seeking to refer the dispute to adjudication, sought a declaration from the court that the UK Act applied to the subcontracts as the enabling and civil works were "construction operations" under the UK Act. Lentjes argued that North Midland's works were excluded from being "construction operations" as they fell within the exclusions under Section 105(2)(c)(i) of the UK Act, which provides that the following acts do not constitute "construction operations", namely

> assembly, installation or demolition of plant or machinery, or erection of steel work for the purposes of supporting or providing access to plant or machinery, on a site where the primary activity is ... (i), nuclear processing, power generation or water or effluent treatment.

Lentjes contended that the works were not construction operations, relying on the *ABB Power*[15] decision which held that on a site which was shown to be for the purposes of one of the exempted operations all works and contracts associated with the site would be exempt. North Midland, however, relied on the *Palmers*[16] decision in which it was held that a narrower interpretation of Section 105(2) of the

14 North Midland Construction Plc v AE & E Lentjes UK Limited [2009] EWHC 1371 (TCC).

15 ABB Power Construction Limited v Norwest Holt Engineering Limited [2000] EWHC 68 (TCC).

16 Palmers Limited v ABB Power Construction Limited [1999] HT 99 0000 90 (TCC).

UK Act was to be preferred and that where an overall site is for the purposes of one of the exempted operations, each individual contract is to be evaluated on the works performed under that contract. Having considered the opposing arguments of the parties, Judge Ramsey held that the narrower interpretation of the scope of the exceptions in Section 105(2)(c)(i) of the UK Act in the *Palmers* case was to be preferred on the basis that:

> the scheme of Section 105 is to set out a very wide definition of "construction operations" and then list specific exclusions. This favours the narrower interpretation of the exclusions. The broader approach requires a strained construction of the words.

4.21. The judge held that, on the facts, the enabling and civil engineering works were clearly within the definition of "construction operations" as they were merely preparatory to carrying out work on the site and could not be described as "assembly, installation or demolition of plant or machinery". The judge held that the intention of the UK Act was to capture a wide range of work carried out in the construction industry, but with some limited specific exclusions, and the enabling and general civil engineering works were not excluded. As a result Judge Ramsey found that the contracts were for the carrying out of construction operations as defined in Section 105(1), and the UK Act applied.

4.22. Prudent parties (and an equally prudent adjudicator) might seek to agree, at the outset, that they accept the adjudicator's jurisdiction to deal with matters within the scope of the Act and also outside the scope of the Act, or which may otherwise be exempted or excluded by the Act. This would minimise the risk of the adjudicator's decision later being held unenforceable,

4.23. Alternatively, the adjudicator could proactively arrive at a non-binding conclusion that he had jurisdiction to decide payment disputes in respect of some, but not all, elements. Where the adjudicator reaches a non-binding conclusion that he has jurisdiction over some matters (which the responding party contests the adjudicator does not) the adjudicator could specifically identify and categorise those specific matters in the decision. This approach allows the contested part of the decision to be severed in the event that a court later determines the adjudicator did not have jurisdiction over the relevant matters and was wrong to adjudicate and decide on those matters.

4.24. In *Cantillon v Urvasco*,[17] HHJ Akenhead set out specific principles on the severability of an adjudicator's decision for the purposes of

17 Cantillon Limited v Urvasco Limited [2008] EWHC 282 (TCC).

enforcement. Urvasco engaged Cantillon to carry out demolition, piling and other works. Cantillon claimed an extension of time under two claims, one for 16 weeks and one for 13 weeks (a variation requiring additional piling work), together with related loss and expense, and these two claims were submitted to adjudication. Cantillon calculated its loss and expense claim by reference to a time period. Urvasco argued the piling took too long, the variation did not cause a delay and there was no loss recoverable for any later period when the piling works were actually done. The adjudicator decided Cantillon was entitled to 9.7 weeks extension of time for the piling variation and assessed loss and expense for the critical delay period, which was a later period than Cantillon had claimed. The adjudicator awarded Cantillon a sum of money, one fifth of which related to 13 weeks' claim. Urvasco refused to pay and Cantillon issued court enforcement proceedings. On the question of severability, if the court found that part of the adjudicator's decision was arrived at improperly, but another part was sound, whether the "bad" part could be severed from the decision, and the "good" part enforced, Judge Akenhead concluded as follows:[18]

(a) The first step must be to ascertain what dispute or disputes has or have been referred to adjudication. One needs to see whether in fact or in effect there is in substance only one dispute or two and what any such dispute comprises.
(b) It is open to a party to an adjudication agreement, as here, to seek to refer more than one dispute or difference to an adjudicator. If there is no objection to that by the other party or if the contract permits it, the adjudicator will have to resolve all referred disputes and differences. If there is objection, the adjudicator can only proceed with resolving more than one dispute or difference if the contract permits him to do so.
(c) If the decision properly addresses more than one dispute or difference, a successful jurisdictional challenge on that part of the decision which deals with one such dispute or difference will not undermine the validity and enforceability of that part of the decision which deals with the other disputes.
(d) The same logic must apply to the case where there is a non-compliance with the rules of natural justice which only affects the disposal of one dispute or difference.
(e) There is a proviso to (c) and (d) above which is that, if the decision as drafted is simply not severable in practice, for instance on the wording, or if the breach of the rules of natural

18 At paragraph 65 of the judgment.

justice is so severe or all pervading that the remainder of the decision is tainted, the decision will not be enforced.

(f) In all cases where there is a decision on one dispute or difference, and the adjudicator acts materially in excess of jurisdiction or in breach of the rules of natural justice, the decision will not be enforced by the court.

4.25. Judge Akenhead held that the rules of natural justice had not been breached and consequently the severability issue did not arise. The judge also held that had the rules of natural justice been breached, he would have severed the decision and given judgment in favour of Cantillon on the other parts of the decision which were unrelated and untainted by any such breaches.

Prohibition on contracting out of the Act

4.26. Section 2(5) of the Act operates as a prohibition to parties seeking to contract out of, limit or exclude the application of the Act. Section 2(5) provides that:

> This Act applies to a construction contract whether or not
>
> (a) the law of the State is otherwise the applicable law in relation to the construction contract, or
>
> (b) the parties to the construction contract purport to limit or exclude its application.

4.27. This provision underpins a primary purpose of the Act, namely, to ensure parties to the construction contract will be entitled to refer payment disputes to adjudication. The provision effectively prevents parties from contracting out, or in any way seeking to limit the application of the Act. Even where the construction contract is governed by a law other than Irish law, if a payment dispute arises the parties will nonetheless be entitled to refer that dispute to adjudication, provided the construction contract is being carried out in the Irish jurisdiction. Where a construction contract to be carried out in Ireland contains a provision providing that the contract be governed by a foreign law, that provision would likely be held inapplicable as regards a party's entitlement to refer payment disputes to adjudication under the Act. For example, if a construction contract is entered into in respect of a project to be carried out in Ireland, but is subject to English law, any payment dispute under the contract would nonetheless be referable to adjudication under the Act, notwithstanding that the law applicable to the substantive issues in dispute would be English law.

4.28. Section 2(5)(b) prohibits the parties agreeing to exclude the whole or certain aspects of the Act. For example, if the parties agree that only payment disputes for less than €1m may be referred to adjudication, and payment disputes for larger sums may not be referred to adjudication, such a provision would be struck down as void, and any payment dispute (irrespective of its value) would be referable to adjudication under the Act.

Beyond adjudication

4.29. Where a losing party fails to comply with an adjudicator's decision the successful party may seek to enforce the decision on an application to court for summary judgment. However, in the intervening period, while the summary judgment proceedings are in process (or where the adjudication otherwise comes to an end without a decision), the parties may refer the payment dispute to whatever further dispute resolution process they have provided for under their contract.

5 The Right to refer payment-related disputes to adjudication

Sections 6(1) and 6(2) of the Act

5.1. Section 6 addresses the core matters concerning the referral of payment disputes to adjudication. It provides for the entitlement to refer payment disputes to adjudication; the timing of the referral; the adjudicator appointment; the adjudicator's time frame to reach a decision; the adjudicator's approach and conduct; the consequences of the decision and other significant elements of the adjudication process and procedure. Sections 6(1) and 6(2) deal with the right to refer payment disputes to adjudication and the notice of intention to refer.

5.2. Section 6(1) provides:

> A party to a construction contract has the right to refer for adjudication in accordance with this section any dispute relating to payment arising under the construction contract (in this Act referred to as a "payment dispute").

5.3. There are a number of issues which arise in respect of the specific wording of Section 6(1), including the following questions:

a) Is the party referring the dispute a party to the "construction contract", as that term is defined in Section 1(1) of the Act?
b) Has the matter being referred to adjudication crystallised as a "dispute"?
c) Does the matter being referred to adjudication constitute a single "dispute" or a number of "disputes"?
d) Is the dispute "relating to payment"?
e) Is the dispute "arising under the construction contract"?

"Party" to a construction contract

5.4. Identifying the correct parties to the construction contract is an obvious precursor to naming those parties in the adjudication referral. Notwithstanding that one would have thought this to be a reasonably straightforward step in pursuing a claim, a number of cases have come before the UK Courts relating to factual errors as to the parties' identities.

5.5. Where issues concerning the correct identity of a party have been canvassed before the adjudicator and the adjudicator has decided on that issue, the court will not entertain any defence in enforcement proceedings which is grounded on the same issue of a party's correct identity.

5.6. In *Brenton v Palmer*,[1] the adjudicator awarded an amount payable to Brenton. Palmer refused to pay, and Brenton issued court proceedings to enforce the adjudicator's decision. In his defence, Mr Palmer argued that his company and not him personally should have been the responding party and consequently the adjudicator did not have jurisdiction. The court held that as Mr Palmer had raised this precise issue in the adjudication, and the adjudicator had decided against him on the issue, the court would not rule on the matter, even if the adjudicator had made a factual error. The court held that an adjudicator's decision on a party's identity, as part of his substantive decision, could not be questioned in enforcement proceedings as it was a decision the adjudicator was empowered to make under the Act. The court enforced the adjudicator's decision.

5.7. Where both parties in an adjudication are aware of a misdescription of a party's name and neither party is misled, the court (in enforcement proceedings) will make a declaration effectively correcting the error in granting summary judgment. However, if there are several companies in the same group and the wrong company is named in the notice of adjudication the court may not be minded to overlook this error.

5.8. In *Total v ABB*,[2] the referring party, Total, was misnamed in the adjudication. In its defence in the enforcement proceedings, ABB contended that the mis-description of Total deprived the adjudicator of jurisdiction. HHJ Wilcox held that both parties were well aware of the contracting party's true identity and nobody had been misled. The judge held that in a case of clear mis-description, where no one was misled, the court could make a declaration that the referring party was the contracting party. Judge Wilcox stated, *obiter*, that where there are similar company names, for instance in a group of companies, or subsidiaries with overlapping management systems and some common directors, a precise description of the referring party would be critical.

5.9. In *Michael John Construction (MJC) v Golledge*,[3] the contractor MJC had entered into a building contract with a rugby club. MJC sought to enforce

1 A J Brenton v Jack Palmer [2001] Unreported, 19 January 2001 (TCC).

2 Total M&E Services Limited v ABB Building Technologies Limited [2002] EWHC 248.

3 Michael John Construction (MJC) v Richard Henry Golledge & Others [2006] EWHC 71 (TCC).

two decisions from an adjudicator. In the first adjudication, the responding party was the fourth defendant in the court proceedings, who at the time of the adjudicator's decision was a club trustee. In the second adjudication, the responding parties were the first, second and third defendants, who were trustees of the rugby club at the time of the contract. All four respondents refused to pay on the basis that the adjudicator did not have any jurisdiction as the proper responding parties should not have been the respondents, but rather all club members. HHJ Coulson QC held that the club was not a separate legal entity, and that only individuals can be pursued and queried as to whether MJC needed to pursue all club members. Judge Coulson held that the issue as to who was liable on behalf of the club was part of the single dispute referred to the adjudicator, and that MJC was simply seeking to be paid by the club by reference to the trustees at the time that the contract was made. Consequently, the judge held that the claim related to a dispute as to the obligations owed by the employer to the contractor and the adjudicator was empowered to decide that question, including as to the identity of the employer.

Has a dispute crystallised?

5.10. It is a statutory precondition to the right to refer a payment dispute to adjudication that there exists a dispute capable of being referred. The Act does not define a "dispute", nor does it refer to language such as "claim". As such, any claim can only be referred to adjudication if that claim has crystallised as a dispute.

5.11. The UK Act provides that for the purposes of Section 108(1) a "dispute" includes any "difference". The Irish Act does not state this. While Section 6(1) of the Act is otherwise similar to Section 108(1) of the UK Act, caution should be exercised when considering English case law in respect of the issue as to whether a "dispute" has arisen, because a "dispute" under the UK Act encompasses "disputes and/or differences". UK Courts have dealt with a considerable number of cases where one of the central issues has been whether a claim or issue has crystallised. Some of these cases have related to arbitration but these decisions are equally applicable to adjudication in the context of considering whether a dispute has crystallised.

5.12. In *Amec*,[4] HHJ Jackson derived seven propositions in determining whether a dispute had arisen:

1) the word "dispute" should be given its ordinary meaning;
2) the mere fact one party notifies the other of a claim doesn't automatically trigger a dispute;

4 Amec Civil Engineering v Secretary of State for Transport [2004] EWHC 2339 (TCC).

3) a dispute doesn't arise unless and until it emerges that the claim is not admitted;
4) there may be an express rejection of the claim, it may be objectively inferred that the claim is not admitted, the respondent may prevaricate or remain silent for a period leading to the inference that he does not admit it;
5) the time for which a respondent may remain silent before a dispute is to be inferred depends heavily upon the facts of the case and the contractual structure. If the gist of the claim is well known and obviously controversial, a short period suffices. If the claim is notified to the respondent's agent who has to consider the claim independently and then give a response, a longer period may be required;
6) if the referring party imposes a deadline for a response that does not automatically reflect a reasonable time, but that will be a relevant factor the court will consider;
7) if the claim is so nebulous and ill-defined that the respondent cannot sensibly respond to it, neither silence nor an express non-admission is likely to lead to a dispute.

5.13. Several cases, decided before and after *Amec*, also assist in determining whether a dispute has arisen.

5.14. In *Fastrack v Morrison*,[5] it was determined that a dispute can only arise when a claim has been notified to the other party and that party has been given time to either accept or reject it. Morrison was the main contractor on a leisure complex and engaged Fastrack as their brickwork subcontractor. Delays occurred during the course of the works and Fastrack applied for an extension of time. Morrison refused the extension of time threatening to take some of Fastrack's work away from them. Fastrack threatened legal action. On 26 February 1999 Fastrack submitted an interim application for payment. Morrison served notice on 2 March stating it was going to monitor Fastrack's work, and on 10 March served a notice of payment reducing several amounts claimed by Fastrack. Morrison also advised that they intended to engage a third party to expedite Fastrack's works. Fastrack treated Morrison's decision in this regard as a repudiation of the contract and withdrew from the site. Morrison then served a notice of determination on Fastrack and subsequently made no payment in respect of Fastrack's application for interim payment by the final date for payment on 20 March. On 17 March, Fastrack submitted Application No. 13 for payment of the gross amount of £383,873.97. Morrison then served a notice of set-off for costs they

5 Fastrack Contractors Limited v Morrison Construction Limited [2000] EWHC 177 (TCC).

had incurred to complete the Fastrack works in the sum of £226,177.00, which exceeded the net sum claimed by Fastrack in Application No. 13. The dispute was referred to adjudication and the adjudicator awarded Fastrack the sum of £85,401.98. Morrison refused to pay arguing that at the time the notice of adjudication was served the only dispute in existence related to the matters set out in Application No. 13. However, the sums claimed in the adjudication were different from those claimed in the application. Morrison therefore argued that the amounts set out in Application No. 13 had been superseded by a new claim, which was not yet in dispute as Morrison had not had the chance to consider the new claim and respond. Morrison argued that there was therefore no dispute and the adjudicator had no jurisdiction. In rejecting these arguments and finding that a dispute had crystallised, Judge Thornton QC held that:

> A "dispute" can only arise once the subject-matter of the claim, issue or other matter has been brought to the attention of the opposing party and that party has had an opportunity of considering and admitting, modifying or rejecting the claim or assertion .... The term [dispute] includes any claim which the opposing party has been notified of which that party has refused to admit or has not paid, whether or not there is any answer to that claim in fact or in law.[6]

5.15. The question as to whether a party has had a reasonable opportunity to consider, admit or reject a claim was examined in *Orange v ABB*.[7] ABB subcontracted mechanical services work to Orange. A dispute arose relating to payment and Orange withdrew from site on 28 May 2002. On 2 December 2002 Orange submitted a final account claim. In the intervening period there had been virtually no communication between the parties other than an exchange of correspondence in early July 2002. ABB had not seen a number of items included in the final account before its receipt and the value of the final account was over twice the value of Orange's previous application for payment in March 2002. On 12 December 2002, ABB advised that it considered a dispute had not yet arisen and needed until 20 January 2003 to consider Orange's claim. ABB also advised that if an agreement was not reached within seven days thereafter ABB would accept that there was a dispute, so that the claim could be referred to adjudication. Orange's representatives said they would take instructions and revert, but did not do so and subsequently on 6 January 2003 Orange served a notice of adjudication

6 At paragraph 27 of the judgment.
7 Orange EBS Limited v ABB Limited [2003] EWHC 1187 (TCC).

on ABB. The adjudicator made a decision in Orange's favour, which ABB failed to comply with, arguing that there was no dispute when the notice of adjudication was served.

5.16. HHJ Kirkham found that a dispute had arisen by the time Orange served notice of adjudication on ABB on 6 January at which stage sufficient time had elapsed for ABB to evaluate Orange's final account claim and enter into any discussion or negotiation. Judge Kirkham noted that in reaching her decision she was mindful of the fact that in December ABB had engaged quantity surveyors and cost consultants who were already familiar with the project. The judge found that ABB had been in possession of the final account for about a month, and as such a dispute had arisen which the adjudicator had jurisdiction to decide, and consequently, Orange were entitled to summary judgment to enforce the adjudicator's decision.

5.17. The *Orange* decision supports the view that a claim does not have to be expressly rejected and can arise through non-acceptance or as a consequence of a period of silence. Whether a dispute will be inferred from such silence depends upon the specific circumstances. The period of silence can be very short, even less than a week. Indeed, in *Beck v UK Flooring*[8] HHJ Akenhead held that:

> In normal circumstances, a gap of five days from the time that a claim or assertion is put forward will often be sufficient to give rise to an inference that it is disputed, particularly, as here, where there had been previous claims and where liability was obviously in dispute for some two months before.[9]

5.18. In the Scottish case of *Stirling v Westminster*[10] the court considered the issue of whether the defender's silence could give rise to a crystallised dispute. The court found that the claim had been asserted and the defenders without explanation failed to pay the sum claimed. Lord Drummond Young held that any disagreement as to the parties' respective rights and obligations should appear in writing and it follows that the mere failure to pay is not enough to effect a dispute or difference. However, the judge found on the facts that a dispute did exist before the adjudication started, because Westminster failed to explain their non-payment whilst disputing that the sum Stirling claimed was due. In the court's opinion, this was sufficient to infer that the defenders disputed the claim. If that weren't so, a party could adopt the tactic of simply failing to respond to repeated invoices, claim letters

8 Beck Interiors Limited v UK Flooring Contractors Limited [2012] EWHC 1808 (TCC).

9 At paragraph 26 of the judgment.

10 John Stirling v Westminster Properties Scotland [2007] Scot EF CSOH 117.

and the like, and thus substantially delay any reference to adjudication. Lord Drummond Young stated that:

> I am conscious that adjudication is a provisional procedure. Nevertheless, one of its major purposes is to ensure contractual cash flow, and that could easily be frustrated if the courts were to impose unduly strict conditions on the raising of adjudication proceedings.[11]

5.19. Lord Drummond Young specifically followed the fifth proposition of Judge Jackson in *Amec*, that where the gist of the claim was well known, a very short period of silence may suffice to trigger the inference that a dispute has arisen. Consequently, where a claim is neither admitted nor rejected and is simply met with a period of silence it will very much depend on the specific circumstances of the case as to whether the period of silence is sufficiently long to constitute rejection of the claim and be construed as disputed. Where a claim is met with silence the referring party should allow the responding party sufficient time within which to consider the claim before construing it disputed and seeking to refer the claim to adjudication.

5.20. It is not uncommon for a responding party to reply to a claim by way of a request for further information or substantiation, including additional documents to support the claim. The question then arises whether the request for further information, either represents evidence that a dispute has arisen, or that a dispute cannot arise until the request is responded to and the claim rejected.

5.21. In *Carillion v Devonport*,[12] it was held that a dispute could not arise whilst a request for further information remained unanswered. Devonport engaged Carillion on a fit-out of a submarine dockyard under a subcontract which provided that Carillion would be paid its actual costs plus accruals and a fee. Carillion's payment entitlements were complex and following completion of the works they claimed over £10 million in a payment application submitted on 16 April 2002. Subsequently, on 6 August 2002 Carillion served a notice of adjudication seeking payment. An adjudicator was subsequently appointed on 12 August 2002, making his decision on 24 September 2002. That decision was not honoured by Devonport, who contended that the decision was not binding on them because the adjudicator had no jurisdiction. Devonport argued, *inter alia*, that no dispute had arisen at the time that Carillion served notice of adjudication because Devonport had repeatedly requested clarification and further information of Carillion's claim so they could understand

11 At paragraph 24 of the judgment.

12 Carillion Construction Limited v Devonport Royal Dockyard Limited [2003] HT-02-395.

the basis of the claim. HHJ Bowsher QC had to consider whether a dispute had arisen in circumstances where Devonport had neither denied nor ignored the claim but rather were seeking clarification as to the grounds on which the claim was being made. Judge Bowsher held that Devonport did not just ignore the payment application but had asked for further information on the amounts claimed and the basis on which they were claimed. The judge acknowledged that this was "an everyday occurrence in the construction industry and if every request for information was regarded as a dispute leading to adjudication there would not be enough adjudicators to go around". Judge Bowsher ruled that the adjudicator lacked jurisdiction as a dispute had not crystallised, and commented that:

> The best approach to clarification that they [Devonport] got was so close in time to the notice of adjudication that they had no opportunity to respond to it. The conduct of the adjudication indicates that, if given a reasonable opportunity to respond, there would have been a dispute, but that is not the point.

5.22. Whether a request for further information or documentation constitutes a genuine request or merely an attempt to obstruct or frustrate a claim very much depends on the circumstances of the case. Where a court considers the request for further information and documentation as not genuine it will construe the request as a rejection of the claim giving rise to a crystallised dispute.

5.23. Another circumstance that can give rise to a dispute not having crystallised is where the referring party seeks to amend or alter the basis on which its claim had previously been made.

5.24. In *Nuttall v Carter*[13] it was held that if the basis of the referring party's claim has changed from the original claim then this represents a new claim which until accepted or rejected is not in dispute. Carter, the main contractor for works on a civic community centre, subcontracted Nuttall in respect of sub-structure and building frame concrete works. Several disputes arose which were referred to adjudication. In the case before the courts the adjudication related to Nuttall's application for payment in May 2001, together with additional costs for delay and disruption, the breakdown of which had been provided in May 2001. Following receipt of the payment application the parties had exchanged correspondences before Nuttall issued a notice of adjudication on 14 December 2001. Nuttall's referral included an expert report in support of its claim which adopted different figures and relied upon several different matters than those

13 Edmund Nuttall Limited v RG Carter Limited [2002] EWHC 400 (TCC).

previously submitted to Carter. Carter objected on the basis that the expert report set out a new claim, which Carter had not seen before, and therefore did not relate to the dispute referred to adjudication. The adjudicator nonetheless continued with the adjudication and made a decision in Nuttall's favour. HHJ Seymour QC concluded that a claim needs to be formulated and put to the other party, and does not become a "dispute" until that other party has had an opportunity to consider the claim and reject it. The judge commented that failure to respond within a reasonable time will amount to a rejection. However, Judge Seymour held that in the circumstances the claim advanced by the expert's report was different to the original claim referred to in the notice of adjudication and therefore the adjudicator's decision was made without jurisdiction and was unenforceable.

5.25. The principle established in *Nuttall* is that where the responding party is confronted with a claim (or the basis on which the claim is being made) for the first time in the notice of referral to adjudication then a dispute will not have crystallised as the responding party had not been given the opportunity to consider the claim and reject it before the dispute was referred to adjudication. While each case will be decided on its own facts, the overriding principle is that the responding party must be given an opportunity to consider the claim being made so as to either accept or reject it. Otherwise the court may find no disagreement has arisen, and a dispute has not crystallised.

5.26. In *Sindall v Abner*[14] it was held that in order for a "dispute" to exist it must be clear that "some point upon which the parties disagreed must have arisen from the process of discussion, and when negotiation has ended that there is something that needs to be decided". HHJ Lloyd held there was no dispute (regarding the extension of time claim) because Sindall provided substantial new information, and Abner's contract administrator did not have sufficient time to consider and reject it before the adjudication notice was served.

5.27. In *Collins v Baltic*,[15] the court accepted Judge Jackson's seven propositions in *Amec*, suggesting that a further proposition be added following the *Sindall* decision, namely that a dispute exists when negotiations or discussions have been concluded and matters remain unresolved. Baltic had engaged Collins under a JCT contract. Collins terminated the contract after Baltic failed to pay an interim certificate, and submitted their final account, which under the terms of the contract required Baltic to pay the full amount properly due within

14 Sindall Limited v Abner Solland [2002] Con LRHT 01/129 (TCC).
15 Collins (Contractors) Limited v Baltic Quay Management Limited [2004] EWCA Civ 1757.

28 days of the final account submission. Collins issued court proceedings for judgment on the interim certificate and payment of the final account, on the basis that no withholding notice had been served. They contended there was no arguable defence to the proceedings in the absence of the service of a withholding notice and therefore no "dispute" capable of being referred to adjudication. Baltic argued that a dispute arose quite simply as a result of a refusal to pay. Lord Justice Clarke cited Judge Jackson's seven propositions in *Amec* endorsing the general approach that "while the mere making of a claim does not amount to a dispute, a dispute will be held to exist once it can reasonably be inferred that a claim is not admitted". Lord Justice Clarke, in dismissing the appeal, held that a dispute had arisen stating that:

> It appears to me that negotiation and discussion are likely to be more consistent with the existence of a dispute, albeit an as yet unresolved dispute, than with an absence of a dispute. It also appears to me that the Court is likely to be willing readily to infer that a claim is not admitted and that a dispute exists so that it can be referred to arbitration or adjudication.[16]

5.28. In *Verry v Furlong*,[17] the court held that a response containing further detail of a claim did not constitute a new claim but, rather, was a fuller explanation of the claim originally made. Furlong engaged Verry to design, supply and erect curtain walling. Contract completion was due in late 2003 but did not take place until July 2004. Furlong granted Verry two extensions of time to 2 February 2004 but Verry maintained they were entitled to a further extension to 24 June 2004. In refusing the extension of time Furlong did not address the detail of the claim and instead sought repayment of an alleged overpayment to Verry in the sum of £143,898.15, referring the matter to adjudication. The notice of adjudication was broadly drafted, anticipating that all issues relating to extensions of time and payment issues under the final account should be resolved within 28 days. In Verry's response to Furlong's referral, Verry included a claim for an extension of time to 27 July 2004 which was for a longer period than they had sought in their previous extension of time application. Furlong objected, contending that this was a new claim which was outside the adjudicator's jurisdiction. The adjudicator, however, ruled that he had jurisdiction and found that Verry was entitled to an extension of time to 27 July 2004. Verry subsequently

16 At paragraph 64 of the judgment.

17 William Verry (Glazing Systems) Limited v Furlong Homes Limited [2005] EWHC 138.

issued court proceedings seeking a declaration that the adjudicator's decision was valid and binding. HHJ Coulson QC found that Verry's response to the referral did not include a new claim but was instead a fuller explanation of the original claim advanced prior to issue of the notice of adjudication. The judge held that Verry's extension of time claim in their adjudication response simply reflected the fact that work continued on site after Verry had first outlined its case. The judge noted that the events relied upon in the response were essentially the same events Verry had relied upon prior to the issue of the notice of adjudication, albeit with longer delay periods ascribed and with further supporting information. Judge Coulson found that even if it was a new claim, the scope of the notice of adjudication was so broadly drafted that the adjudicator would have been entitled to consider it. The judge noted that neither the original notice of adjudication nor the referral did anything to restrict the extension of time claim to be considered by the adjudicator and that Verry were entitled to defend themselves as best they could. In reaching his decision Judge Coulson endorsed Judge Toulmin's decision in *AWG v Rockingham*,[18] where the judge stated that:

> The test in each case is first what dispute did the parties agree to refer to the adjudicator? And, secondly, on what basis? If the basis ... argued in the adjudication is wholly different to that which a defendant has had an opportunity to respond in advance of the adjudication, this may constitute a different dispute not referred to the adjudicator or ... insofar as the adjudicator reaches a decision on the new issues, it is not responsive to the issues referred to him.[19]

5.29. Judge Coulson held that the claim for the extension of time to a later date did not constitute a new claim, a dispute had crystallised, the jurisdictional challenge should fail, and the adjudicator's decision should be enforced.

5.30. The court reached a similar view in *Bovis v The Clinic*,[20] where it was held that once it was clear a crystallised dispute had arisen, it was necessary to differentiate between the substance of the dispute referred to adjudication, and the evidence needed to support or contest that disputed claim. The Clinic contracted Bovis to redevelop a London property as a medical facility. Practical completion was achieved 58 weeks late on 4 August 2006. The architect granted Bovis a four-week extension of time and allowed

18 AWG Construction Services Limited v Rockingham Motor Speedway Limited [2004] EWHC 888.

19 Paragraph 141 of the judgment.

20 Bovis Lend Lease v The Trustees of The London Clinic [2009] EWHC 64 (TCC).

£150,000 prolongation costs on account. In late 2006, Bovis made applications for further extensions of time and engaged an expert to prepare a report on loss and expense for the delays. The Bovis expert produced a preliminary report based on estimated figures, calculating a Bovis entitlement to £3.6 million for loss and expense. Bovis subsequently claimed on the basis of this figure, which the Clinic rejected. In March 2007, the architect prepared a detailed review of Bovis' claims rejecting any further extensions of time. Bovis' experts then produced a second report in June 2008 using actual figures based on additional evidence. This second report was served in July 2008 together with other new expert reports, witness statements and a draft referral notice claiming an extension of time, reimbursement of liquidated damages of just over £1.6 million and almost £3.3 million for loss and expense. The Clinic requested, and was granted by Bovis, extra time to respond to that claim. On 18 August 2008 the Clinic advised that they had considered Bovis' claims, that there was insufficient evidence to support them, requesting further information and stating that Bovis had not made a valid payment application under the contract. Bovis commenced adjudication proceedings issuing a final referral notice substantially identical to the draft referral notice that had been furnished earlier in July 2008. The Clinic argued that the adjudicator had no jurisdiction to decide the loss and expense claim as no dispute on that issue had crystallised. They again argued that they had requested further information which was not forthcoming and that Bovis had not made a valid payment application. The adjudicator decided that Bovis was entitled to approximately £3.8 million including £1.8 million for loss and expense. The Clinic refused to pay on the basis of its original objection to the adjudicator's jurisdiction and a further challenge, raised for the first time, that they had been ambushed in the adjudication and this amounted to a breach of natural justice. Bovis issued court enforcement proceedings. On the jurisdiction issue, the court held that a "dispute or difference" did not arise merely upon notification of a claim or expiry of a deadline set by the claimant, but at a point when the claim was "not admitted", which depends on the surrounding circumstances. The court cast doubt on the line of authority relied on by the Clinic that a request for further information did not amount to non-admission of the claim. Unless the claim was so "nebulous and ill defined" that the respondent could not sensibly respond to it, the court thought that a dispute would usually arise notwithstanding a request for information. The court held that the Clinic had not expressed the request as seeking further information to understand the nature of Bovis' claim, but rather to defend the claim itself. HHJ Akenhead held that once a dispute had crystallised, a distinction should be drawn between the substance of that dispute and the evidence required to support or contest the disputed claim. The judge held that given the Clinic's detailed consideration (and non-acceptance) of the draft referral notice, within a time limit set by the Clinic itself following its request for further time, the whole draft referral was in fact disputed. Judge Akenhead further stated that he did not therefore have to decide whether or not the "new" quantum claim put forward in the draft referral notice was so new that it amounted to a new claim or merely new evidence, although it appeared to suggest that the latter was the case. The judge further held that the issue of whether Bovis had made applications for payment of loss and expense in accordance with the contract was part of the dispute and the adjudication clause should be construed in a purposive and commercial way so that it was not limited to claims established under the contract. The judge enforced the adjudicator's decision dismissing the Clinic's claim that it had been ambushed in the adjudication in a breach of natural justice.

5.31. The common thread throughout the UK case law on whether a dispute has crystallised, particularly where the respondent requests further information or seeks to rely upon a prescribed or agreed timeline, is that each case will essentially be decided upon its own facts. In *All-In-One* ("AIO") *v Makers*,[21] Makers subcontracted AIO in November 2004 on a flat refurbishment project. In July 2005, issues arose regarding an alleged lack of supervision and insufficient labour on site and Makers issued notice of its intention to terminate the subcontract. AIO claimed the subcontract had been repudiated, that a dispute had arisen, and that they would be claiming damages for Makers' breach of contract. Shortly thereafter on 15 August, AIO issued a draft assessment of account to Makers which included a claim for interest and overheads as a result of the alleged repudiatory breach. The claim was not supported by detailed information. On the same date AIO issued a referral notice to adjudication stating that the nature of the dispute was the non-payment of AIO's 15 August application, for which Makers had failed to serve a withholding notice, claiming that Makers was in repudiatory breach of contract. An adjudicator was appointed and decided in AIO's favour awarding a sum of money. Makers did not comply with the decision and AIO commenced enforcement proceedings. Makers challenged the enforcement, claiming there was no dispute at the time of the referral notice because AIO's assessed account was either an interim payment application or akin to a draft final account, and the time allowed under the subcontract for payment of that account had not lapsed by the time of the referral. HHJ Wilcox, citing *Collins*[22] and *Amec*,[23] stated that the proper approach to the question of whether a dispute has arisen is to adopt a rigorous,

21 All-In-One Building & Refurbishments v Makers UK [2005] EWHC 2943 (TCC).
22 [2004] EWCA Civ 1757.
23 [2004] EWHC 2339 (TCC).

common-sense approach and look at the substance of the claims rather than the labels attached to them. The judge held that a claim could be considered a dispute even though the contractual time limit for paying the interim application, or even a final account, had not lapsed. The judge held that a distinction should be made between the date for payment and an entitlement to payment, and that in the claims before the court it was the entitlement to payment that was being denied. Judge Wilcox concluded that it was clear from the outset that both parties accepted the contract had been wrongfully ended, but each blamed the other for that. It was clear that Makers was not going to pay the sums AIO claimed for the alleged financial consequences of the breach, whatever the level of supporting evidence that might have been provided. Judge Wilcox consequently granted summary judgment for AIO enforcing the adjudicator's decision.

5.32. In respect of interim payment applications under construction contracts, Section 4(3) of the Act provides that if a paying party contests any amount claimed, then it must deliver a response to the interim payment claim notice not later than 21 days after the payment claim date. The response must specify the amount the paying party proposes to pay, the reasons for the difference between the amount as proposed and the amount claimed and the basis on which the proposed amount is calculated. A similar provision exists under the UK legislation requiring the paying party to issue a "withholding notice" in respect of interim payment applications where the paying party contests the amount being claimed. The English Courts have considered a number of cases relating to the crystallisation of a dispute in circumstances where withholding notices have not been properly served or where a claiming party has sought to refer a dispute to adjudication in advance of service of a withholding notice.

5.33. In *Working Environment v Greencoat*[24] the court considered whether a dispute over an interim valuation had crystallised prior to Greencoat serving a withholding notice. Greencoat subcontracted Working Environment to carry out mechanical services on an office fit-out works. A dispute arose over Greencoat's valuation, on 2 December 2011, of Working Environment's Interim Application No. 10, plus the value of nine items that Greencoat sought to set off by way of a withholding notice. On 14 December 2011 Working Environment referred the dispute to adjudication, requesting that the adjudicator decide, among other things, the amount to be paid to them when payment fell due on 14 January 2012. Greencoat raised numerous jurisdictional challenges, including that no material dispute had

24 Working Environment Limited v Greencoat Construction Limited [2012] EWHC 1039 (TCC).

crystallised as the date for payment had not occurred. Following the adjudication commencing, and prior to serving its response in the adjudication, Greencoat issued a second withholding notice including 12 items, two of which had not previously been mentioned in its previous withholding notice. The adjudicator rejected Greencoat's jurisdictional argument and determined that Greencoat should pay Working Environment when payment fell due, including the two further items in his determination. Greencoat did not honour the adjudicator's decision and Working Environment issued court enforcement proceedings. Greencoat argued that the dispute had not crystallised because the matter had been referred to the adjudicator before the relevant withholding notice (the second one) had been served by them. HHJ Akenhead rejected this argument on the basis that ten of the 12 matters that had been referred to adjudication had already crystallised as disputes before service of the first withholding notice which had preceded the referral to adjudication. Judge Akenhead held that the court could sever the two matters which had not crystallised as disputes (and which had not in fact been referred to the adjudicator at all) and enforce the other ten valid matters in the adjudicator's decision. Judge Akenhead further held that a dispute over an interim valuation, before the date of payment, was referable to adjudication. In respect of two of the twelve items in the decision, the judge held that the adjudicator lacked jurisdiction, as they had not arisen until some 22 days into the adjudication process when Greencoat had issued its second withholding notice.

5.34. The issue as to whether a dispute has crystallised is essentially an issue as to whether an adjudicator has jurisdiction to decide the claim it is alleged has not crystallised as a dispute. The UK Courts have been willing to sever those matters where it considers a dispute has not crystallised from those matters decided upon by an adjudicator where the court is of the view that a dispute has crystallised. The Irish Courts will likely adopt a similar approach in respect of severing issues in an adjudicator's decision.

5.35. In *Beck v UK Flooring*[25] Beck claimed damages against UK Flooring for repudiation of contract. Beck wrote to UK Flooring after close of business on the day before Good Friday, adding a further claim for LADs. Beck then served the adjudication notice on the first business day after Easter. UK Flooring argued that the adjudicator had no jurisdiction over the LADs claim as no dispute had crystallised as they did not have an opportunity to consider the LADs claim and either accept or reject it. Judge Akenhead agreed with UK Flooring's position for two reasons. Firstly, the judge held that Beck

25 Beck Interiors Limited v UK Flooring Contractors Limited [2012] EWHC 1808 (TCC).

must have known that when they served their letter the day before Good Friday UK Flooring would not have an opportunity to consider the LADs claim prior to Beck serving the adjudication notice the first day after Easter. Secondly, the LADs claim in the adjudication notice was quite different from the claim which had been set out in Beck's letter sent the day before Good Friday. Judge Akenhead specifically stated that:

> It is of course always a matter of fact and degree … as to whether amendments to a disputed claim fall within the umbrella of the original dispute.[26]

5.36. Consequently, the court held that the adjudicator's decision on the LADs claim was unenforceable, and severed the decision on the claim, leaving the remainder of the decision enforceable.

Does Section 6(1) require that only a single "dispute" may be referred?

5.37. Section 6(1) of the Act provides that a "dispute" and not "disputes" may be referred to adjudication. By their very nature construction claims tend to encompass numerous interrelated issues. Thus, a very narrow interpretation of what constitutes the "dispute" could lead to a series of disputes being referred to a series of adjudications, all arising out of the same set of circumstances that may have occurred during the contract and given rise to the various claims in dispute. This issue as to whether a dispute referred is actually a series of separate disputes or is a single dispute with a number of component sub-disputes has come before the UK Courts on many occasions. In each case, the UK Courts have adopted a pragmatic approach on the basis that significant problems could arise if the term "dispute" were too narrowly interpreted.

5.38. In *McLean v Swansea*,[27] HHJ Lloyd QC held, in determining whether an adjudication notice related to one or more than one dispute, that it was appropriate to apply a "sensible" or "benevolent" construction of the term "dispute", bearing in mind that a single dispute might well consist of several discrete elements. McLean contended that it was entitled to an extension of time and Swansea, in contesting the same, gave notice of intention to deduct LADs. McLean issued an adjudication notice for monies due on a payment application which included claims for loss and expense, variations, measured work and

26 At paragraph 29 of the judgment.

27 David McLean Housing Contractors Limited v Swansea Housing Association Limited [2001] EWHC 830 (TCC).

provisional sums adjustments. The adjudication notice also sought an extension of time determination. Swansea argued that McLean had sought to refer a number of disputes to adjudication, and that McLean's payment claim and their claim seeking an extension of time were two distinct disputes. They argued that the UK Act required McLean to separately refer each dispute to adjudication and both disputes could not be decided in one adjudication. The judge held that the dispute related to the amount due to McLean under the payment application, irrespective as to the individual component factors giving rise to the claim for payment. The judge held that a decision as to McLean's entitlement to an extension of time was a necessary and indispensable precursor to loss and expense payable in respect of the same. Judge Lloyd held that the adjudication notice did not refer more than one dispute and that the court should adopt a sensible approach and construe McLean's adjudication notice as referring a single dispute to adjudication, namely "how much should I be paid?".

5.39. This approach was adopted by Judge Thornton in *Fastrack v Morrison*[28] when he held that if the dispute is characterised as "what sum is due?" then that is a single dispute which can encompass a multitude of individual and complex sub-issues.

5.40. Judge Humphrey Lloyd QC reached a similar view in *Sindall v Abner*[29] stating that:

> Where a dispute is referred, there is comprehended within it all constituent elements, including sub-disputes, contentions, issues (some of which might have been referred separately) – in other words all the ingredients which go into the dispute referred.[30]

5.41. In most construction projects the contractor's final account claim often contains a whole series of issues relating to matters in dispute, which can be interrelated and quite complex. In addressing this issue, the UK Courts have adopted the approach that a final account claim effectively represents a claim as regards "what money is due to the contractor?", and considered that addressing this question constitutes a single dispute capable of referral to adjudication.

5.42. In *Witney v Beam*,[31] Beam's notice initiating the adjudication claimed money and time, and a claim that Witney were in repudiatory breach of the contract. The adjudicator was appointed, and Beam served a referral notice setting out full details of its claim. Witney immediately raised an issue that more than one dispute had been

28 Fastrack Contractors Limited v Morrison Construction Limited [2000] EWHC 177 (TCC).

29 Sindall Limited v Abner Solland [2001] 2 TCLR 30.

30 At paragraph 16 of the judgment.

31 Witney Town Council v Beam Construction (Cheltenham) Limited [2011] EWHC 2332 (TCC).

referred to adjudication. Witney argued that, in fact, there were four disputes referred in the single adjudication which included two disputes relating to the draft final account, the final account submitted by Beam, a claim for interest on underpayment of retention and a claim for the payment of the whole retention based on repudiatory breach. The adjudicator made it clear that he did not consider Witney's position as tenable and that they had a poor argument. Witney reserved its position. HHJ Akenhead held that only a single dispute as to what monies were due and owing to Beam had been referred to adjudication. Judge Akenhead's decision included the following conclusions:[32]

a) A dispute arises generally when, and in circumstances in which, a claim or assertion is made by one party and expressly or implicitly challenged or not accepted.
b) A dispute in existence at one time can in time metamorphose into something different to that which it was originally.
c) A dispute can comprise a single issue or any number of issues within it … a dispute … does not necessarily comprise everything which is in issue between them at the time that one party initiates adjudication … everything in issue at that time does not necessarily comprise one dispute, although it may do so.
d) What a dispute in any given case is, will be a question of fact albeit that the facts may require to be interpreted. Courts should not adopt an over legalistic analysis of what the dispute between the parties is, bearing in mind that almost every construction contract is a commercial transaction and parties cannot broadly have contemplated that every issue between the parties would necessarily have to attract a separate reference to adjudication.
e) The Notice of Adjudication and the Referral Notice are not necessarily determinative of what the true dispute is, or as to whether there is more than one dispute. One looks at them but also at the background facts.
f) Where, on a proper analysis, there are two separate and distinct disputes, only one can be referred to one adjudicator unless the parties agree otherwise. An adjudicator who has two disputes referred to him, absent the agreement of the parties, does not have jurisdiction to deal with more than one dispute.

5.43. Determining whether there are one or more disputes referred requires consideration of the facts. If there is a clear link between two or more arguably separate claims or assertions, that link may point to there being

32 Paragraph 38 of the judgment.

just one dispute. If, for example, disputed claim No. 1 cannot be decided without deciding all or parts of disputed claim No. 2, that establishes a clear link pointing to there being only one dispute.

5.44. In UK adjudication cases, where there are several payment claims under the same contract, the UK Courts generally treat them as representing one dispute, namely how much money is owed by one party to the other. Difficulties arise where there are payment disputes and also disputes over other aspects of the contract, as in the *Witney Town Council* case where there was a dispute relating to repudiatory breach, although there were financial consequences of that dispute in the context of release of retention following the repudiatory breach.

5.45. Under the Act, where only payment disputes may be referred to adjudication, Irish Courts are likely to adopt a very similar position to the UK Courts and find that where several payment disputes arise under one contract, those apparently separate disputes may actually be characterised as a single dispute namely what sum is due by one party to the other.

5.46. From the referring party's perspective seeking to refer a single dispute to adjudication, it is critical to carefully draft the adjudication notice and referral to ensure that the payment dispute being referred can be characterised in its totality as "what sums are due" to the referring party. Such an approach reduces the prospects of a challenge that more than one dispute has been referred.

The dispute must be "relating to payment"

5.47. A significant distinction between the Irish and UK Acts is that only payment disputes may be referred to adjudication under the Irish Act whereas any dispute may be referred under the UK Act. There has been speculation that the Irish Courts will interpret the phrase "relating to payment" narrowly, construing the phrase as meaning that only payment disputes arising under Section 4 of the Act (regarding payment claim notices) may be referred to adjudication. It is submitted that this approach is misconceived, and it is more likely that the Irish Courts will adopt a broad interpretation of the phrase "relating to payment". It is anticipated that any dispute relating to payment may be referred to adjudication, irrespective of whether or not such disputes arise out of the contract payment provisions or the circumstances envisaged under Section 4 of the Act.

5.48. It is noteworthy that Section 6(1) uses the phrase "any dispute relating to payment" and not simply "payment dispute". While, for convenience, we use the shorter phrase "payment dispute" in this book, the term "payment dispute" could conceivably be construed more narrowly than the phrase "dispute relating to payment". The latter phrase arguably includes disputes which are not simply payment disputes but also disputes on claims, which when decided the outcome relates to what payment one party to the dispute must make to the other. It is difficult to envisage a claim under a construction contract that could not in some way be characterised as a claim relating to payment or money. Indeed, one would question as to why a referring party would refer a dispute to adjudication where there is no financial or payment implication arising from the adjudicator's decision. Even where a dispute relates to the correct interpretation of a contract provision, or as to whether a standard form has been incorporated into a contract, the consequences of a determination of such matters would likely ultimately culminate in a claim for monies due and owing from one party to the other. However, any claimant seeking to refer to adjudication a dispute over the correct interpretation of a contractual provision would be well advised to include, within its notice to refer, the payment consequences (and preferably a monetary sum) arising from any decision as regards their contended interpretation of the contractual provision.

5.49. A party seeking to refer a dispute involving an issue which may not be immediately apparent as having payment or cost consequences may face a challenge from the opposing party that the referred matter is not a payment dispute, and therefore cannot be referred to adjudication under the Act. Consequently, a prudent referring party should carefully think through the issue being referred, and clearly articulate the payment consequences of the adjudicator's decision on the issue, so as to minimise prospective challenges that the issue referred does not relate to payment.

5.50. Notwithstanding the likelihood that most construction contract disputed claims could be articulated as disputes relating to payment, it is nonetheless regrettable that the Irish legislature chose to limit referrals to adjudication to payment disputes. This is particularly so as it is not uncommon for parties to seek to bifurcate their dispute into liability and quantum, often leading to the benefit of resolving the dispute in a more cost-effective and efficient manner. It would not be possible to bifurcate disputes in this manner under the Act, and consequently the liability and quantum of disputed matters would always have to be referred to, and considered by, the adjudicator at the same time if the dispute were to be properly characterised as a payment dispute.

Disputes arising under the contract

5.51. There is not a substantial amount of Irish case law regarding interpretation of the phrase "arising under the contract" in the context of disputes. Most recently, in *Gulliver v Brady*,[33] Geoghegan

33 [2003] IESC 68; unreported, Supreme Court, Geoghegan J., 19 December 2003.

J. considered the meaning of the phrase "any disputes between the parties ... arising out of or in relation to ..." and approved the following passages from *Commercial Arbitration*[34] setting out Mustill and Boyd's understanding of the meaning of the phrase "arising out of", namely:

> These words have been given a wide meaning. It has been said that they cover every dispute except a dispute as to whether there was ever a contract at all. If the parties to a contract make provision in it as to their rights should certain events occur in the course of the contract, and a dispute arises between them as to their rights following the occurrence of those events, then that dispute as to their rights arises out of the contract.[35]

5.52. The issue as regards the meaning of "arises under the contract" has been exhaustively debated in the UK Courts, and most recently received extensive consideration in the *Fiona Trust* case.[36] The House of Lords considered the scope of an arbitration clause and held that it was to be assumed the parties likely intended that any dispute arising out of the contract, including disputes over the validity of the agreement itself, were to be decided by the arbitrator unless express words in the clause excluded the same.

5.53. However, thereafter in *Camillin v Adelaide*,[37] Judge Akenhead expressed uncertainty as to whether *Fiona Trust* was authority on an adjudicator's jurisdiction and that a narrower approach should be adopted. Irish Courts could possibly choose to adopt Judge Akenhead's approach in *Camillin* and narrowly construe the phrase "arising under". There are a number of reasons why the Irish Courts might be justified in taking this approach.

5.54. Firstly, the House of Lords in *Fiona Trust* was considering the "contract" negotiated and agreed between the parties, and as such the parties' intentions, as regards how disputes would be dealt with (on the face of it by arbitration) in relation to the contract, were extensively considered. However, referral of a payment dispute to adjudication is a statutory entitlement. The courts' interpretation of a statutory provision (in this case payment disputes "arising under the construction contract") might well be dealt with differently given that the Irish Courts have historically adopted the view that the legislature intends statutes to be interpreted on a literal meaning of

34 Mustill and Boyd, *Commercial Arbitration* (London: Butterworths, 1989).

35 At page 120, Mustill & Boyd.

36 Fiona Trust and Holding Company and Others v Yuri Privalov [2007] 4 All ER 951 (HL).

37 Camillin Denny Architects Limited v Adelaide Jones & Company Limited [2009] EWHC 2110 (TCC).

provisions. The courts could well take the view that the legislature specifically intended that the phrase "arising under" be narrowly interpreted, as otherwise the Act could have also included the broader phrases "in connection with", "in relation to" and "out of", but chose not to do so.

5.55. Secondly, given that adjudication is designed as a fast-track procedure where the adjudicator's jurisdiction is strictly interpreted, one would have thought the legislature had this in mind when specifically choosing the phrase "arising under the construction contract". Consequently, the courts could well consider the interpretation of the phrase "arising under" in the context of the speedy nature of adjudication and conclude that the phrase be interpreted narrowly.

5.56. Thirdly, as only payment disputes "arising under the construction contract" may be referred to adjudication, the legislature would be reluctant to see the courts give a wide interpretation to the term on the basis that referral of a payment dispute which related to the contract but not under the contract, could have potential to expand the nature of disputes in a disorderly fashion.

5.57. While the courts may narrowly interpret the phrase "arising under", they may take a wider view as to how the contract is construed, or exists, or where the parties have expanded the provisions of the contract itself. The English Courts have frequently considered the issue of whether disputes arising under side or supplemental agreements to an original construction contract can be referred to adjudication, and as to whether the side or supplemental agreement is an extension of the construction contract, and the dispute thereby arises under the construction contract itself.

5.58. In *Browne v Crosby*,[38] the court held that an adjudication agreement was wide enough to cover disputes about side agreements, as the adjudication provision in the contract specifically referred to disputes arising under, out of, or in connection with, the contract. In *Air Design v Deerglen*,[39] the parties entered into several supplementary agreements following their initial agreement. The court held that the adjudicator had been correct in treating the subsequent agreements as variations of the original contract and that he had jurisdiction to adjudicate on disputes regarding both the contract and its variations, as the variations arose under the contract. In *Amec v Thames Water*,[40] the court held that where the written adjudication provision was contained in a framework agreement, it applied to any disputes arising as regards work packages which formed separate contracts under

38 Browne & Son v Crosby North West Homes [2009] EWHC 3503 (TCC).

39 Air Design (Kent) Limited v Deerglen Limited [2008] EWHC 3047 (TCC).

40 Amec Group v Thames Water Utilities Limited [2010] EWHC 419 (TCC).

the agreement. The issue of whether a dispute arises "under" the construction contract can extend into consideration as to the correct party to a contract, disputes arising after the contract has been rescinded and as to misrepresentation.

5.59. In *Michael John v Golledge*,[41] the court held that a dispute regarding who the correct employer was (and therefore who was liable for payment) concerned the obligations owed "under" the contract by the employer to the contractor, and accordingly the adjudicator had jurisdiction to decide the issue. In *Carter v Nuttal*,[42] the court held that the question as to whether a particular term has been incorporated into an agreement falls within the adjudicator's power to decide. In *Barr v Law Mining*,[43] the responder argued that the contract had been rescinded and the adjudicator could not determine the payment claim for work done as it didn't arise under the contract. The adjudicator declined to answer the rescission issue but awarded sums for the relevant work. The court held that the adjudicator would only have had jurisdiction if he had first decided that there had been no rescission. If the adjudicator decided that a contract had been rescinded, any disputes over matters under the rescinded contract could not come within his jurisdiction.

5.60. In *Hillcrest v Beresford*,[44] it was held that claims involving negligent misstatement and misrepresentation were not disputes arising under the contract but rather under the law of negligent misstatement or misrepresentation, and an adjudicator did not have jurisdiction to deal with them. Beresford agreed to design and construct a residential property for Hillcrest under a contract which provided that any dispute arising under the contract be referred to adjudication under the UK Scheme. The employer's requirements ("ERs") prescribed for the structural engineer's appointment to be novated to Beresford when the parties executed the contract. However, the structural engineer was reluctant to enter into the novation and did not execute the deed of novation until October 2012. By that time, the building works had reached practical completion. Beresford then refused to execute the deed of novation and started adjudication seeking several declarations concerning the novation's validity. The adjudicator found that Hillcrest made a negligent misstatement in the ERs regarding the novation, which induced Beresford to enter into the contract, that the negligent misstatement was a misrepresentation entitling Beresford to damages and that the deed of novation

41 Michael John Construction v Richard Henry Golledge & Others [2006] EWHC 71 (TCC).
42 RG Carter Limited v E Nuttal Limited [2000] HT-00-230 (TCC), 21 June 2000.
43 Barr Limited v Law Mining Limited [2001] Scot CS 152.
44 Hillcrest Homes v Beresford & Curbishley [2014] EWHC 280 (TCC).

executed by the structural engineer was void and his appointment had not been novated. Hillcrest issued court proceedings seeking declarations to the effect that the adjudicator's decision was unenforceable on the grounds that he lacked jurisdiction and breached the rules of natural justice. The court was asked to decide, *inter alia*, whether the claims for misrepresentation and/or negligent misstatement were within the scope of the adjudication clause and consequently whether the adjudicator's decision was enforceable. HHJ Raynor QC held that the adjudicator's decision was unenforceable because the adjudication clause only provided for "disputes under the contract" to be referred to adjudication, which was not wide enough to cover a misrepresentation/negligent misstatement claim. The judge noted that in *Fiona Trust* the House of Lords held, in relation to arbitration clauses, that unless the language of these clauses made it clear that certain questions were intended to be excluded from the arbitral award they should be construed widely to include any dispute arising out of the relationship the parties had entered into. However, Judge Raynor drew a distinction between *Fiona Trust* and the case before him on the basis that adjudication clauses are present or implied by reason of statutory intervention, and whereas the adjudication clause in the case before him concerned disputes arising "under the contract", the arbitration clause in *Fiona Trust* was expressed in much wider terms as relating to "any dispute or difference … of any kind whatsoever arising out of or in connection with this contract". Consequently, Judge Raynor held that the respective draftsmen of these clauses must be taken to have intended that the disputes capable of referral to arbitration were wider than those capable of referral to adjudication.

5.61. Subsequently, in *Murphy v Maher*[45] HHJ Akenhead considered a short, simple but interesting point about an adjudicator's jurisdiction; namely, where there is a dispute as to whether there has been a full and final settlement agreement reached by the parties on the final account, whether the dispute arises "under" the subcontract or under the settlement agreement or both. Judge Akenhead held that

> a dispute as to whether all or some of the alleged entitlements which one contractual party has against the other has been settled in a binding way arises "under" the original contract. That is wholly logical because what is supposedly settled is the alleged entitlement to be paid "under" the original sub-sub-contract … . It would be extraordinary and illogical if the parties here or Parliament had intended that an otherwise properly appointed adjudicator would

45 J Murphy & Sons Limited v W Maher and Sons Limited [2016] EWHC 1148 (TCC).

have jurisdiction if addressing what entitlement a contractor or subcontractor might have to be paid in all circumstances save in relation to where a dispute arises as to whether that entitlement had been settled. If Murphy was right, save by ad hoc agreement, one could never adjudicate in a construction contract on an interim or final account which had been agreed in some binding way; that makes commercial and policy nonsense in circumstances in which such agreements must occur all the time and should be encouraged and supported by retaining the right to adjudicate if one party seeks to challenge the settlement on one basis or another.[46]

A dispute may be referred to adjudication "at any time"

5.62. Section 6(2) of the Act provides that:

> The party may exercise the right by serving on the other person who is party to the construction contract at any time notice of intention to refer the payment dispute for adjudication.

5.63. The key aspect of Section 6(2) is a party's entitlement to exercise its right to refer a payment dispute for adjudication "at any time". UK Courts have expansively interpreted a similar provision under Section 108(2)(a) of the UK Act. It has been held that "at any time" can mean that the notice of intention to refer can be served, *inter alia*, where court or arbitration proceedings have already commenced and are ongoing, where the underlying contract has come to an end (even if repudiated) or where there is a precondition in the contract requiring parties to enter into some process (such as conciliation or mediation) before referring disputes to adjudication.

5.64. The phrase "at any time" literally means that a party's statutory entitlement to refer a dispute to adjudication cannot be limited. Section 2(5)(b) (which prohibits contracting out of the Act) reinforces the parties' statutory entitlement to refer their payment dispute to adjudication, notwithstanding that the contract may include provisions which require the parties to refer their payment disputes to mediation or conciliation prior to adjudication. In the latest version of the RIAI Contract,[47] Clause 38(a)(i) provides that

> should any dispute arise in relation to this contract – and it is not resolved by negotiation – the parties should enter into conciliation to resolve such dispute before recourse to statutory adjudication or arbitration.

46 At paragraph 32 of the judgment.
47 2017 Editions of the Yellow and Blue Forms of RIAI Construction Contracts.

5.65. Whilst the RIAI Contract does not seek to prohibit a party from referring a dispute to adjudication, the choice of language providing that a party "should" refer a dispute to conciliation "before recourse" to statutory adjudication is misleading and could easily be construed (particularly by a non-lawyer) as meaning that a party seeking to refer a dispute to adjudication is obliged, in the first instance, to refer its dispute to conciliation. This interpretation is wrong but would be a consequence of the misleading wording of Clause 38(a)(i) of the RIAI Contract. Despite the intention of the RIAI provisions, a party that wishes to refer a payment dispute to adjudication may do so at any time in advance of, or during, conciliation or even after issue of a conciliator's recommendation. In addition, provided that a dispute has crystallised, a party need not even enter into negotiation as provided for in Clause 38(a)(i) of the RIAI Contracts, nor any other dispute resolution process. Having said that, it is generally advisable that, before a party takes any formal steps in a dispute resolution process, particularly a statutory dispute resolution process such as adjudication, the parties explore the possibility of resolving their dispute through settlement and negotiation, as this will likely elicit a preferable commercial outcome for the parties.

5.66. The UK Courts have repeatedly held that a party's entitlement to refer a dispute to adjudication "at any time" cannot be impaired by any contractual provision imposing a precondition to recourse to mediation or conciliation. Indeed, in *Mowlem v Hydra-Tight*,[48] the court held that a contractual provision that a dispute may not be referred to adjudication until after the disputing party has served notice of dissatisfaction with an engineer's decision conflicts with the party's statutory right to refer a dispute to adjudication "at any time", as such a precondition cannot impede a party's statutory right.

5.67. In *Carter v Nuttall*,[49] the parties disagreed as to the full extent of the terms of, and documents incorporated into, their contractual relationship. Following disputes arising, Nuttall commenced adjudication proceedings, but as soon as the adjudicator was appointed Carter sought a court injunction. Carter argued that without any clear agreement or finding as to the full extent of the contract between the parties Nuttall could not seek either statutory or contractual adjudication. Furthermore, they argued that the contract incorporated a clause requiring mandatory mediation before commencing adjudication. HHJ Thornton QC held that the parties accepted that there was a construction contract between them and that a dispute under the contract existed as to whether a particular document was

48 John Mowlem v Hydra-Tight & Co [2002] 17 Const LJ 358.
49 RG Carter Limited v E Nuttal Limited [2000] HT-00-230 (TCC), 21 June 2000.

a contractual document. Judge Thornton further held that the mandatory mediation provision was inconsistent with a party's unqualified statutory entitlement to refer disputes to adjudication. Consequently, the judge held that Carter was not entitled to injunctive relief, and so long as a dispute had arisen Nuttall may refer it to adjudication.

5.68. A party's entitlement to refer a payment dispute "at any time" also encompasses an entitlement to do so after the contract has been repudiated or terminated, and indeed subsequent to the commencement of court or arbitration proceedings. The UK Courts have dealt with numerous cases upholding a party's entitlement to refer a dispute "at any time", or at the very least determined that right as co-existing with an obligation or entitlement to proceed by way of arbitration or court proceedings.

5.69. In *Herschel v Breen*[50] Judge Dyson held that the principle that a party is prohibited from claiming in two separate proceedings had no application to adjudication. The judge noted that as an adjudicator's decision was not final and binding the relevant dispute could ultimately be determined by a court or arbitrator, and the adjudicator's decision only had a temporary effect. The judge held that it therefore followed that the words "at any time" meant that even if separate court proceedings had been commenced, a party was still entitled to refer the same dispute to adjudication.

5.70. In *Camden v Makers*,[51] HHJ Akenhead refused a court order sought by Camden to prevent Makers from referring a dispute to adjudication in the event that Makers receive an order setting aside an earlier judgment in default, which Camden had obtained against Makers. The judge confirmed a party's statutory right to adjudicate at any time and held that the legislature would have addressed it in the legislation had it intended that a party could not institute adjudication if there was already court or arbitration proceedings initiated. This was even where Camden stressed that there was a threat that Makers would become insolvent during the intervening period, and that if Makers received an adjudicator's decision in their favour then there was no prospect of Camden recovering those costs because of Makers' perilous financial situation.

5.71. In *Connex v MJ Building*[52] the English Court of Appeal held that a dispute may be referred at any time, even after the contract has been repudiated or terminated, or expiry of the limitation period. However, the limitation defence might be relied upon in seeking to

50 Herschel Engineering Limited v Breen Property Limited [2000] EWHC Technology 178.

51 London Borough of Camden v Makers UK Limited [2009] BWHC 605 (TCC).

52 Connex South Eastern Limited v MJ Building Services Group Plc [2005] EWCA Civ 193.

defeat the claim in adjudication. Connex argued that MJ's delay in referring the dispute to adjudication, 15 months after MJ accepted what they believed was a repudiatory breach of contract by Connex (which had the effect of terminating the contract) was an "abuse of process". Dyson LJ held that the phrase "at any time" meant exactly what it said, and had the legislature intended to restrict the time frame for adjudication referral, they would have done so in the UK Act. Dyson LJ held that there was no time limit to a party referring a dispute to adjudication.

5.72. While UK Courts are generally reluctant to enforce adjudication decisions in favour of companies in liquidation, in *Lonsdale v Bresco*,[53] HHJ Fraser considered whether a company in liquidation can refer a dispute to adjudication. The judge found that after liquidation, the only dispute that remains in law is that of "taking the account" under the insolvency rules of the mutual dealings between the parties, and an adjudicator cannot conduct such taking of an account. Consequently, the judge held that:

> A company in liquidation cannot refer a dispute to adjudication when that dispute includes (whether in whole or in part) determination of any claim for further sums, said to be due to the referring party from the responding party.[54]

5.73. Where an insolvent party refers a dispute to adjudication, or where a party that is not insolvent refers a dispute but later becomes insolvent before the adjudication concludes, that party should discontinue such proceedings because adjudication ceases to be a forum for payment disputes from the moment the company goes into liquidation. It follows that should the insolvent company refuse to discontinue the adjudication, then the adjudicator must resign. In this regard Judge Fraser noted in *Lonsdale* that:

> I would be surprised if many, or indeed any, adjudicators would decline to resign if a respondent brought the relevant passages of this case to his/her attention during an adjudication.

Stay of court or arbitration proceedings

5.74. It is also arguable that "at any time" entitles a party, who is referring a payment dispute to adjudication, to seek to stay any court or arbitration proceedings that may have been commenced by the opposing party in respect of the payment dispute. In a construction contract

53 Michael J. Lonsdale (Electrical) Limited v Bresco Electrical Services Limited (in Liquidation) [2018] EWHC 2043 (TCC).

54 At paragraph 76 of the judgment.

which is silent as to the parties' entitlements regarding disputes or which specifically provides that either party is entitled to refer any dispute to arbitration or court, then where either party seeks to refer a payment dispute to court or arbitration, the opposing party would be entitled to nonetheless refer the payment dispute to adjudication. The UK Courts' decisions on this issue will be informative should Irish Courts hear the matter of a party's entitlement to stay such proceedings.

5.75. In *DGT v Cubitt*[55] the court considered whether to order a temporary stay on the court proceedings so as to allow an adjudication to proceed. Judge Coulson determined three relevant principles:

a) The court will not grant an injunction to prevent one party from commencing and pursuing adjudication proceedings even if there is already court or arbitration proceedings in respect of the same dispute.
b) The court has an inherent jurisdiction to stay court proceedings issued in breach of an agreement to adjudicate.
c) The court's discretion whether to grant a stay should be exercised in accordance with the aforementioned principles.

5.76. The contractor Cubitt subcontracted DGT for external cladding works under a subcontract containing an adjudication clause that "Any dispute, question or difference arising under or in connection with the subcontract shall, in the first instance, be submitted to adjudication". The judge found that the word "shall" in the clause meant there was a binding agreement to adjudicate. The judge also held that the court had an inherent jurisdiction to stay any proceeding brought in breach of an agreement to adjudicate. Judge Coulson dismissed the argument that because the UK Act merely provided a party to a construction contract with a right, and not an obligation, to refer a dispute to adjudication, the adjudication clause should be construed in the same way. The judge further noted that the UK Act conferred the right to adjudicate a dispute, not a claim, on any party to a construction contract, and even if the claiming party had started court proceedings, the defending party might be entitled to a stay of the proceedings based on its right to adjudicate being denied. Consequently, Judge Coulson granted a stay of the court proceedings so as to enable the adjudication to proceed. The judge also held *obiter* that as Section 108 of the UK Act provided for an adjudication right, court proceedings should be stayed, unless there was a solid reason not to grant such a stay.

55 DGT Steel & Cladding v Cubitt Building & Interiors Limited [2007] EWHC 1584.

5.77. In the *Cubitt v Richardson*[56] case, Judge Akenhead distinguished the *DGT v Cubitt* decision on the basis that in the latter case the adjudication provision was mandatory, and the arbitration provision was not. However, in *Cubitt v Richardson* the adjudication provision was not mandatory, and the arbitration provision was. Judge Akenhead held that a stay of court or arbitration proceedings will not usually be granted for the sole purpose of allowing adjudication to proceed, as Section 108 of the UK Act did not constitute an obligation to adjudicate, but simply entitled a party a right to refer a dispute to adjudication. Judge Akenhead stated that a party who has started court or arbitration proceedings is entitled to have those proceedings resolved as reasonably expeditiously as can be achieved and as justice demands, and a party should not be forced to have those proceedings delayed or stayed and be forced to adjudicate when it does not want to exercise its right to do so. Cubitt brought the court proceedings seeking, primarily, a declaration that the subcontract did not contain an arbitration agreement which had been properly incorporated into the contract. If the arbitrator found that the arbitration agreement had been incorporated into the contract then the arbitrator would have jurisdiction to deal with the issue whether arbitration proceedings should be stayed in order to facilitate adjudication. Judge Akenhead also considered this issue and held that the arbitration agreement had been incorporated into the contract and accordingly the question of whether there should be a stay was entirely a matter for the arbitrator, not the court.

5.78. It would appear, therefore, that if the parties include a provision in their construction contract which requires mandatory adjudication of payment disputes before either party is entitled to institute court or arbitration proceedings, then if either party were to seek to circumvent that provision a judge or arbitrator may well be disposed to stay the proceedings so as to allow the adjudication to proceed. However, it is more commonly the case that there is no such mandatory adjudication provision, in which case a judge or arbitrator would not stay the proceedings. However, while refusing a stay, the judge or arbitrator would equally not be entitled to prevent the adjudication from proceeding in parallel. Given that adjudication is a fast-track procedure the most practical, sensible approach arbitrators and judges might adopt (having rejected a stay application) would be to set a timetable for the arbitration or litigation to proceed so as to facilitate the timing of the adjudication. While a judge or arbitrator would not be obliged to extend the court or arbitration proceedings timetable, when rejecting

56 Cubitt Building & Interiors Limited v Richardson Roofing (Industry) Limited [2008] EWHC 1020 (TCC).

a stay application, it would nonetheless be a sensible approach to do so as time could be made up at a later stage in the relevant proceedings.

"Adjudicator shopping"

5.79. The right to refer a dispute to adjudication at any time has given rise to what has been termed "adjudicator shopping" in the English Courts. The English Court of Appeal considered this issue in *Lanes v - Galliford.*[57] Galliford issued an adjudication notice following a dispute under their subcontract with Lanes. However, following the nominating body's appointment of an adjudicator, Galliford wrote objecting to the appointment based on its previous poor experience with the adjudicator. The nominating body refused to nominate an alternative adjudicator, so Galliford decided not to serve its referral notice within the time prescribed under the UK Act, thereby allowing the adjudication to elapse. Thereafter Galliford issued a fresh adjudication notice, and a new adjudicator was appointed upon that notice. Lanes sought an injunction restraining Galliford from proceeding with the second adjudication. The court held that Galliford's failure to serve the referral notice within the time prescribed, whilst a breach of contract, did not prevent them from serving a second adjudication notice, and seeking the appointment of a second adjudicator. Lord Justice Jackson held that whilst the situation was not an appealing one, Galliford's conduct was permissible and the "second" adjudicator had jurisdiction to decide the dispute.

Referring the same dispute to adjudication

5.80. The judge's decision in *Galliford* was made on the basis that the first adjudicator never acquired jurisdiction as the referral had not been submitted to him, and therefore he had not made any decision on the dispute. The judge's decision would have been different if the first adjudicator had made a decision on the dispute and then a second adjudicator had been appointed to decide the same issue.

5.81. Where a party is of the view that the dispute being referred to adjudication has been previously referred to an earlier adjudicator and decided upon, then that party would have good grounds to challenge any later appointed adjudicator's decision.

5.82. In *Sherwood v MacKenzie*[58] HHJ Thornton QC held that when there is an allegation that two disputes are substantially the same, the courts may make a limited enquiry to determine whether there is

57 Lanes Group Plc v Galliford Try Infrastructure Limited [2011] EWCA Civ 1617.

58 Sherwood & Casson Limited v MacKenzie Engineering Limited [1999] EWHC 274 (TCC).

a substantial overlap between the two adjudications, but will not consider the merits of the cases. Sherwood was Mackenzie's subcontractor under a contract which did not contain an adjudication clause, and consequently the UK Scheme applied to the adjudication. In a first adjudication, Sherwood's claim sought payment of variations in an interim application, and the adjudicator awarded them a sum of money. Subsequently, Sherwood submitted its final account which contained the same variations, but at a higher valuation, and a claim for loss and expense caused, in part, by those variations. Mackenzie did not pay, and Sherwood referred its final account claim to adjudication. A different adjudicator was appointed, but Mackenzie argued that he should resign because the dispute was the same, or substantially the same, as the dispute previously referred in the first adjudication. The second adjudicator decided otherwise on the basis that an interim payment claim, and a final account claim were different. The adjudicator awarded a higher sum in Sherwood's favour for the variations but dismissed the claim for loss and expense. Mackenzie did not comply with the adjudicator's decision and Sherwood issued court enforcement proceedings in respect of the adjudicator's decision. Mackenzie claimed that the second adjudicator had no jurisdiction and should have resigned. HHJ Thornton QC held that the disputes were different, and the second adjudicator had jurisdiction to make a decision. The judge held that Mackenzie's challenges on the adjudicator's findings of fact were not to be considered by the court in enforcement proceedings on the adjudicator's decision, and that any challenge on the facts should be mounted at a subsequent arbitration commenced after the adjudicator's decision has been rejected. The judge held that the court can examine a jurisdictional challenge on the basis that the dispute is the same or substantially the same as one previously referred, but will only enquire if there are substantial grounds that the adjudicator erred in having decided on that issue in the adjudication.

5.83. If the parties refer to a second adjudicator the issue as to whether the substantive dispute which they wish to refer has been decided on by a previous adjudicator, and that second adjudicator decides upon that issue, then the courts will not interfere with the second adjudicator's decision. That is, unless there is a challenge in respect of the second adjudicator's jurisdiction which demonstrates substantial grounds for showing that the adjudicator erred in the relevant second decision.

5.84. In *Holt v Colt*[59] a dispute over Colt's Interim Application No. 10 was referred to adjudication. Colt requested that the adjudicator order "immediate payment of the balance of the sum due". The

59 Holt Insulation Limited v Colt International Limited [2001] EWHC 451.

adjudicator decided that based on the wording of Colt's referral notice the only decision he could consider making was whether the precise sum Colt claimed was due. The adjudicator considered that his jurisdiction restricted him to consideration of the precise sum Colt claimed. He found that he couldn't agree the total sum was due and therefore his decision had the effect that no money was payable to Colt. Colt served a further adjudication notice and referral contending that it was entitled to the sum claimed as due "or such other sum the adjudicator should find was due". The same adjudicator was appointed, but Holt objected to the further adjudication, arguing that the adjudicator had already been asked to deal with Application No. 10 and he should resign if the dispute was the same or substantially the same as the one previously referred and decided. The adjudicator, however, decided that there was an issue that had not arisen on the earlier referral insofar as the adjudicator had been asked in the second adjudication to decide whether another sum, other than the sum claimed, was due to Colt. The adjudicator awarded Colt a lesser sum than claimed and Holt refused to pay seeking a court order that the second decision be rescinded since the adjudicator had no jurisdiction. Colt counterclaimed seeking judgment enforcing the adjudicator's award. HHJ Mackay concluded that while the disputes referred to the adjudicator related to the same matters arising out of contractual relations between the parties, they did not relate to the same dispute. The referral notices were crucially different and therefore the adjudicator was correct in making the decision in the second adjudication.

5.85. In the context of *Holt* it is crucial that the adjudication and referral notices are accurately drafted as they will be closely examined by any adjudicator appointed so as to establish whether it is a new dispute or the same as a dispute that has previously been referred to adjudication. In addition, *Holt* emphasises the importance of ensuring to draft the adjudication and referral notices so as not to restrict the adjudicator to making a single decision on a precise sum of money, but allow the adjudicator to decide what other sum is owed if the adjudicator decides the precise sum claimed is not due. A prudent referring party should draft their notices sufficiently broadly so as to claim for the precise amount to which they believe they are entitled "or such other sum as the adjudicator may decide".

5.86. In *Quietfield v Vascroft*,[60] HHJ Jackson held that an adjudicator had erred in treating his own decision in an earlier adjudication as conclusive regarding a time extension in a subsequent adjudication. Quietfield engaged Vascroft to carry out renovation works. The

60 Quietfield Limited v Vascroft Contractors Limited [2006] EWHC 174; [2006] EWCA Civ 1737.

works were delayed and were not completed by the contractual completion date. Vascroft made two applications to the architect for time extensions. The first application for an extension of time until 9 June 2005 was made in a letter dated 2 September 2004, identifying 12 matters as the causes of delay. A second application was made on 22 April 2005, seeking an extension of time until 23 September 2005. The architect rejected both applications and did not grant any extension of time. In August 2005, Vascroft issued an adjudication notice referring its claims for an extension of time based on the matters set out in their two applications. The adjudicator rejected Vascroft's claim on the basis that they had failed to discharge the burden of proof to establish the extension of time. Subsequently, Quietfield commenced adjudication seeking LADs for Vascroft's failure to complete the work on time. The same adjudicator was appointed. Vascroft opposed Quietfield's claim contending that they were entitled to an extension of time for the whole period relying on a new document appended to their response setting out numerous causes of delay which they had not previously contended and which also identified the critical path. The adjudicator decided that he was unable to consider the contents of the new document appended to Vascroft's response arguing for a time extension as he was bound by the decision given in the first adjudication on those applications. The adjudicator consequently decided that Vascroft's defence in its response in the Quietfield adjudication was seeking to rely, for a second time, on the same matters that they had unsuccessfully relied on in the first adjudication. The adjudicator decided in Quietfield's favour on their LADs claim, but Vascroft failed to comply with that decision. Quietfield applied for summary judgment to enforce the decision. The court considered the question of whether the adjudicator was correct to ignore the new document Vascroft had relied upon in the Quietfield adjudication. HHJ Jackson refused to enforce the adjudicator's decision on the basis that Clause 25 of the contract permitted the contractor to make successive applications for extensions of time on different grounds. Consequently, the judge held that successive adjudications concerning extensions of time must be permissible provided that each adjudication arises from a separate dispute. As Vascroft's argument for an extension of time was in response to Quietfield's adjudication seeking LADs this was a separate dispute, and consequently Vascroft was entitled to seek a time extension by way of defence in its response to the Quietfield adjudication.

5.87. Quietfield appealed Judge Jackson's judgment to the English Court of Appeal who agreed with the judgment and dismissed the appeal, confirming that the adjudicator's decision should not be enforced. The key points of the Court of Appeal judgment were as follows:

a) More than one adjudication in respect of a dispute is permissible but a second adjudicator cannot decide that which the first adjudicator has already decided.
b) Grounds for an extension of time not individually established in an earlier adjudication could legitimately be submitted in a successive adjudication in conjunction with new grounds due to being on a critical path affected by those new grounds. This is because a composite claim might legitimately be seen as outside the dispute in the earlier adjudication.
c) The key question in the case of successive applications is "what did the adjudicator decide in the first adjudication?". Parties should look at the decision in the first adjudication and the related notice of adjudication to determine what the adjudicator decided. Whether dispute A is substantially the same as dispute B is a question of fact and degree.
d) A contractor cannot make successive applications based on the same material circumstances with the same causes of delay or the same relevant events as previously made under the contract. A contractor must present new supporting material that could reasonably lead the architect to a different conclusion. This does not necessarily mean relying on a different relevant event but requires materially different particulars of the expected effects and/or a different estimate of the expected delay to completion.
e) Where the only difference between disputes relating to claims for extension of time is that the later adjudication makes good shortcomings of the earlier adjudication, the disputes are likely to be considered by an adjudicator as the same.

5.88. The Court of Appeal held that the dispute referred and decided in the first adjudication was Vascroft's entitlement to an extension of time as set out in their two previous applications to the architect. In the Quietfield adjudication, the new document introduced by Vascroft identified new causes of delay not mentioned in their two previous applications for extensions of time and the first adjudication. The court held that the dispute in the second adjudication was therefore substantially different from the first adjudication and the adjudicator was wrong not to consider Vascroft's extension of time claim based on its new document.

5.89. In *HG Construction v Ashwell*,[61] HHJ Ramsey refused to enforce an adjudicator's decision in the third of a series of adjudications stating that it was the same, or substantially the same, as the issue in the first adjudication and therefore the first adjudicator's decision stood unless

61 HG Construction Limited v Ashwell Homes (East Anglia) Limited [2007] EWHC 144 (TCC).

it was overturned in arbitration or litigation. Ashwell engaged HG under a contract for a new housing development. Disputes arose and were referred to adjudication. In Adjudication No. 1 the adjudicator decided that the LADs provision in the contract was valid and enforceable. In Adjudication No. 3 the adjudicator decided that the LADs were not enforceable, and Ashwell should refund monies to HG. HG issued court enforcement proceedings in respect of Adjudication No. 3. Ashwell opposed the proceedings on the basis that the adjudicator decided substantially the same dispute as had already been decided by a different adjudicator in Adjudication No. 1. HHJ Jackson found that, as a matter of contract, the parties accepted that they were bound by an adjudicator's decision. Therefore, to the extent that a subsequent adjudicator purported to decide any dispute already decided by an earlier adjudicator, the later adjudicator's decision would not be valid or enforceable. The judge commented that as a matter of practice an adjudicator should consider whether he is being asked to decide upon a matter which another adjudicator has already made a binding decision on, and if so, he should decline to decide the matter and resign. Judge Ramsey held that the subject matter of Adjudication No. 3 was substantially the same as the dispute the adjudicator had decided in Adjudication No. 1 and the adjudicator's decision in Adjudication No. 3 was therefore not valid. Consequently, Judge Ramsey dismissed the application for summary judgment seeking to enforce the decision in Adjudication No. 3.

The notice of intention to refer

5.90. Under Section 6(2) a party exercises its right to refer a payment dispute to adjudication by serving on the other party to the construction contract a notice of intention to refer. As the notice of intention to refer commences the adjudication process and represents the document from which the adjudicator derives his jurisdiction it is critical that a valid notice is served in accordance with the prescribed procedure in the Act. Any failure to serve a valid notice, or failure to comply with prescribed procedures could potentially culminate in a court finding that an adjudicator did not have the jurisdiction to make a decision, thereby potentially rendering any decision unenforceable.

5.91. An ill-defined or poorly worded notice of intention can result in serious consequences and potentially an unenforceable decision. In addition, a decision by an adjudicator on matters outside those in the notice of intention would exceed his jurisdiction, unless the parties confer additional jurisdiction on him. Irish Courts are likely to adopt a similar approach to the UK Courts regarding the content and service of the notice of intention (referred to as the notice of adjudication in the UK legislation) and treat it as a critical document from which the adjudicator derives his jurisdiction, but which also defines the scope and limits of the referring party's claim in the adjudication. The notice of intention should identify the precise nature and issues in dispute and the financial remedy sought by the referring party. If a referring party fails to adequately address such matters in its notice it could lead to an adjudicator exceeding his jurisdiction or claims in respect of issues which were not contained within the notice. There are numerous UK cases dealing with adjudication notices that did not adequately identify the issues in dispute.

5.92. In *Mecright v Morris*[62] the UK Scheme applied to the subcontract. Morris terminated the subcontract on the basis that Mecright had failed to proceed executing the works in a reasonable and workmanlike manner. Mecright disputed this and Morris referred the dispute to adjudication seeking a declaration that the subcontract was terminated in accordance with its terms, and that Morris was entitled to recover damages arising out of the termination. The adjudicator held in Mecright's favour on the grounds that Morris failed to terminate the subcontract in accordance with its terms. As such the adjudicator found that Morris was not entitled to damages and instead awarded Mecright £30,000, essentially for the value of Mecright's work up to the subcontract termination, and costs incurred from Morris' wrongful repudiation of the subcontract. In the court enforcement proceedings Mecright argued that because the UK Scheme allowed the taking into account of "any other matters which the parties to the dispute agree should be within the scope of the adjudication, or which are matters under the contract which he considers are necessarily connected to the dispute", the adjudicator was entitled to deal with Mecright's claims for payment for work undertaken to the termination date and damages for repudiation. HHJ Seymour held that the adjudicator could determine matters which the parties agree come within the scope of the adjudication, even if they were not originally within such scope and within the adjudication notice. Judge Seymour also held that the adjudicator could decide any matter arising under the contract which he considered necessarily connected with the dispute. However, the judge went on to say that the issue before the adjudicator was whether Morris was entitled to terminate its contract with Mecright and, if so, what amount of money Morris was entitled to as a consequence. Judge Seymour identified that the sources of an adjudicator's jurisdiction were twofold: the adjudicator can decide

62 Mecright Limited v TA Morris Developments Limited [2001] HT01 84.

any matter within the subject of the adjudication notice, and disputes which are aspects of or the resolution of which is necessary, to resolve the dispute which is the subject matter of the adjudication notice Following these principles, Judge Seymour found that the adjudicator only had jurisdiction to decide whether the termination was correct, and if so, what sum Morris was entitled to as damages for wrongful termination. The judge accepted Morris' submission that a dispute as to how much Mecright was entitled to be paid in respect of execution of the works to the date of termination or as a result of wrongful termination was not, on a proper construction of the notice of adjudication, included within the dispute that was referred to adjudication by that notice. The judge agreed with Morris that the adjudicator did not have jurisdiction and consequently refused Mecright's application for summary judgment seeking to enforce the adjudication decision.

5.93. In *Pilon v Breyer,*[63] Breyer engaged Pilon to work on a project which involved two batches of work. The following year, in the intervening period during which Pilon entered into a voluntary arrangement with their creditors, Pilon made an interim payment application relating to the second batch. Pilon was not paid upon the application, so served an adjudication notice which was limited to only the second batch payment application. Breyer's primary defence was that they were entitled to set off an overpayment they alleged was made for the first batch. Breyer had not served withholding notices for the amount they wished to set off but argued that they were not required to do so under the contract terms. The adjudicator decided that the alleged overpayment for the first batch was outside his jurisdiction and the adjudication was limited solely to the second batch payment application. Breyer failed to discharge the adjudicator's decision and Pilon initiated court enforcement proceedings. HHJ Coulson held that the adjudicator's decision on his jurisdiction was incorrect and the jurisdiction extended to consideration of any set-off defence claimed by Breyer. The judge held that in not considering Breyer's set-off the adjudicator's behaviour amounted to a breach of natural justice and the decision should not be enforced. Judge Coulson stated that an adjudicator should think meticulously before ruling out a defence merely because there was no mention of it in the adjudication notice. The judge stated:

> Adjudicators should be aware that the notice of adjudication will ordinarily be confined to the claim being advanced; it will rarely refer to the points that might be raised by way of a defence to

63 Pilon Limited v Breyer Group PLC [2010] EWHC 837 (TCC).

that claim. But, subject to questions of withholding notices and the like, a responding party is entitled to defend himself against a claim for money due by reference to any legitimate available defence (including set off) and thus, such defences will ordinarily be encompassed within the notice of adjudication.[64]

5.94. The principle that a respondent should not be deprived of a defence because of the narrow construction of an adjudication notice cannot, however, be extended so as to include cross-claims that ought to have been the subject of withholding notices but were not. Under Section 4(3) of the Act where a contractor (or a subcontractor) has made an interim payment application by way of a payment claims notice, if the responding paying party contests the amount that is due and payable to the claiming party, the responding party must deliver a notice specifying the amount that it proposes to pay, the reasons for the difference between the amount claimed and the amount to be paid and the basis on which the amount proposed to pay is being calculated. This notice must be served by the responding paying party not later than 21 days after the payment claim date, which is the date on which the claiming party has made its claim. This notice is often referred to as a "withholding notice". A similar "withholding notice" regime exists under the UK legislation and is often incorporated into construction contracts in the UK. There is a line of authority in UK case law that if a responding paying party has not served a valid withholding notice (where one is required to be served under the contract) within which the cross-claim has been identified, the responding paying party is not entitled to raise a cross-claim as a defence to the claiming party's claim in the adjudication.

5.95. In *Letchworth v Sterling*,[65] Letchworth was engaged as Sterling's roofing subcontractor. A dispute arose regarding Letchworth's Interim Payment Application No. 5. Sterling refused to pay the amount raising a cross-claim for delay. Following adjudication, and a decision in their favour, Letchworth initiated court enforcement proceedings. Sterling defended the proceedings on the grounds, *inter alia*, that the adjudicator failed to deal properly or at all with the delay cross-claim and should have taken the amount of that claim into account when awarding Letchworth on its payment application. Sterling argued that a responding party in adjudication is entitled to raise any matter by way of defence which would amount in law, or in fact, to a defence to the claim being pursued. In addressing this issue, Judge Coulson discussed the principles relevant

64 At paragraph 25 of the judgment.
65 Letchworth Roofing Company v Sterling Building Company [2009] EWHC 1119 (TCC).

to the scope of the adjudication notice and held that while a responding party can raise whatever matters it wishes by way of defence for the adjudicator's consideration, that general principle does not permit a responding party to rely on a cross-claim which should have been the subject matter of a withholding notice. Judge Coulson stated that:

> There was nothing in either the adjudication or the referral notice which suggested that the delay cross-claim, regardless of the withholding notice position, was being referred to the adjudicator for decision as part of his consideration of interim application 5.[66]

5.96. Judge Coulson held that although a responding party can seek to raise a cross-claim in an adjudication, if an adjudicator concludes that such a cross-claim required a valid withholding notice, and no such notice has been issued, then the adjudicator is acting entirely properly by not taking this into account when making an award to a claimant. Judge Coulson consequently granted summary judgment for enforcement of the adjudicator's decision in favour of Letchworth.

5.97. If prior to the commencement of an adjudication a claim has been responded to by way of a withholding notice and the claim is subsequently referred to adjudication, the issue as to the validity or otherwise of the withholding notice would also be properly the subject of the adjudicator's decision-making process. If the adjudicator were to decide that no withholding notice was required, or that one was required and it was properly served, then the cross-claim would have to be taken into account by the adjudicator in reaching a decision. If, on the other hand, no withholding notice had been served or the adjudicator determines that a withholding notice should have been issued but was not, or was invalidly served, then the adjudicator is not entitled to take the cross-claim into account in reaching a decision.

5.98. If a responding party raises a cross-claim in the adjudication and the adjudicator rejects such cross-claim as unsuccessful in defeating (or reducing) the referring party's claim, then on the basis that the adjudicator has decided on the merits of the cross-claim it is not open for the responding party to raise the cross-claim again as a set-off in any court proceedings issued by the referring party seeking to enforce the adjudicator's decision.

5.99. In *MI Electrical v Elements*,[67] HHJ Nissen QC enforced an adjudicator's decision in the subcontractor's favour and dismissed the

66 At paragraph 21 of the judgment.

67 MI Electrical Solutions Limited v Elements (Europe) Limited [2018] EWHC 1472 (TCC).

contractor's set-off against the decision on the basis of its cross-claim in respect of the subcontractor's alleged defective work. Elements did not seek to defend the court enforcement proceedings on the basis that the adjudicator lacked jurisdiction, nor had they made any complaint that the adjudicator acted in breach of natural justice. Instead, the sole issue before the court concerned whether Elements were entitled to set-off a cross-claim against the decision. Judge Nissen noted that Elements had already canvassed its set-off claim regarding defects and the resulting cross-claim before the adjudicator who concluded that, whatever the merits of the alleged defects, they could not amount to a defence to a claim in the adjudication. Judge Ramsey concluded that it was simply not open to the defendant to seek to re-run the point, by way of a defence in the enforcement proceedings, that the existence of defects constituted a defence to the claimant's claim. The judge held that the adjudicator determined this point and consequently he gave summary judgment enforcing the adjudicator's decision.

5.100. As previously mentioned, the wording used by the referring party in the notice of intention to refer is critical so as to ensure that the responding party and adjudicator fully understand the nature of the dispute and the referring party's claim.

5.101. In *Griffin v Midas*[68] it was held that the adjudication notice commences the adjudication process and plays an important role in defining the scope of the dispute over which the adjudicator has jurisdiction. Before serving the adjudication notice Griffin's solicitors corresponded with Midas' solicitors regarding Griffin's claim for several unpaid invoices and retention which Griffin considered payable. Griffin's adjudication notice referred back to those letters and stated that as no payment had been received in respect of their claims a dispute existed which should be referred to adjudication. Midas argued that because the notice did not specifically set out full details of the dispute, but rather referred to the correspondence which had passed between solicitors regarding the amounts claimed, it could not be determined from the notice as to which items Griffin intended to refer to adjudication. Consequently, Midas argued that the dispute had been ill-defined in the notice because, in their view, it could not be established as to precisely what matters over which the adjudicator had jurisdiction. Notwithstanding this, Griffin pursued the adjudication and Midas reserved its position on jurisdiction. Judge Humphrey Lloyd QC had to decide whether the dispute referred to adjudication had been precisely described in the notice. Interestingly, Judge Humphrey Lloyd noted that because the

68 Griffin v Midas Homes Limited [2000] EWHC 182 (TCC).

notice had been written by a solicitor it ought not to be given the same latitude as an adjudication notice written without legal advice. Judge Humphrey Lloyd examined what could have been in dispute as at the date of the adjudication notice. The judge found that the notice was invalid to the extent that it could not confer jurisdiction on the adjudicator to make a decision about certain invoices which had only recently been seen by Midas, and in respect of which Midas had not been given a sufficient opportunity to deal with them, and either accept or reject those invoices. The adjudicator's decision encompassed all invoices, but the judge decided to sever the decision on those invoices Midas had only recently seen. The judge, however, granted summary judgment for the other invoices referenced in the notice of adjudication on the basis that the adjudicator had jurisdiction to deal with those. The judge held that although it is possible to give notice of adjudication by reference to other correspondence, a party must ensure that the underlying referenced correspondence is sufficiently clear, and records the dispute with precision. The judge held that it was crucial for a party receiving a notice of adjudication (and the adjudicator) to be able to answer two key questions "the key questions are, what is the brief description of the dispute and what is the nature of the redress sought? These are crucial to the party receiving the notice and to the adjudicator".[69]

5.102. While the importance of a precisely defined notice of intention cannot be overstated it is also vital to recognise that it is the substance of the dispute as defined in the notice which is critical, and not just the label given to it. In other words, while a referring party might characterise its claim as a draft final account claim, if in fact the claim relates to an interim payment application then that is what is important in terms of defining the substance of the claim and ultimately the adjudicator's jurisdiction.

5.103. In *OSC v IDC*[70] HHJ Ramsey held that it was the substance of the dispute that mattered, not the label given to it in the adjudication notice. IDC engaged OSC for drainage and site access works on a medical centre project. Before completing the works OSC submitted Payment Application No. 10 referring to it as their draft final account. Following negotiations regarding the account, IDC responded with their revised final account figure and determined that no further payment was due to OSC. OSC served notice of adjudication stating, the "dispute concerns non-payment by IDC to OSC in respect of OSC's final account" and seeking payment for

69 At paragraph 16, page 5, of the judgment.

70 OSC Building Services Limited v Interior Dimensions Contracts Limited [2009] EWHC 248 (TCC).

the difference between IDC's assessment of the draft final account and the amount OSC considered payable, or such other sum as the adjudicator should decide. In the subsequent referral document OSC sought redress similar to that set out in the notice except for the adjustment of specific sums to a greater value. IDC submitted that under the notice the adjudicator only had jurisdiction to decide on non-payment against IDC's certification and could not determine the value of OSC's final account, which they said represented an expansion of the matters in dispute which had been referred to adjudication under the notice. Judge Ramsey agreed with OSC's submissions that the whole of the notice and referral should be read in context with the relevant documents and that the dispute OSC referred to adjudication was the question of what sum was due to OSC when the notice was issued. The judge held that OSC's reference to IDC's non-payment of OSC's final account, as set out in the notice, encompassed both those sums, which were further identified in the notice and the referral. The judge held that he did not consider that the difference between the sums claimed in the notice and the subsequent referral could sensibly be said to expand the claim when the precise sum was left open by the words "or such sum as the adjudicator may decide". IDC further submitted that the notice did not allow the adjudicator to determine the value of the final account, but merely to decide on the issue of non-payment against IDC's certification. Judge Ramsey held that even though OSC's Application No. 10 was referred to in the notice (albeit inaccurately as the final account) a review of the contemporaneous documents clearly indicated that the application being considered by the adjudicator was an interim payment application. The judge held that the draft final account was not the final account itself which it was envisaged would be provided by OSC at a later date. Judge Ramsey confirmed that the adjudicator had actually only dealt with the Application No. 10, not the final account as IDC argued. As a consequence, Judge Ramsey held the adjudicator's decision did not exceed his jurisdiction and was enforceable.

5.104. While the notice of intention to refer is a critical document in terms of defining the scope and limits of the referring party's claim in the adjudication, it also represents the commencement of the adjudication process.

5.105. In *University of Brighton v Dovehouse*,[71] a dispute arose under a fit-out contract at a university centre after the contract was delayed by three months. Dovehouse claimed an extension of time and loss and expense, with the University claiming for incomplete works

71 University of Brighton v Dovehouse Interiors Limited [2014] EWHC 940 (TCC).

and alleged defects. Six weeks after practical completion the final certificate was issued. The contract provided that the final certificate was conclusive of all matters stated therein save where adjudication, arbitration or other proceedings commenced within 28 days after issue of the final certificate (the "saving proviso"). The parties agreed to extend this 28-day period, as regards the conclusivity of the final certificate, to 66 days in order to seek to negotiate settlement of the dispute. The parties reached no settlement and on day 65 Dovehouse served an adjudication notice. However, the notice identified the wrong nominating body and included an address for service on the University which was different from that specified in the contract. Whilst the flawed notice was received within the 66-day period, the appointed adjudicator resigned a week later following a jurisdictional challenge. Dovehouse then served an amended adjudication notice three days after the resignation. The University argued that the adjudication proceedings could not proceed under the second notice, as the agreed 66-day period concerning the conclusivity of the final certificate had passed and consequently the final certificate became conclusive of all matters. The University instigated court proceedings seeking a declaration that the first notice was ineffective to commence adjudication proceedings because:

a) adjudication did not commence until the referral was served on a properly appointed adjudicator;
b) the notice of adjudication did not identify, and was not served at, the contractually correct address;
c) no adjudication was commenced as Dovehouse applied to the wrong appointing body and the first adjudicator resigned; and
d) the defects in the notice of adjudication and the resignation of the adjudicator could not be cured.

5.106. HHJ Carr held that adjudication proceedings commenced when the notice was given, as Dovehouse argued, and not when the referral was served as the University argued. The judge held as the proceedings "commenced" by service of the first notice, and because Dovehouse was never obliged to identify the appointing authority in that notice, the identification of the wrong appointing authority did not invalidate the first notice. Judge Carr held that the substantive purpose of the adjudication notice, namely the commencement of proceedings, had been achieved and consequently the invalidity of the referral and the first adjudicator's resignation did not negate the sufficiency of the first notice for commencing adjudication proceedings. The judge also considered the issue regarding the extension of the 28-day period to 66 days in respect of the conclusivity of the final certificate and held that the saving proviso (i.e. that the final certificate is conclusive "save" where adjudication proceedings are commenced) was not triggered by any step beyond the commencement of adjudication proceedings. Judge Carr stated that all that was required to engage the saving proviso was the commencement of proceedings and once that had been triggered (irrespective of the flaws within the notice of adjudication) the saving proviso remained in operation during the currency of any subsequent adjudication proceedings. For these reasons Judge Carr dismissed the claim by the University.

5.107. Judge Carr also held that while there was no requirement to identify the nominating body in the notice of adjudication, references to the wrong nominating body did not invalidate the notice. Equally, the judge held that the failure to serve the notice of adjudication on the University's correct address was not fatal because the contract entitled the parties to serve a notice by any effective means and it was not disputed that the University had received the notice. Effectively, the judge took the view that the service of the notice of adjudication commenced the adjudication proceedings and that errors on the face of the notice were not sufficiently material to prevent the notice from achieving its substantive purpose, namely to commence the adjudication proceedings.

6 Adjudicator appointment

Sections 6(3), 6(4), 6(18) and Section 8 of the Act

Appointing the adjudicator

6.1. Section 6(3) of the Act provides that:

> The parties may, within 5 days beginning with the day on which notice under subsection (2) is served, agree to appoint an adjudicator of their own choice or from the panel appointed by the Minister under section 8.

6.2. It is commonly the case, and best practice, for the referring party to nominate (generally within its notice of intention to refer) a number of potential adjudicators that it proposes be appointed. The responding party is under no obligation to agree to any of the adjudicators nominated and may itself nominate potential adjudicators it wishes to appoint. In the event that the parties cannot agree an adjudicator appointment within five days of the date of service of the notice of intention, either party may apply to the Chair of the Minister's panel pursuant to Section 6(4) of the Act.

6.3. Section 6(3) of the Act is similar to Section 108(2)(b) of the UK Act, which has given rise to substantial case law as regards issues surrounding the appointment of the adjudicator.

6.4. In *IDE v Carter*[1] the adjudication referral was subject to the UK Scheme. Paragraph 2(1) of the UK Scheme provides that after giving an adjudication notice (and subject to any agreement between the parties as to who shall act as adjudicator) the referring party shall request the person (if any) named in the contract to act as adjudicator, and that such request must include a copy of the adjudication notice. If nobody is named in the contract or the named person is

1 IDE Contracting Limited v RG Carter Cambridge Limited [2004] EWHC 36 (TCC).

unavailable or unable to act, the referring party shall request the relevant nominating body to select an adjudicator. The UK Scheme further provides that the nominee requested is required to indicate whether he/she is willing to act within two days of receiving the request. If that person is unable or unwilling to act, the referring party may request the nominating body to appoint an adjudicator. The contract between IDE and Carter provided that disputes would be referred (under the UK Act and UK Scheme) to a named adjudicator and in the event that he could not act then the nominating body was to nominate an adjudicator by default. Representing IDE, a chartered quantity surveying firm had telephoned the named adjudicator in early September 2003 to enquire as to whether he would be able to act. However, the adjudicator was unable to accept the appointment due to previous commitments but indicated that he would be available from 2 November onward. Carter received the adjudication notice on 29 September. Simultaneously, IDE made a request to the Chartered Institute of Arbitrators (CIArb) (the nominating body) to appoint an adjudicator. Carter then proposed two alternative names asking IDE to agree on one, but IDE preferred to rely on the nominating body to appoint the adjudicator. Carter reserved its position, stating that the contract required any dispute to be referred to the named adjudicator, and they had not received a copy of an application to him, nor any confirmation that he had been contacted or was unable or unwilling to act. Initially, they participated in the adjudication on a without prejudice basis, but later withdrew completely. The CIArb-nominated adjudicator was appointed and made a decision with which Carter did not comply. IDE issued court enforcement proceedings. Carter sought to defend the proceedings on the grounds, *inter alia*, that the adjudicator had not been duly appointed. HHJ Havery QC held that it was clear that the adjudication notice should be issued first and then followed by a request to the person named in the contract to act as adjudicator. The request to the named adjudicator by IDE had to be in writing and include a copy of the notice of adjudication. The judge found that because IDE's representative had telephoned the named adjudicator enquiring as to his availability a request had not been made which complied with the provisions of the UK Scheme. The judge held that as a result of such non-compliance the adjudicator subsequently nominated by the CIArb, and appointed, did not have any jurisdiction. The judge, in rejecting the claimant's application and refusing to enforce the adjudicator's decision, held that it was important to follow the procedures, because

> it would be open to an intending claimant who did not want the specified person to act as the adjudicator to ascertain,

without the knowledge of the other party, when the specified person would not be available, and to serve the notice of adjudication at that time.[2]

6.5. When referring a dispute to adjudication under Section 6(2) of the Act the notice of intention to refer must come before the parties seek to agree on appointing an adjudicator of their own choice. Any failure to comply with this sequence could potentially deprive an appointed adjudicator of jurisdiction resulting in an unenforceable decision. If, for example, the parties seek to agree an adjudicator appointment before serving the notice of intention, they would nonetheless be required to comply with Section 6(3) (and seek to agree the appointment after the service of the notice of intention) before either party would be entitled to seek an adjudicator appointment by the Chair of the Minister's panel under Section 6(4). Similarly, if either party seeks to procure an appointment from the Chair within the five days provided for agreement on an adjudicator of their choice, again this could lead to a defective appointment.

6.6. Irish Courts would likely adopt a similar approach as the UK Court did in *IDE* if a similar set of circumstances were to arise, because under the Act before applying to the Minister's panel to seek an adjudicator appointment the parties must first serve notice of intention pursuant to Section 6(2) and then (within five days) seek to agree on an adjudicator of their own choice (or agreed from the Minister-appointed panel) under Section 6(3). Where a party seeks to apply to the Chair of the Minister's panel before issuing a notice of intention, or for that matter before the expiry of five days under Section 6(3), a court would likely refuse to enforce a decision subsequently issued by an adjudicator who had not been appointed in compliance with the Sections 6(2) and (3) procedure.

6.7. In order to overcome such difficulties and potential jurisdictional challenges it may be preferable, depending on circumstances, to name, in the contract, an adjudicator to be appointed in the event that a dispute arises. Such an approach could conceivably avoid any issues surrounding the mechanics of an adjudicator appointment. However, until a dispute arises the parties are not best placed to decide on an appropriately qualified person to adjudicate on their payment disputes. However, this problem can be overcome by naming in the contract a panel, consisting of adjudicators with different professional skill sets (e.g. architect, engineer, quantity surveyor, lawyer) from which the appropriate adjudicator may be selected depending on whether the payment dispute relates predominantly to architectural, engineering, legal or quantum issues.

6.8. The timing of the service of the notice of intention and the adjudicator appointment will be critical in ascertaining whether the

2 At paragraph 10 of the judgment.

adjudicator has been validly appointed, either by agreement between the parties or by a nominating body. It will very much depend on the particular circumstances of the case, including any relevant provisions contained in the construction contract between the parties, as to whether a valid adjudicator appointment has been made.

6.9. In *Palmac v Park Lane*,[3] the adjudicator nomination occurred before the adjudication notice was served. Notwithstanding this the court held, on the specific facts of the case, that the adjudicator had the necessary jurisdiction. The parties entered into a contract which provided that disputes could be referred to adjudication in accordance with the contract. Palmac made Payment Application No. 20 while emailing the same to Park Lane. Following Park Lane's failure to pay upon Palmac's application, Palmac notified Park Lane that they intended to refer the dispute to adjudication. Palmac sent an adjudication notice by special delivery to Park Lane's usual place of business. Thereafter, Palmac applied to the Royal Institute of Chartered Surveyors[4] for an adjudicator appointment. The adjudication notice, however, had not reached Park Lane, so Palmac served a second notice and on the same day made a further application to the RICS to re-nominate the same adjudicator. The adjudicator was appointed and issued a decision in Palmac's favour. Park Lane failed to comply with the decision and Palmac issued court enforcement proceedings. In the proceedings, Park Lane argued that there was no dispute at the time of the reference because the service of the payment application by email did not comply with the contract and nor did the service of the adjudication notice. They argued that service by email was not permitted under the contract as the parties agreed to communicate by fax and letter. Furthermore, Park Lane argued that the contract required the adjudication notice to be issued before applying to the RICS for an adjudicator appointment, and as a consequence the adjudicator had acted without jurisdiction. Her Honour Judge Kirkham held that the first issue as to the payment application service by email had been put to the adjudicator and he therefore had jurisdiction to decide the issue and the parties were bound by that decision. The judge held that as the adjudicator had been asked to decide a dispute regarding a payment application, in order to decide that dispute the adjudicator had to decide whether the payment application had been validly made. The judge held that this question was within the adjudicator's jurisdiction. The adjudicator expressly stated in his decision that he decided that the payment application had been validly made. The judge held that as this was a question the

3 Palmac Contracting Limited v Park Lane Estate Limited [2005] EWHC 919; [2005] ELR 301.
4 RICS.

adjudicator was entitled to decide, the court was not permitted to ask whether he had answered it correctly. The judge also held that in any event the parties had an expectation that applications would be made by email and so the service of Application No. 20 was effective. As regards the second ground Park Lane had relied on, Palmac argued that the adjudication notice was served before it applied to the RICS, whilst Park Lane argued that the notice had not been served until the day after Palmac applied to the RICS. The contract stated

> where either party has given notice of his intention to refer a dispute … to adjudication … then … any application to the nominator must be made with the object of securing the appointment … within 7 days of the date of the notice of intention to refer.

Judge Kirkham held that the aforementioned provision

> does not mean that a party must first serve a notice then apply to a nominator … the purpose of that clause is to set out a procedure for appointment of an adjudicator. It does not stipulate that any application for nomination must be made after the notice of adjudication has been served.[5]

The judge noted that Park Lane was not prejudiced and there was no statutory requirement as to the timing of any nomination, only that the contract aimed to secure an appointment within seven days of the adjudication notice.

6.10. However, the UK Scheme did not apply in the *Palmac* case and there were provisions in the construction contract between the parties which did not strictly prohibit an application for an adjudicator appointment by a nominating body prior to serving the adjudication notice, in the same way as set out in the UK Scheme.

6.11. In *Vision v Lancsville*,[6] Vision engaged Lancsville on a residential apartment development under a design and build contract which provided that the UK Scheme apply. Disputes arose regarding dates on which various apartment blocks had been completed by Lancsville, and subsequently Vision sought to determine the contract. Lancsville commenced two adjudications concerning, firstly, whether Vision's contract determination was unlawful and, secondly, in respect of an interim payment valuation. Lancsville subsequently issued a third adjudication notice (Adjudication No. 3) seeking declarations that the nature of the works

5 At paragraph 30 of the judgment.

6 Vision Homes Limited v Lancsville Construction Limited [2009] EWHC 2042 (TCC).

had changed to the extent that LADs no longer applied. On the same day that Lancsville issued the Adjudication No. 3 notice, Vision issued their own adjudication notice (Adjudication No. 4) seeking payment of LADs. On receipt of Adjudication No. 4 notice, Lancsville (1 minute later) requested that the RICS appoint an adjudicator in Adjudication No. 3. Lancsville then issued a modified Adjudication No. 3 notice, with an amendment that Vision be responsible for the adjudicator's fees and expenses (the "Modified Adjudication Notice"). The adjudicator subsequently decided Adjudication No. 3 in Lancsville's favour deciding that LADs were not payable. Vision rejected the adjudicator's decision and contended that it was not binding on the adjudicator in Adjudication No. 4. Lancsville argued the subject matter in Adjudication No. 3 was substantially the same as in Adjudication No. 4 and therefore the adjudicator had no jurisdiction to act any further. The adjudicator indicated that he considered the subject matter of Adjudication No. 3 and No. 4 to be the same, or substantially the same (i.e. whether LADs could be applied), and that the parties should apply to court for a decision on the validity of the previous decision in Adjudication No. 3. Accordingly, Vision issued court proceedings seeking a declaration that as the adjudicator was without jurisdiction his decision was null and void and of no effect. Vision argued that it was clear the adjudicator had acted under the Modified Adjudication Notice because he had ordered Vision to pay his costs and expenses pursuant to the amended item of redress in the Modified Adjudication Notice. Vision submitted that, because the UK Scheme provides that a request for nomination of an adjudicator must come after the adjudication notice, Lancsville's request to the RICS for an adjudicator appointment in Adjudication No. 3 had come after the initial notice of adjudication but prior to the Modified Adjudication Notice. HHJ Clarke agreed with Vision's submission and held that the adjudicator had no jurisdiction as he acted under the Modified Adjudication Notice and, as such, the RICS nomination had preceded the relevant notice. The judge expressed misgivings about this conclusion on the basis that the amendment in the Modified Adjudication Notice was of limited significance in the dispute as a whole but felt this was overwhelmed by the possibility that an amendment introduced in this way could, in other circumstances, be much more significant. The judge did not wish to create a precedent which might later be exploited. Judge Clarke was firmly of the view that it was preferable that any question of jurisdiction should not be decided on by the particular significance of the amendment, but that the principle should remain that any request for nomination must not precede service of the notice of adjudication. Accordingly, Judge Clarke issued a declaration that the adjudicator's decision was null and void and had no effect on the parties.

Agreement on adjudicator appointment

6.12. The precise wording of Section 6(3) of the Act provides that:

> The parties may, within 5 days beginning with the day on which notice under subsection (2) was served, agree to appoint an adjudicator of their own choice.

6.13. It is arguable that parties need not use the full five days allowed for under Section 6(3) in order to agree an adjudicator appointment of their own choice. They could decide after, say two days of seeking to agree, that they are disagreed on the adjudicator appointment and that either party may proceed to apply to the Chair of the Minister's panel for an appointment. The word "may" is potentially problematic insofar as it could facilitate a referring party in making no attempt to agree an adjudicator appointment, and instead proceed immediately under Section 6(4), following the service of the notice of intention, to seek the adjudicator appointment by the Chair of the Minister's panel. Such a party, if it were so minded, could argue that the term "may" infers on either party an option only to seek to agree with the opposing party an adjudicator appointment. Had the term "shall" been used in Section 6(3) instead of "may", then there would have been a strict requirement on the parties to seek to agree the adjudicator appointment within five days of the service of the notice of intention, but that is not the case. In addition, the wording of Section 6(3) could infer that either party could make the unilateral decision, within a shorter time frame than the five days allowed, that the parties were unable to agree an adjudicator appointment. In other words, a party that does not wish to agree an adjudicator appointment and intends to apply to the Chair of the Minister's panel for an appointment, could simply write to the opposing party immediately after receipt of the notice of intention stating that they do not agree with the adjudicator appointment and wish to immediately proceed to the appointment of an adjudicator by the Chair of the Minister's panel.

6.14. However, where the parties have successfully agreed the identity of an adjudicator to be appointed then either party should write to the person agreed, requesting that he accept the appointment. It is good practice to also seek confirmation from the agreed adjudicator that he does not have a conflict of interest (to the best of his knowledge at the time) and is available to act and deliver a decision within the time prescribed under the Act. The process thereafter for finalising the agreed adjudicator's appointment is set out in the Code of Practice.

6.15. Paragraph 9 of the Code of Practice provides that the person who is requested to accept an adjudicator appointment, following the parties' agreement, should within two days of the request, prior to accepting the appointment, write to the parties asking them to disclose any information indicating any potential conflict of interest that may arise from their appointment as adjudicator. The prospective adjudicator is also required to set out in the correspondence their proposed terms and conditions of appointment including the basis for their fees, costs and expenses.

6.16. Paragraph 10 provides that each party shall, within three days of receiving the prospective adjudicator's correspondence, decide if the prospective adjudicator's appointment is to proceed and inform the prospective adjudicator in writing of their decision in this regard. Where a potential conflict of interest is disclosed by a party then, pursuant to paragraph 11 of the Code of Practice, the prospective adjudicator may, subject to the consent of all the parties and on satisfying any professional and/or ethical concerns, accept the appointment.

6.17. Paragraph 12 provides that, if the prospective adjudicator's appointment is to proceed, the prospective adjudicator shall write to each party to accept the appointment. The date of the adjudicator's letter of acceptance is deemed to be the date on which the adjudicator appointment is made for the purposes of Section 6(5)(a) of the Act.

6.18. Section 6(8) of the Act provides that the adjudicator shall comply with the Code of Practice. However, there is no similar legislative provision in the Act which requires the parties to comply with the Code of Practice. If a party fails to comply with the Code of Practice during the course of an adjudication, and the adjudicator proceeds with the adjudication notwithstanding a party's non-compliance, it would be likely that the non-compliance would be a factor in any defence in court enforcement proceedings initiated by the non-compliant party.

Failure to agree on the adjudicator

6.19. If the parties cannot agree on the adjudicator within the five days allowed, then the Chair of the Minister's panel will make the appointment pursuant to Section 6(4) of the Act, which provides that:

> Failing agreement between the parties under subsection (3), the adjudicator shall be appointed by the chair of the panel selected by the Minister under section 8.

6.20. Paragraph 13 of the Code of Practice provides for either party to apply to the Chairperson[7] for an adjudicator appointment from the panel under Section 6(4). Pursuant to paragraph 14 of the Code, a party

7 While the Act uses the term "Chair" the Code of Practice uses the term "Chairperson". The two terms are used interchangeably in this Part 1, where the context dictates.

cannot do so earlier than five days from the date the notice of intention was served including the day of service. Under paragraph 15 the application to the Chair must be in writing and comply with the procedures set out by the Construction Contracts Adjudication Service[8] "from time to time". The application to the Chair should be copied simultaneously, as it is made, to the other parties to the payment dispute and should include the name, address and contact details of each party to the construction contract; relevant details of the payment dispute including the disputed amount (even if it's zero), the nature of the dispute; the site address; a copy of the notice of intention to refer, including any accompanying documents attached thereto; the date when and how the notice was served on the responding parties; and relevant details to identify the contract, which must be attached if in writing, and any supporting information that may assist an adjudicator in understanding the nature of the dispute.

6.21. Pursuant to paragraph 16 the Chair may seek, from the applicant, further information or clarification relevant to the nature of the dispute before appointing an adjudicator. The applicant party should provide such information promptly while copying it to the other parties to the dispute.

6.22. Paragraph 18 provides that the adjudicator appointment from the Minister's panel shall be made by the Chair and notified in writing by the CCAS to the parties "normally within seven days after the receipt of the application to the Chairperson, subject to paragraph 16 of this Code of Practice".

6.23. Presumably this flexibility in the timing for an adjudicator appointment by the Chair is to allow for circumstances where the Chair may be unavailable or is otherwise unable to make the appointment within the "normal" seven-day period. Alternatively, it may be to allow for a situation where a prospective adjudicator appointed by the Chair subsequently discovers that they are conflicted and have to reject the appointment, requiring a further appointment by the Chair.

6.24. While it is not entirely satisfactory that the Code of Practice provides that the CCAS shall "normally" notify the parties in writing within seven days of receipt of the application to the Chair, it is anticipated that the CCAS will endeavour to comply with this seven-day time frame, and experience to date would support that view. Even if the CCAS, for whatever reason, did not comply with the proposed seven-day time frame, it is unlikely that their failure to do so would exceed beyond a short period and any delay in the proceedings would likely be negligible.

8 Construction Contracts Adjudication Service of the Department of Jobs Enterprise and Innovation (the "CCAS").

6.25. The date of the CCAS' letter informing the parties of the adjudicator appointment by the Chair is deemed to be the date of the adjudicator's appointment for the purposes of Section 6(5)(a).

6.26. Under paragraph 19 an adjudicator appointed by the Chair shall write to the parties within two days of the appointment requesting them to disclose any potential conflict of interest in writing. The adjudicator should provide the parties with his terms and conditions including the basis for his fees, costs and expenses. If the information the parties disclose indicates a potential conflict of interest then, under paragraph 20, the adjudicator may only proceed with the adjudication where he is satisfied that the disclosures are frivolous or vexatious, that no professional or ethical concerns arise and that no actual conflict exists.

Revocation of adjudicator appointment

6.27. Section 6(18) of the Act provides that:

> The parties to a dispute may at any time agree to revoke the appointment of the adjudicator and the parties shall be jointly and severally liable for the payment of the reasonable fees, costs and expenses incurred by the adjudicator up to the date of the revocation.

6.28. Section 6(18) entitles the parties to revoke the adjudicator's appointment, by agreement, in advance of the adjudicator reaching a decision, whereas Section 6(17) (which is addressed later in this chapter) entitles an adjudicator to resign at any time. Clearly, if the adjudicator has reached a decision then the adjudicator is *functus officio* and consequently the question of revocation does not arise. The most common situation in which an adjudicator's appointment might be revoked arises where the parties, having referred their payment dispute to adjudication, settle the matter prior to the adjudicator making a decision. However, there may be circumstances where the adjudicator's appointment is revoked due to the adjudicator having misconducted the adjudication, and the parties agree that the adjudicator's appointment should be revoked.

6.29. Section 6(18) of the Act is similar to paragraph 11 of the UK Scheme, save in one significant respect. Under paragraph 11(2) of the UK Scheme, where the revocation of the adjudicator's appointment is due to the adjudicator's default or misconduct, the parties are not liable to pay the adjudicator's fees and expenses. No such provision exists under the Irish Act. Consequently, it is arguable that, even where an adjudicator is in default or has misconducted the adjudication or himself, the parties may be liable to pay the adjudicator's fees should his appointment be revoked. It will be interesting to observe how the Irish Courts interpret Section 6(18) in circumstances where an adjudicator's appointment is revoked due to an adjudicator's default or misconduct. If an adjudicator's default or misconduct gives rise to circumstances where the adjudicator is found to have no jurisdiction then any adjudicator's decision would be unenforceable, and as he had no jurisdiction he would not be entitled to his fees. If, on the other hand, the adjudicator had been seized with jurisdiction, but thereafter is in default or misconducts the adjudication, and the parties agree the adjudicator's appointment should be revoked, then in those circumstances it appears that, under Section 6(18), the parties would nonetheless be liable for the adjudicator's fees incurred to the date of revocation.

Panel appointments of adjudicators

6.30. Section 8 of the Act provides for the mechanism by which the Minister appoints the panel of adjudicators. The Minister is required, from time to time, to select persons to be members of the panel, having regard to their experience in dispute resolution under construction contracts, to act as adjudicators on payment disputes and select one of them to chair the panel. A person may not be selected as a member unless the person is a registered professional as defined in Section 2 of the Building Control Act 2007; a chartered member of the Institution of Engineers of Ireland; a barrister/solicitor; a fellow of the CIArb; or has a qualification equivalent to any of the foregoing duly obtained in a EU member state. The term of office for panel members is five years commencing on the date of their selection and members shall be eligible for re-selection. The Minister may, for good and sufficient reason, remove a panel member, who may also at any time resign by giving written notice to the Minister.

Challenges to the adjudicator's jurisdiction

6.31. Challenges to an adjudicator's jurisdiction often form the grounds on which a party opposing enforcement of an adjudicator's decision seeks to argue that the decision is unenforceable.

6.32. Under the Act an adjudicator only has jurisdiction over a payment dispute arising under a construction contract. If one party argues that a construction contract has not been entered into then it follows that the party contesting the existence of a construction contract is entitled to contest that any adjudicator purportedly appointed does not have jurisdiction, as that adjudicator's jurisdiction derives from the existence of a construction contract. The issue as regards an adjudicator's jurisdiction has been before the UK Courts on numerous occasions.

6.33. If, during adjudication, a party has any objection to the adjudicator's jurisdiction the objection should be raised without delay. Failure to object at the time may constitute acceptance of the procedure and the courts may reject any later objection. The UK Courts have adopted this approach to jurisdictional challenges and Irish Courts will likely adopt a similar approach.

6.34. In *Cowlin v CFW*,[9] Cowlin engaged CFW as their architect on a design and build housing project for the UK Ministry of Defence. CFW sought to terminate its contract on the basis of alleged outstanding payments. Before CFW could refer the payment dispute to adjudication Cowlin submitted an adjudication notice ("the First Adjudication") seeking a declaration that the form of contract under which CFW were engaged was the UK RIBA form of appointment for architects (the "RIBA Contract"). CFW issued a "counter-notice", which mirrored the Cowlin adjudication notice but in reverse, seeking a declaration that the form of contract was "SFA/99 together with DB2/99", an alternative form of architect's appointment (the "SFA Contract"). The adjudicator was appointed on the nomination of RIBA. Two weeks after CFW served their counter-notice, they wrote to Cowlin questioning the validity of the adjudicator's appointment arguing that the RIBA Contract had not been concluded between the parties, and that the adjudicator did not have jurisdiction to decide which form of contract applied. The adjudicator nonetheless proceeded with the Cowlin adjudication finding that the contract entered into was the SFA Contract and not the RIBA Contract. Subsequently, Cowlin claimed CFW had delayed Cowlin's commencement of the works by 14 weeks and that they had suffered losses arising from this delay, which they claimed against CFW. A dispute arose which Cowlin referred to adjudication and applied to the RIBA for an adjudicator appointment (the "Second Adjudication"). RIBA appointed an adjudicator in the Second Adjudication but CFW challenged the second adjudicator's jurisdiction on the grounds that there was no contract in writing for the purposes of the legislation. The adjudicator proceeded to make a decision that CFW pay Cowlin a sum of money. CFW refused to pay Cowlin on foot of the adjudicator's decision and Cowlin issued court enforcement proceedings in respect of the adjudicator's decision. CFW argued that the adjudicator in the Second Adjudication did not have jurisdiction to determine the dispute as there was no written contract between the parties and that there was therefore no dispute capable

9 Cowlin Construction Limited v CFW Architects [2003] EWHC 60.

of being referred to adjudication. Cowlin argued that it had been decided in the First Adjudication that there was a contract. CFW argued that they had not agreed that the adjudicator in the First Adjudication had jurisdiction to determine whether a contract had been concluded and consequently argued that because the adjudicator in the Second Adjudication took the adjudicator's decision in the First Adjudication as the basis for her award, this compromised the second adjudicator's jurisdiction.

6.35. HHJ Kirkham disagreed with CFW's contentions and held that by submitting counter-notice in the First Adjudication CFW accepted the adjudicator's jurisdiction to determine the form of contract between the parties. The judge held that CFW expressly accepted, in the First Adjudication, the adjudicator's jurisdiction to decide the issues set out in the counter-notice, which included the issue as to the correct form of contract. CFW were therefore bound by that acceptance and waived their right to object to the adjudicator's decision in the First Adjudication. The judge held that the first thing that comes to a responding party's mind when an adjudication notice is served is jurisdictional issues and whether there is a construction contract. Judge Kirkham concluded that CFW had sufficient knowledge to accept the adjudicator's jurisdiction at the time of service of their counter-notice and that any challenge to jurisdiction should have been made at the earliest opportunity. Judge Kirkham concluded that where CFW had "elected to affirm the adjudicator's jurisdiction and expressly sought decisions by the adjudicator … CFW waived its right to object to the jurisdiction of the adjudicator". Consequently, a party accepts that an adjudicator has jurisdiction over a dispute they cannot resile from that position in a subsequent adjudication.

6.36. In *Maymac v Farraday*,[10] Farraday engaged Maymac as a mechanical and electrical subcontractor. Maymac received only part payment on a payment application for works done, and thus referred the dispute to adjudication. The adjudicator awarded the payment claimed together with interest, but Farraday failed to comply with the decision. Maymac applied for summary judgment to enforce the adjudicator's decision in its favour. In its defence, Farraday claimed, in a witness statement from its solicitors, that the adjudicator had no jurisdiction to make the award because no construction contract ever came into existence. Farraday argued, on the documents before the adjudicator, that it may well have appeared there was a contract concluded between the parties, evidenced by a purchase order, but in fact there was no contract. They argued that there was a letter dated

10 Maymac Environmental Services Limited v Farraday Building Services Limited [2001] 75 Com LRHT 00/222.

6 December 1999 from a Maymac director to a Farraday director which stated, under the heading "Contract Conditions", that Maymac was unaware of the terms of the head contract between Farraday and the employer and until Maymac received a copy of the same they could not accept or be bound by the relevant conditions, nor could they be aware of how those conditions impacted upon the subcontract between Farraday and Maymac. Farraday argued that this letter was decisive in demonstrating non-existence of any contract between the parties, or at least not one which was subject to an adjudication referral. Farraday conceded that the substance of the works would have been capable of being covered by a construction contract and therefore within the terms of the UK Act, but that the parties had proceeded on the basis of a shared mistake, namely that they had thought they had a contract when they had no contract. Farraday also claimed the adjudication should be set aside if the terms of the contract turned out to be different from the contract put before the adjudicator. Farraday nonetheless agreed to take part in the adjudication but had failed to bring the letter relied upon to the adjudicator's attention.

6.37. HHJ Toulmin QC held that nowhere in the correspondence put before the adjudicator, and which had been brought to the court's attention, was there reference to the letter of 6 December 1999, and that the parties had a common understanding that there was a concluded contract in writing on agreed terms, and that this was one to which the UK Act applied. The judge held that the construction contract was accepted by Farraday as one to which the UK Act and UK Scheme applied and that Farraday explicitly sought to limit the adjudicator's jurisdiction to Maymac's entitlement to payment under their application for payment. Judge Toulmin also held that Farraday had accepted the adjudicator's jurisdiction to determine the dispute which Maymac had referred, and on which the adjudicator had made his decision. The judge held that the adjudicator had clearly adjudicated on the dispute which had been put before him, that Farraday had consented to submit to the adjudication and admitted that there was a contract to which the UK Act and UK Scheme applied, and the adjudication was conducted on that basis. Accordingly, Judge Toulmin held that Farraday were estopped, by representation and convention, from arguing that the UK Act and UK Scheme did not apply, and that the adjudicator was not entitled to make a decision which would be binding until the final determination of the dispute. Judge Toulmin held that, even if no contract existed and the referral was not under the UK Act, it was nonetheless made by the parties on the basis that the adjudication took place by agreement between the parties on the same terms as the UK Act and UK Scheme and that such an agreement between the parties was enforceable on the same basis as if the UK Act had applied. Judge Toulmin ordered summary judgment in favour of Maymac.

6.38. The *Maymac* case highlights the significance of a party raising a jurisdictional challenge during the adjudication. If a party seeking to oppose the enforcement of an adjudicator's decision happens upon an argument that the adjudicator did not have jurisdiction to make the decision, because of circumstances which arose either in advance of or during the adjudication, but such circumstances were not drawn to the adjudicator's attention at the time, then that party may have significant difficulties in convincing a court not to enforce the adjudicator's decision. The UK Courts, in general, have been loath to reject the enforcement of adjudicator decisions on grounds of lack of jurisdiction, though they have done so in several cases.

6.39. In *Carillion v Devonport,*[11] HHJ Jackson reviewed the UK case law on jurisdiction and breach of natural justice and outlined four basic principles: (1) the adjudication procedure does not involve the final determination of anybody's rights (unless all the parties so wish); (2) adjudicator's decisions must be enforced, even if they result from errors of procedure, fact or law; (3) where an adjudicator has acted in excess of jurisdiction or in serious breach of the rules of natural justice, the court will not enforce his decision; and (4) errors of law, fact or procedure by an adjudicator must be examined critically before the court accepts that such errors constitute excess of jurisdiction or serious breaches of the rules of natural justice.

6.40. Judge Jackson's decision was reaffirmed when the UK Court of Appeal heard the *Carillion* case on appeal.[12] The court held that an adjudicator's decision is to be treated as enforceable, and will only be interfered with in rare situations, even if the adjudicator has made obvious or highly arguable errors of fact or law. Lord Justice Chadwick stated that:

> The objective which underlies the Act and the statutory scheme requires the Courts to respect and enforce the adjudicator's decision unless it is plain that the question which he has decided was not the question referred to him or the manner in which he has gone about his task is obviously unfair. It should be only in rare circumstances that the Courts will interfere with the decision of an adjudicator.[13]

11 Carillion Construction Limited v Devonport Royal Dockyard Limited [2005] EWHC 778 (TCC).

12 [2005] EWCA Civ 1358.

13 At paragraph 85 of the judgment.

> … It is only too easy in a complex case for a party who is dissatisfied with the decision of an adjudicator to comb through the adjudicator's reasons and identify points upon which to present a challenge under the labels "excess of jurisdiction" or "breach of natural justice". … [T]he task of the adjudicator is not to act as arbitrator or judge. The time constraints within which he is expected to operate are proof of that. The task of the adjudicator is to find an interim solution which meets the needs of the case … The need to have the "right" answer has been subordinated to the need to have an answer quickly.[14]

6.41. Consequently, whilst jurisdictional challenges are not uncommon it is apparent that the UK Courts' preference is to respect adjudicators' decisions and enforce them.

Options when challenging an adjudicator's jurisdiction

6.42. A responding party seeking to challenge an adjudicator's jurisdiction has the following options:

(1) It can agree to refer the jurisdictional dispute to the same adjudicator and be bound by that adjudicator's decision in relation to jurisdiction.

(2) The parties can seek to refer the dispute on jurisdiction to a second adjudicator and be bound by the second adjudicator's decision. However, in these circumstances such action would not halt the first adjudication since that first adjudication has to be completed within the 28-day statutory time period. Consequently, a referring party rarely pursues this option.

(3) The responding party can seek a court declaration that the proposed adjudicator lacks jurisdiction. While this process needs to be initiated straight away it has not been that common in the UK Courts and requires a certain degree of cooperation between the parties to the adjudication and by the proposed adjudicator himself.

(4) The most common option is for the responding party, who is challenging the adjudicator's jurisdiction, to clearly set out the grounds of its objection and thereafter proceed with the adjudication, reserving its position on the lack of the adjudicator's jurisdiction. If the referring party is successful in both its jurisdictional and substantive case and seeks to enforce the decision the responding party, having properly reserved its position, can seek to challenge the adjudicator's decision in any court enforcement proceedings.

14 At paragraph 86 of the judgment.

6.43. The Act does not empower an adjudicator to determine his own jurisdiction. However, the parties can expressly agree to refer any jurisdictional dispute to the same adjudicator to whom the payment dispute has been referred. In such circumstances, the parties would be bound by that adjudicator's decision on the matter or jurisdiction. Circumstances may arise where the parties, by an implied agreement or inadvertently, empower the adjudicator to determine his own jurisdiction. In *Maymac* the parties conceded that the adjudicator determine whether the parties had concluded a contract, and as regards the terms and conditions of that contract. Consequently, the parties had thereby conceded that the adjudicator had been determining his own jurisdiction on the basis that the adjudicator's jurisdiction (in the particular circumstances) arose out of the contract. However, where one of the parties objects to the adjudicator's jurisdiction there cannot be implied an agreement to confer the determination of a jurisdictional dispute on that adjudicator. In such circumstances, if an objecting party wishes to later seek to rely on its jurisdictional objection, the objection should be raised at the earliest opportunity either before or during the adjudication.

The adjudicator's non-binding conclusion on jurisdiction

6.44. Any challenge by a party to an adjudicator's jurisdiction should be accompanied by a reservation of the right to maintain that objection. This is the most common scenario following a jurisdictional challenge. The objecting party should clearly set out the grounds of its objections, reserve its position and allow the adjudicator to proceed with the adjudication. This allows the adjudicator to reach a non-binding conclusion on his own jurisdiction, and if the adjudicator's non-binding conclusion is that he considers he has jurisdiction then the adjudicator can proceed to make a decision on the substantive dispute referred to adjudication. In the event that the adjudicator decides he does not have jurisdiction then the adjudicator would be obliged to withdraw from the adjudication. Such withdrawal would not, however, be construed as a resignation because where an adjudicator reaches a non-binding conclusion that he did not have jurisdiction then he was *prima facie* not appointed adjudicator and therefore had no role from which to resign.

6.45. In *Christiani v Lowry*,[15] HHJ Thornton QC held that an adjudicator could not decide his own jurisdiction because the UK Act only gave the adjudicator jurisdiction to decide disputes arising under the

15 Christiani & Nielsen Limited v The Lowry Centre Development Company Limited [2000] HT 001/59 (TCC), 26 June 2000.

contract. Judge Thornton stated that an adjudicator could ignore the jurisdictional challenge and proceed thereby allowing the court to consider the question of jurisdiction at the time of enforcement. The judge went on to say that alternatively an adjudicator could investigate the question of his own jurisdiction and reach his own "conclusion". Having reached his conclusion the adjudicator could then either withdraw or proceed with the adjudication depending on whether his conclusion was that he had jurisdiction to proceed. The judge held that while an adjudicator might reach such a conclusion, his decision on jurisdiction is non-binding as the adjudicator's jurisdiction would be a matter solely for the court in any enforcement proceedings.

6.46. On the basis of the *Christiani* ruling, an adjudicator's decision as to the existence or scope of his jurisdiction is therefore, in principle, a non-binding decision and merely a conclusion. However, where the jurisdiction issue overlaps with a substantive issue on which the adjudicator makes a decision then the adjudicator in deciding the substantive issue has, effectively empowered by default, decided the jurisdiction issue as opposed to having reached a non-binding conclusion.

6.47. If a party raises a jurisdictional objection but nonetheless allows the adjudicator to proceed, and thereby consider and decide on jurisdiction, it is crucial that the objecting party does not give the adjudicator the implied power to decide the jurisdictional issue. In such circumstances, it is imperative that the objecting party reserve the right to maintain its jurisdictional objection at a later date, generally after the adjudicator has issued a decision (during the course of enforcement proceedings) and to make clear that any purported decision by the adjudicator on jurisdiction may only give rise to a non-binding conclusion by the adjudicator and not a decision. If an objecting party fails to make clear that they are only facilitating an opportunity for the adjudicator to reach a non-binding conclusion, then the objecting party would be restricting itself from later maintaining the jurisdictional challenge in seeking to defend any enforcement proceedings on the adjudicator's decision. It is therefore critical that the objecting party makes it clear that, despite asking the adjudicator to consider his own jurisdiction and reach a non-binding conclusion in that regard, they do not agree to the adjudicator having power to make a binding decision on jurisdiction, or indeed on the substantive dispute.

6.48. In *Project Consultancy v Gray*,[16] HHJ Dyson held that the adjudicator himself had no power to rule on his own jurisdiction but was

16 The Project Consultancy Group v The Trustees of The Gray Trust [1999] HT/99/29 (TCC).

entitled to make enquiries into the question of whether he had jurisdiction. The judge held that the non-binding conclusion was made by the adjudicator so that the adjudicator could decide whether to proceed with the adjudication or withdraw. Project Consultancy's claim was for its fees arising out of works carried out for Gray in respect of a nursing home. The UK Act does not apply to construction contracts entered into before 1 May 1998. Gray argued that if there was a construction contract it was concluded prior to 1 May 1998 and therefore there was no right to refer the dispute to adjudication under the Act. Gray argued before the adjudicator that no contract was ever concluded but continued to participate in the adjudication reserving its position on the issue of jurisdiction. The adjudicator reached a decision that a contract had been concluded on 10 July 1998 (i.e. after the UK Act's operative date) and awarded Project Consultancy £65,000.00. Gray didn't pay upon the decision, so Project Consultancy applied for summary judgment. Project Consultancy argued before the court that a decision whose validity was challenged was nonetheless a decision within the meaning of the UK Act and sought to rely on the *Macob* decision.[17] Judge Dyson differentiated his decision in *Macob* on the basis that the alleged invalidity in that case arose from what was claimed by the defendant to have been a procedural error which amounted to a breach of natural justice. Judge Dyson's view was that different considerations apply where the adjudicator purports to make a decision which he is not empowered by the Act to make. Consequently, Judge Dyson rejected Project Consultancy's argument in this regard. Judge Dyson then considered the issue whether Gray had properly reserved its position on the adjudicator's jurisdiction, while continuing to participate in the adjudication, and held that Gray had stated in the clearest terms that it contested the adjudicator's decision and would not comply with any decision, while reserving its right to participate without prejudice to that contention. Judge Dyson quoted the actual words used by Gray's lawyers that:

> the Act cannot apply and your Notice of Reference to adjudication is invalid. We suggest that in the circumstances adjudication is inappropriate and enquire whether you intend to withdraw the reference. If however your client proceeds with adjudication, our client shall dispute the adjudicator's jurisdiction. If the adjudicator makes a decision notwithstanding the objection to jurisdiction, our clients will not comply with any award made on the basis that it was made without jurisdiction. These issues will be

17 Macob Civil Engineering v Morrison Construction [1999] EWHC Technology 254.

> placed before the Court should your client issue any application for enforcement of the adjudicator's award. Without prejudice to the above, if you proceed with the adjudication, we reserve our clients' rights generally, and in particular to appear and present their case to the adjudicator.[18]

6.49. Judge Dyson noted that Gray's solicitors' letter had been sent well in advance of the adjudicator's decision. Judge Dyson was strongly of the view that Gray's position and approach did not prevent their subsequent challenge to the adjudicator's decision in the enforcement proceedings and, consequently, upheld the challenge dismissing Project Consultancy's application for enforcement of the adjudicator's decision.

6.50. Parties may also, expressly or impliedly, give the adjudicator power to make a binding decision on his own jurisdiction. In *Whiteways v - Impresa*,[19] HHJ Bowsher QC held that the adjudicator's jurisdictional decision resulted from the parties having given the adjudicator power to determine his own jurisdiction, and that consequently the adjudicator was entitled to determine his own jurisdiction and the parties were bound by that determination. Impresa engaged Whiteways as a plastering subcontractor for hotel renovation works under a UK standard-form subcontract DOM/1, which provided for disputes or differences to be referred to adjudication. A dispute arose which Whiteways referred to adjudication and the adjudicator made a decision ordering Impresa to pay monies to Whiteways. Impresa did not pay Whiteways, who then issued court enforcement proceedings in respect of the decision. In the court proceedings Impresa asserted that the adjudicator acted outside his jurisdiction and the decision should not be enforced. Impresa argued that the adjudicator's jurisdiction was limited to the disputes indicated in the adjudication notice and the adjudicator should not have dealt with additional points made by Whiteways in their subsequent referral document, but not indicated in the notice. On receipt of the referral Impresa's solicitors had raised the jurisdiction objection by letter to the adjudicator stating

> We invite you to decide on this issue as a matter of urgency as our response to Whiteway's Notice of Referral will depend on your decision. Our client does not wish to incur costs on matters which, in our view fall outside the jurisdiction of the adjudication.[20]

18 At paragraph 11 of the judgment.

19 Whiteways Contractors (Sussex) Limited v Impresa Castelli Construction UK Limited [2000] EWHC Technology 67 (TCC).

20 At paragraph 16 of the judgment.

The adjudicator replied maintaining that after a detailed reasoned analysis on each of the disputed items, "I do have jurisdiction to consider this matter". Subsequently, in the adjudicator's decision, he further stated

> although I consider that an adjudicator has no power under Clause 38A of DOM/1 to determine his or her jurisdiction, I conclude from the parties' submissions, both from leading firms of Solicitors, that the parties have extended my jurisdiction to determine the matter.[21]

6.51. Judge Bowscher held that Impresa's solicitors' letter raising the jurisdictional challenge had not in fact enquired into the adjudicator's jurisdiction but rather had actually invited the adjudicator to decide his own jurisdiction. Judge Bowsher held that the adjudicator's decision, which included his decision on matters which were subject to the jurisdictional challenge, was enforceable and the judge gave summary judgment in Whiteways' favour enforcing the adjudicator's decision.

6.52. Circumstances can arise where it is not so easy to distinguish between a substantive matter which has been referred to adjudication and a jurisdictional issue. Where the two issues overlap or are entwined, the situation can become quite complex. In *Air Design v Deerglen*[22] it was held that where a substantive matter within the adjudicator's jurisdiction overlaps with a question of jurisdiction itself, an error in the decision relating to the substantive matter can result in a jurisdictional challenge. Deerglen, as main contractor, engaged Air Design as a mechanical services subcontractor on an office block project. Matters were complicated by the fact that there was a "base build" contract followed by a "supplementary contract" and two other separate arrangements. A payment dispute arose, and Air Design served an adjudication notice. The adjudicator issued a decision that Deerglen pay monies to Air Design. Deerglen did not comply with the adjudicator's decision and Air Design issued court enforcement proceedings. Deerglen argued that the adjudicator had no jurisdiction in respect of the separate contracts between the parties and no jurisdiction over disputes arising out of those contracts. Air Design responded by arguing that the adjudicator had jurisdiction as the supplementary contract, and the separate arrangements, were simply variation agreements on the "base build" contract. The adjudicator agreed with Air Design's submissions and gave his "non-binding" view that he had jurisdiction. Judge Akenhead agreed with the adjudicator's jurisdictional decision as the substantive dispute

21 At paragraph 22 of the judgment.

22 Air Design (Kent) Limited v Deerglen (Jersey) Limited [2008] EWHC 3047 (TCC).

put before the adjudicator required him to consider whether there was more than one contract before reaching a decision on the substantive dispute. In those circumstances, Judge Akenhead held that it was therefore within the adjudicator's jurisdiction to decide that there was one contract varied by an agreement. The judge held in favour of Air Design enforcing the adjudicator's decision.

7 Referral to the adjudicator

Section 6(5) of the Act

Referral notice

7.1. Pursuant to Section 6(5), following the adjudicator appointment the referring party must, within seven days, beginning on the day on which the appointment was made, refer the payment dispute to the adjudicator ("the referral"). Simultaneously, the referring party must also provide a copy of the referral, together with all accompanying documents, to the responding party.

The seven-day period

7.2. Section 6(5)(a) of the Act is similar to Section 108(2)(a) of the UK Act and paragraph 7(1) of the UK Scheme, which also impose a seven-day time frame on the service of the referral. The UK Courts have repeatedly held that failure to serve the referral notice within the seven days prescribed renders any adjudicator's subsequent decision a nullity. The seven-day time frame for service of the referral under Section 6(5)(a) will likely be construed as mandatory by the Irish Courts, and that has been the experience in the UK under the UK Act and the UK Scheme.

7.3. Irish Courts are likely to consider UK case law persuasive on the interpretation of the seven-day time frame within which the referral notice must be served, and the UK legislative requirement in this regard has been construed rigidly by the UK Courts. Section 108(2)(b) of the UK Act provides that a construction contract shall include a provision in writing so as to provide for a timetable with the object of securing the appointment of the adjudicator and referral of the dispute to him within seven days of the adjudication notice. Where the construction contract does not contain such a provision the UK Scheme automatically applies. For the purposes of understanding the UK case law on the seven-day period for service of the referral it is important to note the distinction between the wording of Section 108(2)(b) of the UK Act and paragraph 7(1) of the UK Scheme. Section 108(2)(b) provides that the "referral of the dispute" to the adjudicator shall be served "within seven days of" the notice of intention to refer a dispute to adjudication. The wording of paragraph 7(1) of the UK Scheme is far more mandatory insofar as it provides that "the referring party shall, not later than seven days from the date of the notice of adjudication, refer the dispute in writing … to the adjudicator".

7.4. Section 6(5)(a) of the Act prescribes the seven-day period more rigidly than Section 108(2)(b) of the UK Act, insofar as Section 6(5)(a) provides that a party "shall refer the payment dispute to the adjudicator within seven days beginning with the day on which the appointment is made". Section 6(5)(a) therefore leaves no scope for the seven-day period to be extended either by the adjudicator on his own initiative with the agreement of the referring party, or by agreement between the parties. This is because it is a statutory time frame within which the referral should be served and the parties cannot, pursuant to Section 2(5)(b), purport to limit the application of the Act including seeking to limit any statutory time frames prescribed in the Act. Consequently, the seven-day period for service of the referral is a mandatory period, which if not complied with by the referring party and the adjudication nonetheless proceeds, the adjudicator's decision would be a nullity.

7.5. Whereas the wording of Section 108(2)(b) of the of the UK Act is more directive in nature, paragraph 7(1) of the UK Scheme is more mandatory and is therefore more comparable to Section 6(5)(a) of the Act. In *Hart v Fiddler*[1] Hart engaged Larchpark (the second defendant in the court proceedings) to carry out construction work on Hart's behalf. Fiddler was the engineer. A wall of the property collapsed and when Larchpark sought to claim payment for the work that it carried out, both before and after the collapse, Hart refused to pay. Larchpark issued a notice of intention to refer the payment dispute to adjudication and an adjudicator was appointed. However, Larchpark failed to serve the referral notice in accordance with the UK Scheme (which applied) serving the referral notice eight days after the notice of intention to refer. Hart immediately adopted the position that the adjudicator had no jurisdiction to deal with the dispute. However, Larchpark carried on with the adjudication. The adjudicator dismissed the suggestion that Larchpark were responsible for the collapse of the wall and made a decision awarding Larchpark payment for the works carried out. Hart failed to pay the sums and

1 Hart Investments v Fiddler & Another [2006] EWHC 2857 (TCC).

Larchpark issued court enforcement proceedings in respect of the adjudicator's decision. Hart opposed the enforcement proceedings on two grounds, one of which was that the referral had not been served in accordance with the UK Scheme, was out of time and the adjudicator's decision was therefore a nullity.

7.6. HHJ Coulson agreed with Hart's argument, stating that although his initial reaction was that it might be harsh to conclude that a one-day delay in service of the referral notice could render the adjudication a nullity, all sorts of difficult questions arose if the failure to comply with the seven-day period in paragraph 7(1) of the UK Scheme was ignored. Judge Coulson held that the adjudicator's decision (following the referral notice having been served late) was a nullity and could not be enforced because the referral notice was served outside the seven days stipulated by paragraph 7(1) of the UK Scheme. The judge repeatedly stated in his judgment that the whole purpose of adjudication was that speed was given precedence over accuracy and it was important that the adjudicator's decision be given within the prescribed time frame, be a quick decision and need not necessarily be a correct one. Larchpark had argued that the adjudicator could extend the seven-day time limit without consent under paragraph 13 of the UK Scheme, but Judge Coulson rejected this argument. Paragraph 13 of the UK Scheme permits the adjudicator to decide the procedure including giving directions on the timetable for the ajudication. The judge stated that everything done pursuant to the UK Scheme, including the 28-day period for the adjudication itself, flows from the date of the referral notice. The adjudicator is not seized of the adjudication until the referral notice is provided and the 28-day period starts to run. The adjudicator therefore has no power until he gets the referral notice. Consequently, the judge held that the adjudicator has no power to extend the seven-day period which occurs before his jurisdiction begins. In any event, the judge determined that the adjudicator could not extend time retrospectively which is what he considered the adjudicator had attempted to do.[2] As Judge Coulson stated, if the adjudicator extended the timetable without consent there was a significant danger that the statutory time frames would become eroded and the core objective of eliciting a quick decision in adjudication would be undermined. Significantly, Judge Coulson held that the word "shall" (in paragraph 7(1)) imposed a mandatory obligation on the referring party to serve the referral notice on the adjudicator within the seven-day period from the adjudication notice date.

2 At paragraph 52 of the judgment.

7.7. Irish Courts are likely to find that the seven-day period to serve the referral is mandatory given of the specific use of the word "shall" in Section 6(5)(a) of the Act.

7.8. In *KNS v Sindall*,[3] HHJ Humphrey Lloyd QC confirmed Ramsey J's view, *obiter*, that the adjudication notice determines an adjudicator's jurisdiction which commences on the service of the referral notice. The judge held that the adjudicator's jurisdiction does not derive from other "accompanying" documents, despite the fact that these documents are likely to help the adjudicator determine what is to be decided.

7.9. The *KNS* and *Hart* decisions are persuasive, in an Irish context, to support the following principles:

(1) While the adjudicator derives his jurisdiction from the notice of intention to refer, the adjudicator is not seized of the adjudication until the referral notice is served.
(2) The adjudicator's jurisdiction does not commence until the referral notice is served and therefore the adjudicator has no power until he receives the referral notice.
(3) The adjudicator does not have jurisdiction to extend the period of time within which a referral notice is served as he is only empowered on receipt of the referral notice.
(4) If the referral notice is not served within the seven-day period then it is not valid, and any subsequent adjudicator's decision is a nullity.

7.10. The adjudicator's jurisdiction is defined by and derived from the notice of intention to refer. However, the adjudicator's jurisdiction to actually proceed with the adjudication referral leading to a decision cannot commence until the adjudicator is in receipt of the referral notice. As a consequence, if the referral notice is not served within the seven-day time period then it is not valid with the effect that the dispute has not been referred to the adjudicator. If the adjudicator then proceeds to consider and decide the dispute his decision is a nullity and unenforceable.

7.11. While Section 6(5)(a) of the Act and paragraph 7(1) of the UK Scheme are similar, insofar as they both contain the mandatory word "shall" in respect of service of the referral notice, they differ in a significant aspect. Paragraph 7(1) states that the referring party shall "not later than seven days from the date of the notice of adjudication" serve the referral notice. However, Section 6(5)(a) provides that the referral notice be served on the adjudicator "within 7 days".

3 KNS Industrial Services (Birmingham) Limited v Sindall Limited [2001] 75 Con LR.

While it is a matter of semantics as to whether "not later than 7 days" has the same meaning as "within 7 days", it will be interesting to see whether the Irish Courts interpret "within 7 days" similarly to how the UK Courts have interpreted "not later than 7 days".

7.12. It should also be noted that the Act differs from the UK Act in respect of the date from which the seven-day period commences. Under Section 6(5)(a) the seven-day period commences from the day on which the adjudicator's appointment is made. Paragraph 7(1) of the UK Scheme provides that the seven days run from the date of the adjudication notice. This is an important difference. Where the parties have agreed the adjudicator appointment the prospective adjudicator, pursuant to paragraph 12 of the Code of Practice, shall write to each party to accept the appointment. The date of the adjudicator's letter of acceptance to the parties is deemed to be the date on which the adjudicator's appointment is made for the purposes of Section 6(5)(a). Where the parties are working to a tight time frame, and particularly where the referring party is required to serve the referral within seven days of the date of the adjudicator appointment, if the adjudicator's acceptance letter is not received by the referring party for a number of days after the adjudicator has sent it to the parties then this could lead to difficulties. In such circumstances, the referring party's period of time within which to prepare its referral notice for service (within the seven days beginning on the day of the adjudicator's appointment) will have been reduced through no fault of the referring party. This is not an entirely satisfactory situation. While a referring party would be advised to have prepared its referral notice in advance of serving its notice of intention to refer, the referring party is nonetheless entitled to avail of the full seven days between the adjudicator's appointment and service of the referral notice. The issue is further complicated by virtue of the fact that, as previously discussed, Section 6(8) provides that the adjudicator shall comply with the Code of Practice but does not similarly provide that the parties are required to comply with the Code of Practice.

7.13. Difficulties may also arise where the parties cannot agree an adjudicator appointment and the Chair of the Minister's panel makes the appointment. Paragraph 18 of the Code states that the date of the CCAS letter to the parties shall be deemed to be the date on which the appointment of the adjudicator is made for the purposes of Section 6(5)(a) of the Act. Consequently, if the CCAS letter is delayed in reaching the referring party, then the seven-day period for serving the referral notice will be shortened.

7.14. Clearly, the fact that the seven-day period for service of the referral notice begins with the date on which the adjudicator's appointment is made could result in a referring party finding itself unable or ill prepared to serve its referral notice within the prescribed time under Section 6(5)(a). Under the UK legislation the seven-day period runs from the adjudication notice date. While the start date for the seven-day period under the Act is therefore different to that under the UK legislation, the principle that the seven-day period is mandatory applies under the Act and UK legislation.

7.15. In *Cubitt v Fleetglade*,[4] Fleetglade engaged Cubitt as its contractor under an amended UK standard form JCT 1998 contract. The architects under the contract issued a final certificate on 20 September 2006, and on the same day Cubitt served an adjudication notice. The following day Cubitt wrote to the RICS seeking the appointment of an adjudicator. However, the adjudicator was not appointed until 27 September 2006. The parties' respective solicitors spoke over the telephone late on 27 September 2006 and Cubitt's solicitor offered Fleetglade's solicitor the referral notice, but without the accompanying documents. Fleetglade's solicitors refused the offer. Cubitt served the referral notice on the adjudicator the following day. Fleetglade's solicitor immediately took the point in writing that the referral was not served within seven days of the adjudication notice and that the adjudicator therefore had no jurisdiction. The adjudicator made a decision in Cubitt's favour but Fleetglade argued that the decision was reached out of time and was a nullity on a number of grounds, including the ground that the referral was served outside the seven-day period rendering the adjudicator without jurisdiction. HHJ Coulson QC found that the wording of the relevant contract clause (Clause 41A of the JCT contract) relating to the seven-day period was mandatory rather than directory, unlike the relevant provisions of the UK Act, which have been held to be directory. However, the judge held that the clause had to be operated in a sensible and commercial way. The judge noted that Clause 41A expressly recognised that an adjudicator may not be appointed until after the seven-day period has expired and he considered that the words of Clause 41A were mandatory. The JCT contract provided for what should happen where the appointment of the adjudicator occurred within the seven days (Scenario 1), and what should happen where the appointment occurred after the seven-day period (Scenario 2). In Scenario 1 (where the adjudicator is appointed within the seven-day period), the referral had to be served within seven days of the adjudication notice. In Scenario 2 (where the adjudicator is appointed after the seven-day period), the referral had to be served "as soon as possible thereafter". While Judge Coulson found that the effective date of the adjudication notice was 20 September 2006, he did

4 Cubitt Building & Interiors v Fleetglade Limited [2006] EWHC 3413(TCC).

not consider the referral notice a nullity, having been served on the adjudicator on 28 September 2006. Judge Coulson held that although Clause 41A did not expressly provide for the situation where, through no fault of the referring party, the adjudicator appointment did not occur until very late on the seventh day, a sensible interpretation of Clause 41A would be to deal with such circumstances as if it had occurred after the seven-day period (i.e. deal with the situation as if it were a Scenario 2 circumstance). The judge held that it would be nonsensical for the referring party to be better off if the appointment occurred on day eight (a Scenario 2 circumstance which would mean the referring party could serve the referral "as soon as possible thereafter") rather than late on day seven (a Scenario 1 circumstance where the referring party would have to serve the referral on the seventh day). The judge noted that the contract provided for circumstances where the appointment was only confirmed after expiry of the seven-day period and that in such circumstances the referral notice must be served "as soon as possible thereafter" in order to constitute good service. However, the contract did not provide for circumstances where the appointment of the adjudicator occurred late on the seventh day (namely at 5.35 p.m. on Wednesday 27 September 2006) and consequently the judge held that service of the referral notice the following day constituted service "as soon as possible thereafter", and in such circumstances constituted good service. The judge held, therefore, that the referral notice was validly served in accordance with the contract and the adjudicator had the necessary jurisdiction.

7.16. Clearly, the *Cubitt* decision was reached on the merits of the specific circumstances pertaining, and in particular the wording of the JCT contract that had been entered into between the parties. However, the *Cubitt* decision highlights the difficulties that can arise where the seven-day period for service of the referral notice runs from the date of the adjudicator's appointment in circumstances where neither party will be aware of that date until they are in receipt of the letter sent by either the adjudicator, pursuant to paragraph 12 of the Code of Practice, or by the CCAS pursuant to paragraph 18 of the Code of Practice.

7.17. Where a party finds itself unable or ill prepared to serve its referral notice within the statutory seven-day period it is open to that party to abandon the referral to adjudication and restart the process. Clearly, in such circumstances, the referring party may face a claim for any costs the adjudicator or the responding party incur in the aborted adjudication, though it is arguable that such costs might not be recoverable. This could arise where a referring party fails to serve its referral notice within the seven-day period because it only became aware of the date of the adjudicator's appointment on receipt of the adjudicator's letter after the expiry of the seven-day period. The referring party's exposure to a claim for wasted costs might be greater if the referring party failed to serve its referral notice within the seven-day period because it was ill prepared or because it deliberately took the decision not to proceed with the adjudication on the basis that it did not like the adjudicator appointed. The wasted costs incurred by both the adjudicator and the opposing party might well be recoverable in such a scenario.

7.18. In *Lanes v Galliford*,[5] the English Court of Appeal held that the referring party's failure to refer the dispute within a seven-day time period did not prohibit them from starting the entire adjudication process again. The court had to consider two appeals from the decisions of the English High Court in relation to an adjudicator's jurisdiction and enforcement of the adjudicator's decision. Lanes agreed to carry out works for Galliford at a maintenance depot under a subcontract which incorporated the adjudication procedure of the Institute of Chartered Engineers (ICE) (the "ICE Adjudication Procedure"). The subcontract was terminated by Galliford following delays which Lanes claimed had arisen due to, *inter alia*, Galliford's extensive variations. Lanes claimed a substantial extension of time and wrongful termination of the subcontract. Lanes' claimed damages of circa £2 million. Galliford claimed approximately £2.7 million in damages arising out of Lanes' failure to proceed with the works with due diligence. When the dispute arose Galliford commenced adjudication proceedings and applied to the ICE to nominate an adjudicator. The ICE nominated a Mr Klein as adjudicator. However, Galliford decided not to send the referral notice to Mr Klein. Galliford's failure to serve the referral notice was contrary to Clause 4.1 of the ICE Adjudication Procedure. Galliford's solicitor's decision not to send the referral to Mr Klein arose because of difficulties they had with him in a previous case, where Galliford perceived that Mr Klein was biased. Instead of serving the referral notice, Galliford's solicitors served a fresh notice of adjudication and sought a fresh appointment from the ICE, who appointed a Mr Atkinson as adjudicator. Lanes' solicitors protested that Mr Atkinson did not have jurisdiction as Mr Klein had been appointed on the dispute. Lanes commenced High Court proceedings to restrain Mr Atkinson from proceeding with the adjudication on the grounds that Galliford's failure to serve the referral notice was a repudiatory breach of the agreement to refer disputes to adjudication, and that Lanes had accepted the repudiatory breach. The effect of acceptance by Lanes of Galliford's alleged

5 Lanes Group Plc v Galliford Try Infrastructure Limited [2011] EWHC 1679 (TCC).

repudiatory breach would have been to render the subcontract determined and *void ab initio*, which appears to be the outcome that Lanes desired so as to facilitate an argument that as no contract existed no further disputes could arise that could be referred to adjudication. Lanes were unsuccessful in their court application and the adjudication before Mr Atkinson proceeded. Mr Atkinson issued a decision in the adjudication and Lanes sought to challenge it on the basis that, *inter alia*, Mr Atkinson did not have jurisdiction where Mr Klein had been validly appointed. The High Court found against Lanes and the Court of Appeal upheld that finding. In addition, the Court of Appeal agreed that the concept of a repudiation was not applicable to adjudications and in any event Galliford could not have forfeited its rights to have other disputes adjudicated, so the contractual provisions for adjudication had to remain in force. As regards the validity of Mr Atkinson's appointment, the Court of Appeal was not sympathetic to the proposition that a claimant in an adjudication could simply let an adjudication lapse if it did not approve of the appointed adjudicator. Nevertheless, the court held that Galliford's attempt to appoint a different adjudicator, Mr Atkinson, was permissible by the contract and was not an abuse of process. The court was persuaded that the contract between the parties, the ICE Adjudication Procedure and the UK Scheme recognised the right to restart an adjudication in a variety of circumstances. Accordingly, the court held that a failure to serve the referral notice did not preclude a party from starting an adjudication afresh. The court commented, *obiter*, that while the failure to serve the referral in time represented a breach of the UK Scheme (albeit a non-repudiatory breach) it did not prevent the referring party from referring again, albeit that such action might entitle the opposing party to claim damages arising from that breach. In other words, the wasted costs associated with the aborted adjudication arising from the failed referral might be recoverable by the responding party and the adjudicator. The Court of Appeal held that there was no basis for implying an absolute or qualified bar to restarting an adjudication where there had been a previous failure to refer that same dispute and there was no prejudice arising from any second process that may be initiated.

All accompanying documents

7.19. Section 6(5)(b) of the Act obliges the party referring the dispute to the adjudicator to simultaneously "provide a copy of the referral and all accompanying documents to the person who is party to the construction contract". While the UK Courts have strictly construed the seven-day period within which the referral must be served, they have adopted a more flexible approach in their interpretation of paragraph 7(2) of the UK Scheme in respect of the requirement that the referral be accompanied by copies of "all accompanying documents". Paragraph 7(2) of the UK Scheme is similarly worded to Section 6(5)(b) of the Act and it is likely that the Irish Courts will adopt a similar approach as the UK Courts have adopted to paragraph 7(2) of the UK Scheme.

7.20. Paragraph 7(2) of the UK Scheme provides that the referral notice, when served on the adjudicator, shall be "accompanied by copies of, or relevant extracts from, the construction contract and such other documents as the referring party intends to rely upon".

7.21. While Section 6(5)(b) provides that the referring party supply "a copy of the referral and all accompanying documents" to the opposing party it does not stipulate that the referring party furnish those accompanying documents to the adjudicator. However, paragraph 22 of the Code of Practice seeks to remedy this particular lacuna, insofar as it provides for the referral of the payment dispute to the adjudicator to include specific details, including (as provided for at paragraph 22(vi) of the Code of Practice) "any supporting information that may assist an Adjudicator in understanding the nature of the payment dispute". While the two provisions are broadly similar insofar as they require the referring party to serve the referral with accompanying documents, there is a difference between any "supporting information" that may assist an adjudicator in understanding the nature of the payment dispute and all "accompanying documents" which must be copied to the opposing party. Consequently, it is arguable that the referring party could potentially serve on the adjudicator (pursuant to paragraph 22(iv) of the Code of Practice) different documents from those that it serves on the opposing party pursuant to Section 6(5)(b). While it is likely that a referring party will serve the same set of documents on the adjudicator as it does on the opposing party, there nonetheless could potentially arise a situation where the documents are not the same and that could lead to all sorts of difficulties arising, in the course of the adjudication and in respect of the adjudicator's decision.

Timing of service of all accompanying documents

7.22. The 28-day period within which the adjudicator must reach a decision commences when the adjudicator receives the referral notice. It is commonly the case in the UK that the referral and accompanying documents are served electronically by email and that is certainly the experience to date in adjudication under the Act. Ideally, the referral notice should include and be accompanied by all relevant documents. However, in reality there are situations where some documents may not be capable of being served electronically, for example drawings, photographic images or large PDF documents unsuitable for electronic transfer. Where such circumstances arise the adjudicator may receive certain documents electronically within the prescribed time (the seven-day period) and later receive by hard copy (after the seven-day period) those documents which could not be sent electronically. In such circumstances the question arises whether the documents which the adjudicator has received on time are sufficient to "assist an Adjudicator in understanding the nature of the payment dispute", as is provided for under paragraph 22(vi) of the Code of Practice. However, if the referring party has experienced the same difficulty in electronically submitting documents to the opposing party the question also arises as to whether the referring party has complied with its obligations under Section 6(5)(b) of the Act in providing "all accompanying documents" with the referral notice. The first issue, as regards proper referral of the payment dispute to the adjudicator, is whether the adjudicator has received the referral notice in its entirety and is therefore seized with jurisdiction. The second issue is whether the 28-day time period for the decision has started to run. The UK Courts have approached these issues on two separate bases, namely as to when the clock starts to run in relation to the 28-day period; and secondly as to whether the adjudicator is seized with jurisdiction despite the fact that he may not be in possession of all accompanying documents.

7.23. In *PTB v ROK*,[6] a jurisdiction challenge was raised on the basis that the construction contract (which would generally always be construed as an important accompanying document and which is to accompany the referral pursuant to paragraph 7(2) of the UK Scheme) had been provided by PTB the day after the referral notice was served. ROK argued that serving the referral without the construction contract was contrary to paragraph 7(2) of the UK Scheme which required that the contract accompany the referral. HHJ Ramsey rejected the jurisdictional challenge differentiating between the decision in *Hart Investments*, in relation to a breach of paragraph 7(1) of the UK Scheme, and the purported breach of paragraph 7(2) of the UK Scheme. The judge held that on the basis that service of the referral notice within seven days was a fundamental provision in the adjudication process, the service of accompanying documents with the referral notice was simply an associated procedural requirement. The judge held that while a breach of paragraph 7(1) could deprive the adjudicator of jurisdiction, a breach of paragraph 7(2) amounted to no more than a trivial breach of the terms of the UK

6 PT Building Services Limited v ROK Build Limited [2008] EWHC 3434 (TCC).

Scheme, which should not be seized upon to impeach the adjudication process. Judge Ramsey held that where there had been late service of documents supporting the referral notice, that in itself does not affect the validity of the service of the referral and the adjudication process, unless the supporting documents are so significant that without them the referral itself is deficient to the extent that it cannot be decided upon by the adjudicator. The judge held that the validity of the referral will always be a question of fact and will depend on the degree of the deficiency in providing the supporting documents that the referring party intends to rely upon.

7.24. Section 6(5)(b) requires the referring party to provide a copy of the referral and "all accompanying documents" to the opposing party at the same time (i.e. within the seven-day statutory period) as the referring party serves the referral on the adjudicator. Section 6(5)(b) does not provide for "all accompanying documents" being served with the referral on the adjudicator and this is left to the Code of Practice to address. Paragraph 22 of the Code of Practice sets out what must be included in the referral to the adjudicator. As previously discussed that includes, at paragraph 22(vi), "any supporting information that may assist an Adjudicator in understanding the nature of the payment dispute". Despite the fact that this is inconsistent with the term "all accompanying documents" in the Act it is also problematic insofar as there is no provision of the Act which requires the parties to comply with the Code of Practice.

7.25. The question arises as to whether a referring party could be said to have failed to comply with Section 6(5)(a) of the Act in serving the referral within seven days in circumstances where the referring party has failed to include "any supporting information" in the referral to the adjudicator, and therefore having not complied with paragraph 22(vi) of the Code of Practice. While non-compliance by a party with the Code of Practice may not place that party in breach of the Act, it is likely that the courts, when considering whether a referral has been properly served on an adjudicator within the statutory seven-day period, will examine what accompanying documents and supporting information was provided to the adjudicator with the referral so as to comply with paragraph 22 of the Code of Practice. A referring party that does not comply with paragraph 22 of the Code of Practice as to the contents of the referral to be furnished to the adjudicator, would be at risk of facing an argument that it failed to comply with its obligations to successfully refer the payment dispute to the adjudicator within the statutory seven-day period provided for in Clause 6(5)(a) of the Act. The Code of Practice is arguably a procedural code as opposed to a statutory requirement, particularly as regards a party's obligations to comply with it. If the Irish Courts adopt a similar view to the UK Courts, in relation to service with the referral of any supporting information or accompanying documents, then one could conceivably envisage the Irish Courts taking the view that if the referral is served within the statutory seven-day period but without all the supporting information and documents, then the latter would merely represent a procedural non-compliance by the referring party and would not result in the adjudicator being deprived of jurisdiction. The critical issue to be determined will be what "supporting information" is sufficient to assist the adjudicator in understanding the nature of the payment dispute. If the necessary documents required to assist the adjudicator in understanding the nature of the dispute are not served with the referral that might lead to a situation where it is more than a procedural non-compliance and constitutes a breach of Section 6(5)(a), depriving the adjudicator of jurisdiction as he has not received the referral because it was not accompanied by the necessary documents to assist the adjudicator in understanding the nature of the payment dispute. Clearly, the issue could come down to a debate as to what supporting information is required, in the context of the particular circumstances of the payment dispute that has been referred to the adjudicator, in order for that particular adjudicator to understand the nature of the payment dispute. This is not an entirely satisfactory situation and it may well result in challenges that the referral itself has not been served within the seven-day period as the referring party failed to accompany the referral with sufficient supporting information to assist the adjudicator in understanding the nature of the payment dispute.

7.26. In *Linnett v Halliwells*,[7] the court held that a failure to serve supporting documents within the time limit for service of the referral is a procedural non-compliance and the court should be slow to find that such a failure deprives the adjudicator of jurisdiction. Mr Linnett was appointed as adjudicator in a dispute between ISG Interior Exterior Plc and Halliwells in relation to a contract to carry out fit-out works at Halliwells' Manchester offices. The referral had been served by ISG on Mr Linnett by fax (without the accompanying documents which were sent by post) on 29 May 2008 and served on Halliwells the following day. Mr Linnett did not receive the original referral (and therefore had also not received the accompanying documents) which seemingly were lost in the post, but received a duplicate copy five days later. Halliwells challenged the adjudicator's jurisdiction. Halliwells' argument was, *inter alia*, that the referral had been served out of time because the adjudicator's copy of the attachments had been lost in the post and was not received within

7 Christopher Michael Linnett v Halliwells LLP [2009] EWHC 319 (TCC).

seven days of the adjudication notice. Mr Linnett sent his terms of engagement to both parties but Halliwells refused to sign them on the basis of their jurisdictional challenge. Instead, Halliwells invited the adjudicator to resign or alternatively to extend the date for service of their response as a result of the late referral. Mr Linnett decided that he had jurisdiction. Halliwells participated in the adjudication, without prejudice to their jurisdictional objection, reserving their position in this regard. While the central issue before the court was the effect of a jurisdictional challenge on the liability of a party to pay the adjudicator's fees and expenses, the court also had to consider, as an integral part of its consideration of the fees issue, whether the adjudicator had jurisdiction.

7.27. HHJ Ramsey held that although the accompanying documents were not served within the seven-day period, and this was a breach of Clause 41A.4 of the contract, it was not a breach sufficient to invalidate the adjudication or nullify the decision itself. The judge considered the *Cubitt* decision where it was held that Clause 41A must be operated in a sensible and commercial way and also considered the specific wording of Clause 41A 5.6 that

> any failure by either party to enter into the JCT adjudication Agreement or to comply with any requirement of the Adjudicator under Clause 41A 5.5 or with any other provision in or requirement under Clause 41A shall not invalidate the decision of the Adjudicator.

Having considered those issues Judge Ramsey held that in the present circumstances, Clause 41A covered the failure to serve the accompanying documents on the adjudicator within the seven-day period and thus the adjudicator retained jurisdiction, despite the lateness of the accompanying documents being served.

Form and content of the referral

7.28. The referral is the document on which the adjudicator's jurisdiction commences. Consequently, the requirements as to the contents of the referral or what documents need to accompany the referral, and whether those documents are delivered to the adjudicator and/or the opposing party, are critical issues. The Irish Courts are likely to adopt a similar approach to the UK Courts in considering issues surrounding supporting documents to be furnished with the referral, namely documents which assist the adjudicator to determine the dispute to be decided but do not, *prima facie*, deprive him of jurisdiction to proceed with the adjudication.

7.29. The preferred position is to ensure that all relevant and accompanying documents are served at the same time as the referral. However, this is not always possible and indeed there will be occasions where this does not happen for whatever reasons, be they logistic reasons or otherwise. In such circumstances, the referring party would be wise to ensure that the referral document itself is robustly drafted so as to be construed, *prima facie*, as sufficient to enable the adjudicator to understand the nature of the payment dispute. If the referral document is adequately drafted so as to inform the adjudicator sufficiently in understanding the nature of the payment dispute, then the referring party will have complied with Section 6(5)(a) of the Act, and the adjudicator will be seized with jurisdiction of the adjudication on receipt of the referral (without the accompanying documents and provided the referral is served within the seven-day period). In such circumstances, were any issue to arise as to when or whether the accompanying documents have been furnished such matters would unlikely lead to a situation where the adjudicator would be deprived of jurisdiction. Instead, it is likely that failure to timely deliver all accompanying documents would merely be construed as a procedural breach, in respect of which the adjudicator could seek to have the breach addressed within the timing of the adjudication or in his decision.

7.30. The Act does not provide for any particular form or content of the referral notice. These matters are addressed in the Code of Practice. Paragraph 22 of the Code of Practice provides that the referral of the payment dispute to the adjudicator shall include:

(i) the name, address and contact details of each party to the construction contract;
(ii) relevant details of the payment dispute to include the amount in dispute (even if the amount is zero), the nature of the payment dispute, and the site address;
(iii) a copy of the Notice of Intention to Refer including any accompanying documents attached to that Notice;
(iv) the date when the Notice of Intention to Refer was served on the Responding Party/Parties and how this was done;
(v) the contentions on which the Referring Party intends to rely upon to support their case; and
(vi) relevant details to identify the construction contract and any supporting information that may assist an Adjudicator in understanding the nature of the payment dispute. Where a written construction contract exists, this must be attached.

It is imperative that the referral be consistent with the notice of intention to refer. While the adjudicator's jurisdiction commences on receipt of the referral, the adjudicator's jurisdiction is derived from the notice of intention. The referring party should take great care to ensure that the referral does not expand or materially change the dispute which has been set out in the notice of intention. If new issues or matters are introduced in the referral, which have not been included in the notice of intention, this could give rise to the responding party arguing that the referral relates to disputes which have not yet arisen and in respect of which the adjudicator has no jurisdiction. In a worst-case scenario this situation could even have the effect of entirely depriving the adjudicator of jurisdiction.

8 Timing of the adjudicator's decision

Sections 6(6) and 6(7) of the Act

8.1. The date the referral[1] is made is a critical date in demonstrating compliance with the statutory seven-day period under Section 6(5)(a). It is also an equally critical date in starting time running for the adjudicator to reach a decision within 28 days of that date, as required by Section 6(6) which provides that:

> The adjudicator shall reach a decision within 28 days beginning with the day on which the referral is made or such longer period as is agreed by the parties after the payment dispute has been referred.

When does time start to run?

8.2. The 28-day time limit in Section 6(6) runs from the date on which the referral "is made". Section 6(5)(a) simply states that the referring party shall refer the payment dispute to the adjudicator but does not identify as to when that referral is "made".

8.3. Paragraph 25 of the Code of Practice, however, provides that the "date of receipt [of the referral] shall be regarded as the date on which the referral of the payment dispute to the Adjudicator has been made for the purposes of Section 6(6) of the Act." Paragraph 25 also requires the adjudicator, upon receipt of the referral, to inform the parties in writing of the date he received it. As the adjudicator must comply with the Code of Practice (pursuant to Section 6(8) of the Act) his obligation to inform the parties of the referral receipt date is effectively a statutory obligation.

8.4. Paragraph 25 of the Code of Practice is similar to paragraph 7(3) of the UK Scheme which includes the words "upon receipt of the

1 The terms "referral" and "referral notice" are used interchangeably in Part 1 of this book and share the same meaning.

referral notice the adjudicator must inform every party to the dispute of the date that it was received". In *Aveat v Jerram*[2] the court held that the 28-day period commenced from the date the adjudicator received the referral on the basis that the period was already a very short deadline and therefore it should not be reduced any further.

8.5. While the parties can extend the 28-day limit by agreement under Section 6(6) of the Act, or with the referring party's consent under Section 6(7), the time frame within which an adjudicator must reach a decision is very short, reflecting the overriding objective of adjudication to elicit a quick decision so as to facilitate cash flow under the contract.

Adjudication timetable

8.6. The Act does not set out the procedure to be followed by the parties and the adjudicator during the 28-day period, nor does it specifically require the responding party to serve a response to the referral notice prior to the adjudicator's decision. It is good practice, however, for an adjudicator to issue directions on the intended conduct of the adjudication in his first correspondence with the parties after receiving the referral. The minimum procedural stages an adjudicator should direct require a response to the referral from the responding party and where such response raises matters which the referring party might want to address then the referring party should be entitled to serve a reply. Having to prepare, review and respond to these documents within the 28-day time frame may put significant pressure on the parties' and the adjudicator's time and resources, but such is the nature of adjudication.

8.7. Paragraphs 26, 27 and 28 of the Code of Practice set out certain procedures which the adjudicator is entitled to direct and adopt in order to progress the adjudication within the 28-day period. Such procedures could include, for example, setting out a timetable for parties' submissions. The UK experience has seen that adjudicators will often, in their first letter to the parties, direct and set out a timetable for the adjudication including the date on which the responding party will respond to the referral, a date for the referring party to reply to the responding party's response, a date for any meeting to be confirmed or vacated during the adjudication, and the date on which the decision will be made. The latter date will, of course, be that which the adjudicator has calculated to be 28 days from the date of his receipt of the referral.

8.8. Section 6(6) of the Act is a similar provision to Section 108(2)(c) of the UK Act insofar as both provisions deal with the 28-day time period within which the adjudicator must reach a decision. Paragraphs 19(1) and 19(3) of the UK Scheme address the period within which

2 Aveat Heating Limited v Jerram Falkus Construction Limited [2007] EWHC 131 (TCC).

the adjudicator must reach a decision and when that decision must be delivered to the parties. The Irish Act does not contain a provision regarding delivery of the decision after it is reached. This lacuna may prove problematic in determining when an adjudicator should deliver his decision after reaching it.

The time for reaching the decision is mandatory

8.9. The 28-day period requiring an adjudicator to reach a decision is mandatory. The UK Courts have strictly adhered to the 28-day period and it is likely the Irish Courts will adopt a similar approach.

8.10. In *Cubitt v Fleetglade*,[3] the JCT contract provided for an adjudicator's decision to be delivered "forthwith". A dispute arose which was referred to adjudication. The adjudicator sent his decision to the parties 12½ hours late. HHJ Coulson had to consider whether the decision was out of time and, therefore, a nullity. The judge held that if adjudicators failed to provide a decision in the relevant period, then the decision may well be a nullity. Judge Coulson agreed with Cubitt's submission that an adjudicator's decision involved a two-stage process. Stage 1 is the completion, or reaching, of the decision, as stipulated in Section 108(2)(c) of the UK Act and paragraph 19(1) of the UK Scheme. Stage 2 is the communication, or delivery, of that decision to the parties as provided for by paragraph 19(3) of the UK Scheme, which requires the decision to be delivered "as soon as possible". Cubitt submitted that the adjudicator's decision was reached on 23 November 2006 and was finalised the following day. Cubitt further submitted that the adjudicator's decision was available for transmission at 10.45 p.m. on 24 November and transmitted electronically in the early hours of 25 November, at 12.21 a.m. in the morning. Cubitt argued that the adjudicator's decision had been delivered "forthwith" for the purposes of the contractual provision. Fleetglade argued that the decision had not been reached by midnight on 24 November and that, consequently, the decision was out of time. In analysing what was meant by the word "forthwith" Judge Coulson cited, *obiter*, comments made by Judge Toumlin in the *Bloor* case[4] where the judge held that:

> the word "forthwith" … meant what it said and required that the process of communication of the decision should have started immediately after the decision had been reached.[5]

3 Cubitt Building & Interiors Limited v Fleetglade Limited [2006] EWHC 3413 (TCC).

4 Bloor Construction (UK) Limited V Bowmer & Kirkland (London) Limited [2000] EWHC 183 (TCC).

5 At paragraph 71 of the Cubitt v Fleetglade judgment.

8.11. Judge Coulson also cited *Barnes v Taylor*[6] in which HHJ Humphrey Lloyd QC dealt with a situation where the decision was signed on the agreed date but was put into the document exchange so that it was not received until at least the following day. In that case Judge Humphrey Lloyd QC held that:

> a decision arrived at in time is in principle authorised and valid, and in my judgment does not become unauthorised and invalid because by an error by the adjudicator in dispatching the decision it does not reach the parties within the time limit. However, I should emphasise that this tolerance does not extend to any period (unless perhaps the parties had agreed to a very long duration), nor does it entitle an adjudicator not to complete the decision within the time allowed.[7]

8.12. Judge Coulson held that a decision which is not communicated until after the expiry of the 28-day period "will be valid, provided always that it can be shown that the decision was communicated forthwith" and as the adjudicator's decision was made within the agreed extended period, and was communicated to the parties the following day, it was therefore not a nullity.

8.13. The UK Courts have repeatedly held that the stage 1 time limit, requiring the adjudicator to reach his decision within the statutory 28-day time limit (or as may be extended by consent or agreement), may not be otherwise extended by any contract conditions. The reasoning is that were the contract to include a provision seeking to extend the statutory time limits for the adjudicator to reach a decision, such provision would contravene Section 108(2)(c) of the UK Act. However, the English Courts have accepted that (provided the decision has been reached within the statutory time limits) a provision which allows for the decision to then be delivered thereafter within a certain time frame (i.e. stage 2 of the two-stage test) does not contravene the mandatory provisions of Section 108(2)(c) of the UK Act.

8.14. Under Section 6(6) of the Act the 28-day period within which the adjudicator must reach his decision is mandatory and can only be extended by 14 days by the consent of the referring party or for such longer period as agreed by both parties.

8.15. In *Epping v Briggs*,[8] the 28-day period expired on 1 November 2006. The referring party, Epping, consented to an initial period extending the date for the adjudicator's decision. Subsequently the adjudicator sought a further extension to the date for issue of a decision to

6 Barnes & Elliott v Taylor Woodrow Holdings Limited [2003] EWHC 3100 (TCC).

7 At paragraph 71 of the Cubitt v Fleetglade judgment.

8 Epping Electrical Company v Briggs and Forrester [2007] EWHC 4 (TCC).

21 November 2006. Briggs agreed to consent to this extension on condition that the decision was issued by 21 November 2006. The adjudicator completed his decision on 21 November but only issued it on 23 November. HHJ Havery QC held that the statutory time periods were mandatory and could only be extended in the first instance by 14 days with consent of the referring party, and both parties consent thereafter. The judge held that Briggs' consent to the second extension of time to 21 November was conditional upon the decision being issued by that date. The judge found that while the adjudicator had reached his decision by 21 November, he did not issue the decision forthwith thereafter, which meant that the condition imposed by Briggs in consenting to extend the period for issue of the decision had not been met and thus the decision was late. Judge Havery held that as the adjudicator's decision was reached late, it was out of time and unenforceable.

Delivering the decision

8.16. The Act states that the adjudicator shall reach a decision within 28 days,[9] but does not provide for when the adjudicator is required to deliver or communicate that decision to the parties. Paragraph 19(3) of the UK Scheme provides for the delivery of an adjudicator's decision as soon as possible (i.e. the second stage) after it is reached. There is no equivalent second stage provided for in the Irish Act. Furthermore, there is no provision in the Code of Practice indicating when the adjudicator should deliver or communicate his decision to the parties. One could envisage circumstances where an adjudicator reaches a decision within 28 days, in accordance with Section 6(6), but delivers or communicates that decision after the 28-day period. It will be interesting to observe how the Irish Courts approach the issue and whether the courts imply an obligation on the adjudicator to deliver his decision within the actual 28-day period or "forthwith", "as soon as possible" or within "a reasonable period of time", following having reached his decision within the 28-day period. It may come down to an analysis as to whether "as soon as possible" and "a reasonable period of time" have the same meaning, or whether the latter could be construed as a longer period. Such analysis will likely be carried out on a case-by-case basis.

8.17. The fact that neither the Act nor the Code of Practice provide for the timing of an adjudicator's delivery or communication of his decision to the parties creates uncertainty as regards an adjudicator's

9 As provided for pursuant to Section 6(6) (or such longer period) as agreed consent of the referring party or as agreed between the parties.

obligations in this respect. While the *Cubitt* decision was informed on the basis of the contract provisions and the *Epping* decision was decided on the specific facts, the principles established in those cases will be informative in an Irish context when considering the timing of delivery of the adjudicator's decision. In addition, there have been further UK Court decisions which may also be relevant in appreciating the adjudicator's obligations in delivering a decision where the adjudicator has reached a decision before the expiry of the 28-day time limit but seeks to withhold the decision until after his fees have been discharged.

8.18. In *Mott McDonald v London & Regional Properties*,[10] HHJ Thornton QC held that even if a decision is reached before the final date for delivery, it should be delivered as soon as possible. The adjudicator reached his decision five days before an agreed extended date for the adjudicator's decision. However, the adjudicator withheld his decision pending receipt of his fees. The adjudicator's fees were paid on the final day of the extended period and on the same day he sent his decision to the parties by post and did not send it by email or facsimile. The parties received the decision the following day, one day later than the final day of the agreed extended period. Judge Thornton QC held that the adjudicator had not complied with his statutory obligation (under the UK Scheme) to deliver his decision "as soon as possible" after it had been reached. Judge Thornton also held that an adjudicator may not impose a lien on his decision (or reasons) and may not withhold delivering it pending the payment of his fees, because in seeking to do so the adjudicator was restricting himself from complying with his statutory obligation to deliver the decision "as soon as possible" after it had been reached. In the particular circumstances of the case, the adjudicator had made it a condition of his appointment that his fees would first have to be paid by the referring party before he delivered his decision to the parties. Judge Thornton held that this condition appeared to lack impartiality particularly as the adjudicator, having reached his decision, had appeared to seek to enforce that pre-condition in his appointment. The judge held that the adjudicator should not be, or appear to be, financially beholden to one party, particularly the referring party, or place himself in the position in which he might appear to be more partial to one side than the other.

8.19. The UK Courts have held that the adjudicator cannot withhold delivering his decision until his fees are discharged because of the adjudicator's overriding obligation to deliver his decision as soon as possible after reaching the decision. Any attempt by an adjudicator to

10 Mott MacDonald Limited v London & Regional Properties Limited [2007] EWHC 1055 (TCC).

include a provision in his terms and conditions of appointment with the parties which provides that the adjudicator will not deliver his decision until his fees have been discharged, would almost certainly be construed by the Irish Courts as contravening the Act. An adjudicator's decision would likely be considered a nullity if the adjudicator refused to deliver or communicate his decision (having reached the decision in accordance with Section 6(6)) until after his fees are paid, even if the adjudicator had included such provision in his terms and conditions of appointment with the parties.

Extending the 28-day period

8.20. Section 6(7) of the Act provides that "The adjudicator may extend the period of 28 days by up to 14 days, with the consent of the party by whom the payment dispute was referred." Section 6(7) is a similar provision to Section 108(2)(d) of the UK Act (and paragraph 19(1)(b) of the UK Scheme).

8.21. Construction disputes, particularly those in respect of payment matters, can be notoriously complex, often involving wide-ranging issues and voluminous documentation. In addition, the adjudication process can often be quite adversarial, and it is common for parties to refuse to extend the already challenging 28-day period by agreement. Indeed, it is equally common for a referring party to refuse to consent to a 14-day extension, particularly if they are unconcerned as to how such refusal might impact on the adjudicator. There have been numerous decisions of the UK Courts relating to the extending of time for the adjudicator's decision in the context of the complex and detailed nature of construction disputes.

8.22. In *CIB v Birse*,[11] Birse claimed adjudication was an inappropriate forum for determining the dispute because of its complex nature. Birse also argued it had been prejudiced by CIB's actions generally and in particular in their adjudication notice, all of which they argued put pressure on Birse's time for responding, timescales generally, expert evidence and documentation generally. HHJ Toulmin QC held that Section 108 of the UK Act conferred a general right to adjudication which applies irrespective of the complexity of the issues in dispute. The judge held that the test to be applied in this regard concerns the ability of the adjudication process to deliver a fair outcome for the parties. The judge noted that while the timetable was tight the adjudicator was not unfair in imposing it. The judge also noted that the adjudicator had specifically stated in his decision that he believed he had sufficient time within which to

11 CIB Properties Limited v Birse Construction Limited [2004] EWHC 2365 (TCC).

consider all the issues and make a decision which was fair. The judge enforced the adjudicator's award against Birse.

8.23. If, having received and considered the referral, an adjudicator believes he has insufficient time available within which to make a decision which is fair to both parties, and fair generally, the adjudicator may resign under Section 6(17) by giving written notice to the parties. Where the adjudicator proceeds and seeks the consent of the referring party for a further 14 days, or seeks both parties' consent for a longer period, a non-consenting party should unequivocally communicate its non-consent, as silence could be construed as consent particularly where that party continues to participate in the adjudication. In *Yule v Speedwell,*[12] the 28-day period was extended to 42 days. On day 42, the adjudicator asked for a further extension of two days, which the referring party, Yule, consented to. Speedwell remained silent but continued to participate in the adjudication by making submissions in accordance with the proposed extended timetable including providing the adjudicator with further documentation during the extended two-day period. Subsequently, the adjudicator made a decision in Yule's favour. Following Speedwell's failure to pay, Yule issued court enforcement proceedings in respect of the adjudicator's decision. Speedwell argued the adjudicator's decision was late. HHJ Coulson QC rejected Speedwell's argument and held that by its conduct Speedwell was taken as having agreed to the proposed two-day extension agreed by Yule. Judge Coulson concluded that the adjudicator's relevant obligation (in paragraph 19(1)(b) of the UK Scheme) was to reach his decision within 28 days and/or any agreed extended period, in order for it to be valid. Judge Coulson held Speedwell was taken to have accepted the two-day extension and therefore the adjudicator's decision was enforceable for the following reasons:

a) Adjudicators are under time pressure to provide their decision within a strict timetable. If they are to do their job properly, simultaneously allowing parties to provide them with late information, it would be reasonable for them to request further days to reach their decision. If either party did not accept the request, their silence was not adequate to demonstrate the refusal. If a party did not accept a request, when agreement of both parties was required then the refusing party is under a duty to communicate that refusal to the adjudicator.
b) Speedwell did more than acquiesce in the extension of time by silence. They provided further information and documentation

12 AC Yule & Son Limited v Speedwell Roofing and Cladding Limited [2007] EWHC 1360 (TCC).

to the adjudicator during the extended 42-day period which also further delayed his decision. By continuing to participate and causing delay, Speedwell could not argue they had not agreed to the further two-day extension of time.

8.24. Judge Coulson held that the necessary ingredients of an estoppel were in place stating that:

> there can be no question that a reasonable man would have expected Speedwell, acting honestly and reasonably, to be under a clear duty to make clear, if that was their position, that they objected to the extension requested by the adjudicator. In all the circumstances, I also find that, to the extent necessary, there was the required unconscionability in Speedwell's acquiescence and conduct, in order to found an estoppel. If they did not agree to the extension, they had to say so, and by acting in a way that was only consistent with their having agreed to such an extension, Speedwell must have known that both Yule and the adjudicator mistakenly believed that the relevant completion date had been extended to 5th April. On that basis, they were taking advantage of the mistaken belief that both the other party and the adjudicator plainly had. For all those reasons, therefore, it seems to me that the necessary ingredients of an equitable estoppel are in place.[13]

8.25. While endorsing the principle that certainty as regards the 28-day period is of paramount importance, Judge Coulson held that because he found that Speedwell agreed (by their silence) to Yule's consent to extend the period by two days, the adjudicator's decision was enforceable.

8.26. Silence may not be the only circumstance in which agreement is construed. A party may inadvertently implicitly agree to extend the 28-day period. In *Letchworth v Sterling*[14] Judge Coulson held that Letchworth implicitly agreed to extend time for the adjudicator's decision. Sterling had subcontracted Letchworth for roofing work. A dispute arose as to Letchworth's Interim Application No. 5, which Sterling refused to pay because it claimed it was entitled to raise a cross-claim for delay. Letchworth issued its referral to adjudication on 9 December 2008 specifically stating, given the impending Christmas period, that they anticipated that a time extension may be necessary for the adjudicator to reach a decision. Subsequently, Sterling applied for a time extension to serve its response. The adjudicator approved it and simultaneously also extended the time for the decision. The adjudicator set out, in multiple

13 At paragraph 20 of the judgment.

14 Letchworth Roofing Company v Sterling Building Company [2009] EWHC 1119 (TCC).

letters to the parties, his understanding of the proposed timetable which incorporated a seven-day extension to the 28-day statutory limit. He also reserved the right to a further week if it proved necessary. Neither Sterling nor Letchworth raised any queries or challenges to the adjudicator's understanding of the time limits and his proposed extensions. The adjudicator made an award in Letchworth's favour which Sterling failed to comply with. Letchworth issued court enforcement proceedings seeking to enforce the decision. Sterling resisted the enforcement on the grounds, *inter alia*, that the decision was issued out of time and therefore a nullity. HHJ Coulson held that, in the absence of an agreed time extension, the adjudicator is obliged to complete his decision within 28 days or if extended by the referring party, 42 days. If the adjudicator fails to do so he has no remaining jurisdiction and any subsequent decision that he might purport to provide is a nullity. Judge Coulson found that the decision was provided by the adjudicator within the 42-day period. The judge held that because neither party suggested that the adjudicator's understanding of the timetable was incorrect, nor raised any point about the proposed extended date for the decision, there was an implicit agreement to extend the 28-day time period to 42 days. Consequently, the judge held that the adjudicator's decision was not reached out of time and was enforceable.

8.27. In *KNN v City Holdings*,[15] KNN initiated adjudication against City Holdings for payment of liquidated and ascertained damages. The adjudicator, in acknowledging the receipt of the referral notice and setting out the timetable leading to a decision, miscalculated the date for issue of his decision. He indicated that his decision would be made on 1 March 2013, as opposed to 28 February 2013, which would have been the correct period pursuant to paragraph 19(1) of the UK Scheme. City Holdings did not raise an objection to the adjudicator's timetable and the miscalculated decision date. However, when the adjudicator ultimately issued his decision on 1 March 2013, as he had indicated, City Holdings challenged it on the grounds that the referral notice had been issued on 31 January 2013 and that consequently the 28-day period for the adjudicator's decision had expired on 28 February 2013. Mr Justice Stuart-Smith rejected City Holdings' argument and held that City Holdings had acquiesced in the implied extension of time for delivery of the decision to 1 March by reason of their silence and failure to raise any objection during the intervening period. Consequently, the judge held that the adjudicator's decision was enforceable.

15 KNN Coburn LLP v GD City Holdings Limited [2013] EWHC 2879 (TCC).

9 The adjudicator's role, responsibilities and powers

Sections 6(8), 6(9), 6(14) and 6(17) of the Act

9.1. The extent of the adjudicator's role, responsibilities and powers are set out in a number of provisions of the Act and also within the Code of Practice. Section 6(8) provides that the adjudicator shall act impartially and comply with the Code of Practice. One of the adjudicator's most wide-ranging powers arises under Section 6(9) which entitles the adjudicator to take the initiative in ascertaining the facts and law in deciding a payment dispute. In addition, paragraph 24 of the Code of Practice empowers the adjudicator to, amongst other things, request documents, invite written submissions and evidence from the parties, meet jointly with the parties and question them, make site visits and inspections, appoint experts, assessors or legal advisors, give directions as to the timetable, or conduct of the adjudication. In exercising these powers, carrying out the role, and conducting the adjudication, the adjudicator must at all times act impartially. Pursuant to Section 6(14) the adjudicator is immune from liability in respect of the role provided that he does not act in bad faith. Section 6(17) provides that the adjudicator may resign from the role at any time.

Adjudicators shall act impartially

9.2. Section 6(8) of the Act provides that "The adjudicator shall act impartially in the conduct of the adjudication and shall comply with the code of practice published by the Minister under section 9, whether or not the adjudicator is a person who is a member of the panel selected by the Minister under section 8".

9.3. There are two important aspects of Section 6(8). Firstly, it requires the adjudicator to act impartially in the conduct of the adjudication. Secondly, it requires the adjudicator to comply with the Code of Practice, irrespective of whether the adjudicator is appointed from the Minister's panel or not. The first part, which requires adjudicators to act impartially, is similar to Section 108(2)(e) of the UK Act. The UK Courts have considered this obligation in several cases with the consensus that the adjudicator's responsibilities as to impartiality are not dissimilar to those of judges and arbitrators. In considering the issue of adjudicator impartiality the UK Courts have also looked to those cases which have examined the issue in the context of an arbitrator's duty to act impartially. Both the Irish and UK Courts have construed an arbitrator's obligation to act impartially as an obligation to act without bias towards any party. Indeed, the terms "fairness", "impartiality", "without bias", "in accordance with the principles of natural justice" and many variations on those terms have all been opined and decided upon by the courts.

9.4. An adjudicator's failure to act impartially, showing bias towards either party, would represent grounds for challenging the validity of his decision, and if successful render the adjudicator's decision unenforceable.

9.5. In addition to the adjudicator's obligation to act impartially, the common law principles of natural justice require that the parties to the adjudication are dealt with by the adjudicator in a fair manner. While neither the Act nor the Code of Practice specifically state that the adjudicator must apply the principles of natural justice, paragraph 23 of the Code of Practice states explicitly that the adjudicator shall "be impartial, independent … shall observe the principles of procedural fairness, which shall include giving each party a reasonable opportunity to put their case and to respond to the other party's case". Consequently, any consideration of Section 6(8) regarding the adjudicator's responsibility to act impartially requires an examination whether the adjudicator has also applied the principles of natural justice and procedural fairness, and not displayed any bias toward either party. These principles of natural justice, procedural fairness and bias in the context of adjudication have been considered by the UK Courts.

9.6. In *Glencot v Barrett*,[1] HHJ Humphrey Lloyd QC considered the issue of actual and apparent bias in an adjudication under the UK Act and took the opportunity to review the law on natural justice, impartiality and bias generally. Judge Humphrey Lloyd summarised the law in relation to bias as follows:

a) The adjudicator must conduct the proceedings in accordance with the rules of natural justice or as fairly as the limitations imposed by statute permit.

1 Glencot Development and Design Co v Ben Barrett & Son (Contractors) Limited [2001] EWHC 15 (TCC).

b) Regardless of any adjudication rules which apply, the concept of "impartiality" must be given its usual meaning, which is the same (and latest) meaning as at common law, namely entitling everyone to a fair hearing by an independent and impartial tribunal.

c) The test of bias is an objective one:

The views of the person involved are either irrelevant or not determinative. The test is whether the circumstances would lead to a fair minded and informed observer [having considered the facts] to conclude that there was a real possibility or a real danger, the two being the same, that the tribunal was biased.[2]

9.7. In the Scottish case *Costain v Strathclyde*,[3] Lord Young held that the test of fairness was whether the adjudicator was seen to have acted fairly and dealt equally with the parties. Costain referred to adjudication dispute seeking, *inter alia*, an order that Strathclyde pay Costain monies which had been withheld as LADs. The adjudicator decided in Costain's favour, ordering Strathclyde to pay the withheld monies. Strathclyde did not comply with the adjudicator's decision and Costain issued court enforcement proceedings in respect of the adjudicator's decision. Strathclyde argued that the adjudicator's decision should not be enforced as it had been vitiated by a breach of the principles of natural justice. Strathclyde contended that three days before the adjudicator was due to issue his decision he wrote to both parties asking Costain to grant a four-day extension required to reach his decision. The adjudicator explained that he was seeking the extension because he wished "to discuss one point in particular with his appointed legal advisor". As the referring party, Costain was entitled to grant the adjudicator the extension and duly did so. The adjudicator then issued his decision within that extended time limit. The adjudicator did not disclose to the parties the result of his discussion with the legal advisor and neither party made any request to be told the terms of the discussion or the results thereof. In addition, the adjudicator did not invite either party to comment or make submissions on his legal advisor's advice and neither party requested an opportunity to do so. Strathclyde reasoned that the legal advisor's advice to the adjudicator was material to which the adjudicator was minded and probably had significance in the decision reached. Consequently, Strathclyde argued that the adjudicator's failure to disclose the substance of the legal advice and invite comments or submissions on it before arriving at his decision was a breach of the principles of natural justice. They claimed that the legal advice may have

2 At paragraph 21 of the judgment.
3 Costain Limited v Strathclyde Builders Limited [2003] Scot CS 316.

influenced the decision and the advice may have been "erroneous, incomplete, irrelevant or otherwise exceptionable".[4] Strathclyde further argued that in any event the parties had no opportunity to counter or correct the legal advice. Lord Drummond Young accepted Strathclyde's argument and found that the adjudicator had breached the principles of natural justice. The court held that it was not necessary that actual injustice be shown to have occurred but was sufficient if a party could demonstrate that there was a mere possibility of injustice. Nobody knew whether the point under discussion was important or crucial and on that basis the possibility of injustice could not be excluded. The judge held that in the absence of knowledge about what the adjudicator and his legal advisor discussed it was impossible to know whether the breach of the principles of natural justice was substantial and relevant, and consequently the court refused enforcement of the adjudicator's decision. In delivering his decision Lord Drummond Young set out some guidelines for adjudicators so that they might ensure they exercise fairness between the parties:

a) The adjudicator should give the other party an opportunity to comment on any supplemental information even if such opportunity is to be exercised within a tight time frame.
b) The adjudicator can use his own knowledge, but the parties must be told what it is and have an opportunity to give their comments.
c) Any technical and legal advice obtained by the adjudicator must be disclosed to each party to allow them to comment.

9.8. The *Costain* decision is in line with the approach of the Irish Courts in cases in relation to assertions as regards a breach of the rules of natural justice. An adjudicator should be conscious to ensure that any input that the adjudicator receives from any professional advisors (be they experts or legal advisors) is properly put before the parties. The adjudicator should also ensure that the parties have an opportunity to comment on the input of those experts or legal advisors.

9.9. In *RSL v Stansell*,[5] HHJ Seymour QC held that it is appropriate for the adjudicator to tell the parties what he plans to do, and necessary that he inform them of any findings or reports he intends to rely upon in reaching his decision. Stansell engaged RSL under a subcontract which included adjudication provisions. A dispute arose on the final account and specifically regarding RSL's extension of time claim and associated loss and expense. The dispute was referred to adjudication. As the primary issue before the adjudicator related to delay events and the

4 At paragraph 5 of the judgment.
5 RSL (South West) v Stansell [2003] EWHC 1390 (TCC).

extension of time claim, both parties agreed the adjudicator could appoint a programming expert. However, Stansell requested a copy of the adjudicator's letter of instruction to the programming expert together with his response and copies of any expert report prepared. The adjudicator received the expert's preliminary advices which he duly forwarded to the parties. Stansell did not consider that a response was required because the preliminary advice appeared to show that the adjudicator's appointed programming expert did not support RSL's position. The adjudicator then issued his decision, in which he specifically referred to a "final report" prepared by the programming expert and awarded RSL a 55-working day extension of time together with a sum of money for its loss and expense claim. Stansell refused to pay the award and RSL issued court enforcement proceedings in respect of the adjudicator's decision. Stansell argued that the adjudicator's decision was unenforceable because of a breach of the principles of natural justice contending that the adjudicator had failed to comply with the basis upon which Stansell had agreed to the appointment of the programming expert.

9.10. HHJ Seymour QC considered that the main grounds on which Stansell could rely was the failure of having an opportunity to comment on the programming expert's final report. Stansell had been given the opportunity to comment on the programming expert's preliminary advice but had not done so because it appeared to be in their favour. However, as the adjudicator's decision was based upon the programming expert's "final report", which neither Stansell nor RSL had received, Stansell had not had the opportunity to comment upon that final report. Judge Seymour held that the parties must have an opportunity to respond to, and comment on, reports and their findings, or any report the adjudicator intends to rely upon in reaching his decision. Judge Seymour held that it is absolutely essential for an adjudicator, in observing the rules of natural justice, to give the parties the opportunity to comment upon any materials or reports, from whatever source, including the adjudicator's own knowledge and experience, and on which the adjudicator intends to rely or attribute some significance to in reaching his decision. Consequently, Judge Seymour held that the adjudicator's decision did not comply with principles of natural justice and should not be enforced.

9.11. The principle of natural justice dictates that an adjudicator should ensure the parties have an opportunity to comment on the input of any expert or legal advisor that the adjudicator has relied upon in reaching his decision and that extends to circumstances where the adjudicator might seek to rely upon his own expertise in reaching his decision.

9.12. In *Balfour Beatty v Lambeth*[6] the courts did not enforce the adjudicator's decision because the adjudicator had not given either party the

6 Balfour Beatty Construction Northern v London Borough of Lambeth [2002] EWHC 597 (TCC).

opportunity to comment on his use of the critical path analysis which the adjudicator had himself produced based on as-built information provided by the parties. Lambeth engaged Balfour Beatty under a JCT contract for a property refurbishment. The works were delayed, and the architect awarded Balfour Beatty a series of extensions of time but also issued a certificate of non-completion which resulted in Lambeth deducting LADs. Balfour Beatty disputed the deducted LADs and referred that dispute to adjudication. Balfour Beatty argued that they should have received a further extension of time from the architect, and that the architect should not have issued a certificate of non-completion entitling Lambeth to deduct LADs. Balfour Beatty's case centred on an as-built programme and analysis to demonstrate the extension of time to which they contended they were entitled. The adjudicator did not consider Balfour Beatty's programme adequate in that it did not identify each relevant delay event, the date of the occurrence of the event, the activity directly affected by the event, and the nature of that effect on the completion date. The adjudicator was partially successful in obtaining the relevant information in respect of the delay events from the parties but was dissatisfied that he had sufficient information. Consequently, the adjudicator decided to prepare his own critical path analysis and then proceeded to make his decision based upon this. The adjudicator subsequently made a decision in favour of Balfour Beatty. However, in reaching that decision the adjudicator had not given either party the opportunity to look at the critical path analysis and assessment upon which his decision was based. Lambeth refused to pay the amount awarded by the adjudicator and Balfour Beatty subsequently issued court enforcement proceedings in respect of the adjudicator's decision. Lambeth argued that in reaching his decision the adjudicator had not acted impartially and had not complied with the principles of natural justice, in that he had not given the parties the opportunity to consider the argument upon which his decision was based. Balfour Beatty argued in response that the adjudicator had merely "set his own procedure" and taken the initiative in ascertaining the facts and law as he considered necessary and which he was expressly permitted to do under the UK Act, and indeed the JCT contract. HHJ Humphrey Lloyd QC held that the adjudicator's actions constituted bias as he had not just taken the initiative in ascertaining the facts but had gone further by effectively preparing part of the case for one of the parties. In addition, the adjudicator had not given the parties the opportunity to review and comment upon the critical path analysis produced by the adjudicator. In reaching his ruling that the adjudicator's decision was unenforceable, Judge Humphrey Lloyd extensively considered the acts and omissions of the adjudicator in the course of the adjudication and in his decision. The judge commented that simply because an adjudicator's decision did not expressly refer to a particular matter which had been put to the adjudicator during the course of the adjudication, this did not infer that the adjudicator had not dealt with that matter, and such an inference would be wrong. The judge stated that there is a presumption that the adjudicator has complied with the obligation to consider all information submitted by both parties unless the content and nature of the decision, and the reasons on which the decision is derived, suggests that the adjudicator did not comply with such an obligation.

9.13. The *Balfour Beatty* decision is not only informative as regards the adjudicator's obligations to ensure that the parties have an opportunity to comment upon any document the adjudicator might himself generate based on his own expertise, but also informative on how the court interprets the adjudicator's power to take the initiative in ascertaining facts and the law. In this regard, an adjudicator who decides to take the initiative and ascertain the facts or the law should be conscious not to go so far that his actions might be construed as effectively helping to prepare part of the case for one of the parties.

9.14. The principles of natural justice also dictate that a party is entitled to know the case being pleaded against it and have sufficient opportunity to answer and respond to that case. In *LAP v Waterman*,[7] the court refused enforcement of the adjudicator's decision for a perceived lack of procedural fairness and impartiality of the adjudicator. LAP engaged Waterman as structural and civil engineers for a shopping centre development. LAP maintained that Waterman failed to release substantial design information by specific dates, thereby causing critical delays to the works. LAP claimed that Waterman were professionally negligent resulting in LAP suffering loss. The matter was referred to adjudication. At a very late stage in the adjudication timetable LAP provided the adjudicator (and Waterman) with a supplemental statement in order to deal with deficiencies in its original quantum statement. The adjudicator decided in LAP's favour that Waterman pay a substantial sum. Waterman applied for a court declaration that the adjudicator did not have, or had exceeded his, jurisdiction and simultaneously LAP applied for enforcement of the adjudicator's decision. Waterman argued that the adjudicator lacked jurisdiction for several reasons including, as they perceived it, the fact that within the time required Waterman did not have any reasonable opportunity of properly responding to the case, and in particular the supplemental quantum

7 London & Amsterdam Properties v Waterman Partnership Limited [2003] EWHC 3059 (TCC).

statement LAP furnished at a very late stage in the adjudication. Deputy Judge Mr Acton-Davis QC considered LAP's late provision of a supplemental statement to be an evidential "ambush" that was "clearly deliberate". Deputy Judge Acton-Davis held that "a mere ambush 'however unattractive' didn't amount to procedural unfairness". However, he found, on the facts, that LAP could have adduced the quantum evidence (in its supplemental statement) at a much earlier stage of the adjudication but did not. The deputy judge held that the material and information in the supplemental quantum statement were central to the adjudicator's decision and on receipt of same, at such a late stage in the adjudication, the adjudicator should have sought a time extension from LAP, or by agreement of both parties, to allow Waterman a reasonable opportunity to respond to the supplemental quantum information. If LAP refused to consent or agree to a time extension and there was insufficient time for Waterman to consider and address the supplemental quantum evidence then the adjudicator should have excluded the late material from his consideration. The adjudicator clearly did not appear to appreciate these alternatives and in the judge's words this was "a substantial relevant breach of the rules of natural justice". As a result, Deputy Judge Acton-Davis held that Waterman demonstrated a live and triable issue and consequently he refused enforcement of the adjudicator's decision.

Unilateral communications with the adjudicator

9.15. It is not uncommon for an uninformed (or mischievous) party to seek to communicate directly with an adjudicator in the other party's absence. Such circumstances could, potentially, trigger allegations of bias against the adjudicator. The adjudicator should avoid communicating unilaterally with either party and where he does so, either inadvertently or if contacted directly by a party, he should fully disclose to the other party not only the fact that the contact occurred but also full and precise details of the contact.

9.16. In *Discain v Opecprime*[8] a payment dispute arose which Discain referred to adjudication. During the adjudication Discain's employee telephoned the adjudicator twice and discussed substantive issues in the adjudication. The conversations were not recorded and communicated to Opecprime. The adjudicator notified Opecprime by facsimile that he had communicated with Discain's representative, but the notification did not disclose materially important aspects of the discussions. The adjudicator subsequently made a decision in Discain's favour. HHJ

8 Discain Project Services v Opecprime Developments Limited [2001] EWHC 450 (TCC).

Seymour QC held that there was a very serious risk of bias because the adjudicator failed to consult with one party on important submissions the other party made. The judge held that it is not sufficient to merely inform a party of the nature of a telephone conversation with the other and the adjudicator should have provided the other party with full particulars of his telephone conversation. The judge held that the adjudicator's actions indicated apparent bias rendering his decision unenforceable.

The adjudicator shall comply with the Code of Practice

9.17. Section 6(8) specifically states that the adjudicator "shall comply with the Code of Practice". Paragraphs 23 to 28 and paragraph 32 of the Code of Practice address the roles and responsibilities of the adjudicator in relation to the adjudication procedure.

9.18. Paragraph 23 essentially reiterates the first part of Section 6(8) of the Act, providing that the adjudicator shall act impartially and independently, but adds that the adjudicator shall "only adjudicate where satisfied that no actual conflict of interest exists". Paragraph 23 further provides that the adjudicator shall observe the principles of procedural fairness giving each party a reasonable opportunity to put their case and to respond to the other party's case.

9.19. Paragraph 24 sets out the adjudicator's powers to

(i) request any reasonable supporting or supplementing documents pertaining to the payment dispute detailed in the notice of intention and/or in the referral;
(ii) take the initiative in ascertaining the facts and matters required for a decision and make use of his own specialist knowledge where relevant and disclose that to the parties as appropriate;
(iii) appoint experts, assessors or legal advisors while notifying the parties of their identity and terms of reference;
(iv) make site visits and inspections or conduct tests subject to prior notification to the parties and obtaining any necessary consent from a third party or parties;
(v) invite written submissions/representations and evidence from the parties if appropriate;
(vi) meet jointly with the parties and their representatives to enable further investigation;
(vii) hold a teleconference with the parties with their consent; and
(viii) hold an oral hearing where appropriate.

9.20. Pursuant to paragraph 24(i), if either party fails to furnish the adjudicator with supporting or supplementary documents when requested by the adjudicator, then pursuant to paragraph 32, the adjudicator may draw adverse inferences from such failure or refusal to provide those documents.

9.21. With respect to paragraph 24(ii), where an adjudicator takes the initiative in ascertaining the facts and matters, such as making use of his own specialist knowledge, regard should be had to the *RSL* and *Balfour Beatty*[9] decisions. The adjudicator should advise and inform the parties of the fact that he is availing of his own specialist knowledge or that of an expert advisor in making a decision, so as to allow the parties the opportunity to comment on, or make submissions, in respect of the accuracy or relevance of the particular knowledge or advice that the adjudicator is seeking to rely upon.

9.22. Paragraph 25 of the Code of Practice provides that the date of receipt of the referral by the adjudicator shall be regarded as the date on which the referral has been made for the purposes of Section 6(6) of the Act. Consequently, the 28-day period within which the adjudicator is required to make a decision runs from that date. The adjudicator should write to the parties on receipt of the referral informing them of the receipt, thereby making it crystal clear to the parties as to the precise date from which the 28-day period shall run.

9.23. Paragraph 26 provides for the adjudicator's general entitlement to determine the procedure he intends to apply during the adjudication process and further expands the provisions of paragraph 25 and Section 6(9) of the Act. The adjudicator should furnish all relevant details as to this intended procedure in his letter to the parties confirming the date the referral was received. This will generally be the first communication from the adjudicator to the parties.

9.24. Neither the Act nor the Code of Practice address the issue as regards the submission by the responding party of a response to the referral, or as regards any aspect of the procedure that should follow, during the 28-day period, after the receipt of the referral by the adjudicator. Consequently, it is a matter for the adjudicator to decide what procedure is appropriate. In the UK and other jurisdictions that have adopted statutory adjudication, the normal practice is for the adjudicator to allow the responding party to deliver its response within an appropriate time frame following receipt of the referral, and thereafter allow the referring party to reply to that response. Following this exchange of documents, a hearing may or may not take place (and in most cases a hearing does not take place) but in any event (unless there is an agreement to extend

9 Balfour Beatty Construction Northern Ireland v London Borough of Lambeth [2002] EWHC 597 (TCC).

time) the adjudicator's decision must be made in the statutory 28-day period. However, the procedure can be adapted depending on the nature and complexity of the payment dispute, and it is entirely within the adjudicator's discretion to determine the timetable, notwithstanding the parties' views on the same.

9.25. While a number of the Irish standard-form construction contracts have been amended to take account of statutory adjudication under the Act,[10] none of them currently include any provisions prescribing a procedure that the parties might adopt in adjudicating their payment dispute adjudication.

9.26. The UK standard-form contracts, on the other hand, have embraced adjudication and many UK standard-form contracts contain outline procedures for the conduct of an adjudication. It is also not uncommon in the UK for construction contracts to be amended to include procedural steps to be taken by the parties during the course of an adjudication.

9.27. In *CJP v Verry*,[11] HHJ Akenhead held that an adjudicator's decision was a nullity and unenforceable where he had erroneously interpreted the applicable adjudication procedure under the subcontract determining that he did not have discretion to consider a response that had been served late. The parties entered into a UK standard-form JCT DOM/2 subcontract, pursuant to which Clause 38A provided that the response in the adjudication was to be served within seven days of the referral. Verry requested an extension of time to serve the response. The adjudicator stated that he had no power to circumvent Clause 38A and that Verry was therefore obliged to serve the response in accordance with the timetable, unless the parties agreed otherwise. CJP agreed an extension of time for service of the response to 12.00 p.m. on 14 May 2008, but Verry did not serve the response document until 5.30 p.m. on that day. CJP submitted that the adjudicator could not consider the response because it had not been served within the extended time frame agreed between the parties. The adjudicator informed the parties that he had no discretion under the contractual adjudication procedure to extend time for service of the response and would therefore not consider the response in making his decision. The adjudicator subsequently issued his decision awarding CJP the full value of their interim payment application. Following Verry's refusal to comply with the adjudicator's decision, CJP commenced court enforcement proceedings. Verry argued that there had been a breach of natural justice as the adjudicator had not considered

10 Both the RIAI and Public Works contract suites have been amended to acknowledge the introduction of statutory adjudication but have not set out any procedures in respect of same.

11 CJP Builders Limited v William Verry Limited [2008] EWHC 2025 (TCC).

Verry's response. Judge Akenhead held that, as a matter of contractual construction of the adjudication procedure, since the adjudicator was given the power to "set his own procedure" and had "absolute discretion" in ascertaining the facts and the law, he therefore had the power to grant an appropriate extension of time for the service of the response. By not doing so the judge held that the adjudicator was wrong to disallow the late response and thereby failed to apply the rule of natural justice that each party has a right to be heard and to have its evidence and arguments considered by the tribunal. Judge Akenhead dismissed CJP's application for enforcement of the adjudicator's decision, and concluded that:

> One of the entitlements of parties to an adjudication is a right to be heard, that being the rule of natural justice *Audi Alterem Partes*. There is thus a reasonable expectation of parties to an adjudication that, within reason and within the constraints of the overall requirement to secure the giving of a decision within the requisite time period, each party's submissions and evidence will be considered by the adjudicator. It is a draconian arrangement (to which the parties are of course free expressly to agree) that a party is denied its right to be heard unless it has been given a fair and clear opportunity to put its case. Very clear wording would be required to ensure that such a right was to be denied.[12]

9.28. Paragraph 27 of the Code of Practice obliges the adjudicator to adopt a procedure commensurate with the nature and the value of the payment dispute and provides that he should be mindful of whether or not an oral hearing is required having regard to all relevant matters including whether there is a conflict of fact. While the adjudicator must consider any relevant information submitted to him by any of the parties to the dispute, the adjudicator's overriding obligation is to reach his decision within the time limit. The adjudicator does not have to consider late submissions in any detail. It is enough to look quickly at any document, to ascertain its general nature and whether it contains anything of real significance.

9.29. Paragraph 28 of the Code provides that the adjudicator use reasonable endeavours to process the payment dispute in the shortest time, at the lowest cost and promptly notify the parties of any matter that will slow down or increase the cost of making a determination. Clearly, this paragraph needs to be read in conjunction with other

12 At paragraph 79(f) of the judgment.

paragraphs regarding the procedure the adjudicator is to adopt so as to ensure his efforts to arrive at a decision as quickly and cost-effectively as possible are commensurate with the nature and value of the dispute. A prudent adjudicator will err on the side of caution considering payment disputes within the shortest time and at the lowest cost but would be well advised to use the full 28 days in reaching a decision. However, this objective must be balanced against the adjudicator's obligations to comply with natural justice principles and ensure that he considers any relevant documents or information the parties submitted, provided of course he complies with the 28-day deadline to make his decision.

9.30. Paragraphs 32(i) to (iv) of the Code of Practice set out circumstances where an adjudicator may exercise his options as to how the adjudication proceeds. Where a party (i) fails to attend a meeting; (ii) fails to comply with the adjudicator's directions; (iii) fails to disclose information indicating a potential conflict of interest; or (iv) fails to produce a document requested by the adjudicator, then the adjudicator may exercise one or more of the options in the following paragraphs 32(a) to (e). In all of the aforementioned circumstances the adjudicator may proceed with the adjudication (on the basis set out at paragraphs 32(a) to (e) of the Code of Practice), including continuing the adjudication in the absence of a party or without the document or written statement requested, and draw such inferences from the failure of a party to comply. The adjudicator may also make a decision based on the material properly provided and apportion fees, costs and expenses as appropriate, taking into account the relevant party's failings.

9.31. Depending on the nature of the payment dispute it may be appropriate, albeit unusual, for the adjudicator to convene a meeting with the parties. Generally if a meeting is necessary it will likely take place by telephone or video conference. In rare cases, a meeting might be scheduled at the outset to discuss the timetable and procedural matters. However, in the majority of adjudications, there is no time for meetings because of the tight deadlines involved.

9.32. In the UK most adjudications are decided on a document-only basis. It is rare for the adjudicator to require oral evidence principally due to time constraints but also on the basis that the referral and response documents should be supported fully by all relevant documents and substantiating information, which would likely include, *inter alia*, contemporaneous correspondences, emails, site diaries and minutes of meetings and possibly sworn witness statements. On rare occasions oral evidence may be necessary where matters of fact are disputed.

Ascertaining the facts and the law

9.33. Section 6(9) provides that "The adjudicator may take the initiative in ascertaining the facts and the law in relation to the payment dispute and may deal at the same time with several payment disputes arising under the same construction contract or related construction contracts".

9.34. There is nothing to prohibit an adjudicator from making a decision based on his own knowledge of the facts or the law. However, should the adjudicator seek to do so then, as held in *Balfour Beatty*, this fact must be brought to the parties' attention to allow them the opportunity to make submissions in this regard. The adjudicator should not seek to rely on his own knowledge or experience to the extent that he makes an argument for one of the parties which has not been put forward by that party.

9.35. In *Herbosh–Kiere v Dover,*[13] Herbosh engaged Dover on a contact for the removal of a cargo ship and debris from the seabed. The contract provided for a time extension entitlement if Herbosh was delayed because of specific events, and also for a pain/gain share of time-related costs designed to financially reward Herbosh for finishing earlier and the opposite for finishing late due to culpable delay. The contract provided for any dispute to be referred to adjudication. The project overran by three months and disputes arose on the final account. The parties exchanged written correspondences regarding delay events and the financial calculations appropriate to ascertain Herbosh's entitlement to additional payment, but were unable to reach agreement and Herbosh referred the dispute to adjudication seeking a determination that the final account's correct value was the value they contended. Dover responded with their overall valuation of the work done, allowing for a deduction for alleged inefficiency on Herbosh's part and provided a breakdown of their valuations calculating the amount due to delay and disruption in effect on exactly the same basis as Herbosh had, albeit for a lesser amount. In reaching his decision in Herbosh's favour, the adjudicator assessed the value of the pain deduction based upon a daily resource rate which he derived by relying on a particular formula that neither party had put to him. He also assessed the extension of time due and then applied the same daily resource rate to reach the amount payable to Herbosh for the delays without separately assessing the various resources. Dover refused to honour the adjudicator's decision and Herbosh issued court enforcement proceedings in respect of the adjudicator's decision. Dover argued that the adjudicator exceeded his jurisdiction by not

13 Herbosh–Kiere Marine Contractors Limited v Dover Harbour Board [2012] EWHC 84 (TCC).

basing his decision on the dispute as outlined between the parties, reasoning that the parties were in dispute regarding the periods of delay for specific resources and were not proceeding based on the application of a composite overall resource rate to the overall delay. Dover argued, in the alternative, that the adjudicator's actions were materially unfair as he did not advise the parties that he intended to apply a composite resource rate to the overall delay. Herbosh argued that the fundamental issue in dispute was the amount owed in the final account and even if the adjudicator erred in his method of calculating the sums owed a mistake should not undermine his decision.

9.36. HHJ Akenhead found that the adjudicator breached the rules of natural justice as he had not allowed the parties a chance to consider and make submissions on the daily resource rate formula which he relied on in reaching his decision. Judge Akenhead's view was that the adjudicator had made his decision without jurisdiction. While each of the parties put forward slightly differing assessment methods, the parties had no dispute as to the correct method. It was clear from the notice of dispute initiating the adjudication that the assessment method formed no part of the dispute referred, yet the adjudicator availed himself of a different method. Consequently, Judge Akenhead held that the adjudicator exceeded his jurisdiction and breached the rules of natural justice and, accordingly, the adjudicator's decision could not be enforced.

9.37. In taking the initiative to ascertain the facts and the law the adjudicator is effectively empowered to conduct the adjudication in an inquisitorial fashion. This is in contrast to the established conduct of proceedings in litigation and arbitration under common law systems where, while a judge or arbitrator may question parties on specific aspects of their case during the course of the hearing, they do not generally behave in an inquisitorial fashion. Should an adjudicator choose to take the initiative in ascertaining the facts and the law, and conduct the proceedings in an inquisitorial fashion, the adjudicator needs to be careful to act fairly and impartially between the parties and also avoid introducing matters which do not form part of the dispute and are therefore outside the adjudicator's jurisdiction.

Multiple disputes

9.38. The second part of Section 6(9) of the Act provides that the adjudicator "may deal at the same time with several payment disputes arising under the same construction contract or related construction contracts". The issue as to what constitutes a dispute in the context of the correct interpretation of the word "dispute" under Section 6(1) needs to be distinguished from the intention of the provisions of Section 6(9) of the Act. The use of the term "dispute" in the singular in Section 6(1) appears to prohibit a multitude of distinct payment disputes being referred in the one referral to adjudication. Under Section 6(9) the same adjudicator would be empowered to deal with several payment disputes that may have been separately referred to adjudication, and to do so at the same time. While it could be argued that there is an inconsistency between Section 6(1) and Section 6(9), there would certainly be a counter-argument that in the interests of cost efficiency and time it makes sense that the same adjudicator deal with separate payment disputes under the same contract. In practice, it may not be feasible for either the adjudicator or parties to conduct multiple payment disputes "at the same time", but pursuant to Section 6(9) that is what the adjudicator would be empowered to do if he so wished.

9.39. In addition, Section 6(9) empowers an adjudicator to deal with several payment disputes arising under related construction contracts. This provision would facilitate the same adjudicator dealing with a payment dispute under a main contract between a client and main contractor and, at the same time, a connected payment dispute between the contractor and subcontractor under a subcontract. Such a situation would clearly be advantageous from the main contractor's perspective as it could expect to receive consistent adjudicator decisions across the two adjudications. If two separate adjudicators were to be appointed in the aforementioned scenario, then there arises the possibility of conflicting decisions on the same issue in dispute across the two separate adjudications. Such an outcome could result in the main contractor losing the adjudication against the employer and at the same time, in relation to the same issue in dispute, losing the adjudication against the subcontractor. This would be an entirely unsatisfactory outcome, particularly for the main contractor. Where disputes under separate contracts are referred to the same adjudicator the parties could reasonably expect the adjudicator's decisions to be consistent across the two adjudications, making it highly unlikely a party would lose in both forums.

Adjudicator immunity

9.40. Section 6(14) provides that:

> The adjudicator is not liable for anything done or omitted in the discharge or purported discharge of his or her functions as adjudicator unless the act or omission is in bad faith, and any employee or agent of the adjudicator is similarly protected from liability.

This Section 6(14) effectively creates immunity for the adjudicator in relation to anything done or omitted in the discharge or purported discharge of his functions as adjudicator. However, where the adjudicator has not reached a decision within the prescribed and/or statutory time frame it can be argued that he cannot avail of the immunity afforded by Section 6(14).

9.41. In *Cubitt v Fleetglade*[14] HHJ Coulson QC stated that if it was "concluded that he [the adjudicator] had not completed his decision in time, the decision would probably have been a nullity". In such a case, where an adjudicator's decision is found to be a nullity, the adjudicator's protection in respect of anything done in the discharge of his functions would not have protected him. This is because the failure of the adjudicator to reach his decision within the prescribed time represents a complete failure on his part to discharge his functions as an adjudicator. Consequently, the adjudicator loses any immunity from a prosecution that he might have otherwise been protected from pursuant to Section 6(14). Section 6(14) is similarly worded to Section 108(4) of the UK Act and paragraph 26 of the UK Scheme.

9.42. The adjudicator's immunity from prosecution is also lost where the relevant act or omission of the adjudicator is in bad faith. This is to be contrasted with a similar immunity which is granted to an arbitrator under the Arbitration Act 2010, at Section 22(1), which provides that an arbitrator shall not be liable in any proceedings for anything done or omitted in the discharge or purported discharge of his functions. However, Section 22(1) of the Arbitration Act 2010 does not make any reference to whether the arbitrator acted in good faith or not. Consequently, under the Arbitration Act 2010, an arbitrator could act in bad faith and still be immune from prosecution. However, an adjudicator would not have a similar protection under the Act and were it to be shown that the adjudicator acted in bad faith then the adjudicator's immunity from prosecution would be lost.

9.43. In *Fenice v Jerram*,[15] Judge Waksman QC held that immunity from prosecution had nothing to do with the adjudicator's entitlement to fees. The judge held that the test to assess the adjudicator's entitlement to fees was reasonableness, which was not inconsistent with the proposition that the adjudicator's fee could only be challenged on the grounds of bad faith. Fenice engaged Jerram under a design and build contract. A dispute arose and was referred to adjudication. The adjudicator made a decision in Fenice's favour which included

14 Cubitt Building & Interiors Limited v Fleetglade Limited [2006] EWHC 3413 (TCC).
15 Fenice Investments Inc v Jerram Falkus Construction Limited [2011] EWHC 1678.

a decision directing that Jerram pay the adjudicator's fees. Jerram refused to pay the adjudicator's fees stating that they were excessive, and subsequently paid a small amount which they considered as reasonable. As the parties were jointly and severally liable for the fees the adjudicator claimed the remainder from Fenice who duly accepted the fees being reasonable and paid the balance. Fenice then commenced proceedings against Jerram for the adjudicator's fees together with its own legal costs. Fenice claimed that because the fees were ordered to be paid as part of the adjudicator's decision, the amount should be summarily enforced as would any other valid adjudicator's decision. In addition, Fenice argued that the adjudicator's fees can only be challenged on the grounds of bad faith by the adjudicator, which Jerram was not contending. Fenice claimed that in any event the adjudicator's fees were reasonable. Judge Waksman QC held that the adjudicator's fees were reasonable and ordered Jerram to repay the balance of the fees to Fenice. As to Fenice's first argument, the parties' adjudication agreement either expressly or impliedly incorporated the provisions of the UK Scheme and therefore it was a term that the adjudicator should state who is liable for his fees. Judge Waksman held that the claim for fees was not part of the decision in the same way as the substantive ruling in respect of the dispute which is referred to adjudication. The judge held that under the UK Scheme, which was applicable as regards the adjudicator's entitlement to determine which party should discharge his fees, the adjudicator was entitled to be paid "reasonable fees" and it could not be the case that the adjudicator was entitled to whatever he decided those fees were. Consequently, the judge rejected the proposition that the adjudicator's decision on payment of his fees was part of his decision on the substantive issue and that consequently his fees should be paid regardless of whether they are reasonable. The judge held that the adjudicator's fees are open to assessment of the reasonableness of the fees. The judge also disagreed with Fenice's second contention as bad faith is not the only grounds on which to challenge an adjudicator's fees. The judge held that a party could also challenge an adjudicator's fees on the reasonableness of the fees, and this right also arose out of paragraph 25 of the UK Scheme. The judge acknowledged that whilst a party had the right to challenge an adjudicator's fees on the grounds of reasonableness, it would be hard for a party to do so successfully. However, the overall burden of proving reasonableness rests upon the adjudicator. The nature of adjudication is that it is fast paced, and an adjudicator needs to analyse the issues and reach a decision in a short period of time. The judge held that whilst initially the adjudicator has to provide a level of justification for his fees, such as the rate, the hours worked and a short narrative, the evidential burden will then be on the party challenging the fees to show that they are unreasonable. If a party fails to acknowledge an adjudicator's invitation to agree his fees, later complaining that the rate is excessive, this would be unlikely to provoke sympathy. The judge held that Fenice, as a party who was jointly and severally liable, had a right to recoup payment for the adjudicator's fees that it had discharged. The judge analysed the fees incurred as against the shifting nature of the parties' contentions, particularly those of Jerram, and considered that the adjudicator had only analysed and decided on issues that had been properly raised by the parties. The judge held that the adjudicator's fees were reasonable and Fenice was entitled to recover from Jerram the fees they had paid to the adjudicator.

Adjudicator's resignation and entitlement to fees

9.44. Section 6(17) provides that "An adjudicator may resign at any time … and the parties shall be jointly and severally liable for the payment of the reasonable fees, costs and expenses incurred by the adjudicator up to the date of resignation".

9.45. When a party challenges the adjudicator's jurisdiction, it is sometimes the case that as part of that challenge application the challenging party invites the adjudicator to resign in the event that the adjudicator decides that he does not have jurisdiction to make a decision. Such a request by the challenging party that the adjudicator resign due to lack of jurisdiction is misperceived because if the adjudicator, on consideration of such an application, reaches the decision that he does not have jurisdiction and that the jurisdictional challenge is valid, then the adjudicator was never appointed by the parties and therefore has no appointment to resign from. Adjudicators should be mindful as to whether they have been properly appointed so as to enjoy this threshold jurisdiction, and if they decide that they do not have jurisdiction then they were never appointed adjudicator. Consequently, an adjudicator can decide to withdraw, but it is incorrect to say the adjudicator should or can resign. Where the adjudicator is never appointed he would have no entitlement to fees under the Act or the Code of Practice, unless the parties had agreed to discharge his fees for his time spent considering the jurisdictional challenge. Section 6(17) is therefore not designed to provide for an adjudicator resigning due to lack of jurisdiction. Instead, its principal purpose is to facilitate circumstances where an adjudicator is of the view that, for example, there is insufficient time for him to fairly make a decision due to the complexity and extent of the payment dispute referred.

9.46. While it is not an ideal scenario, there is nothing to stop an adjudicator resigning if, for example, having considered the referral, he has genuinely formed the view that he will have insufficient time available to reach a decision within 28 days, or even an agreed extended period. In such circumstances, an adjudicator could advise the parties that he is of the view that the decision cannot be reached within the 28-day period, and invite the parties to consent to an agreed extended period which the adjudicator might propose. If, however, the parties cannot reach agreement to extend the 28-day period (or the referring party refuses to consent to a 14-day extension) the adjudicator could choose to resign. It would be most unsatisfactory, however, if an adjudicator threatened resignation if the parties did not agree to his proposed extension but were he to make such a threat and subsequently resign, it is regrettable that there is no recourse for the parties, and no repercussions for the adjudicator, under the Act.

9.47. Where an appointed adjudicator considers (or one of the parties contends) that the payment dispute put to him has already been subject to a decision in a previous adjudication between the parties, the adjudicator may resign. In such circumstances, the previous adjudicator's decision is binding on the parties until the matter is arbitrated or litigated and there would be no dispute to refer to adjudication again. Consequently, the appropriate action by the subsequent adjudicator would be to resign. In such case, he would be entitled to payment of his fees under the terms of his appointment.[16]

16 Christopher Michael Linnett v Halliwells LLP [2009] EWHC 319 (TCC).

10 The content and enforceability of the adjudicator's decision

Sections 6(10) to 6(13) and Section 7 of the Act

10.1. Section 6(10) and 6(12) provide that:

> 6(10) The decision of the adjudicator shall be binding until the payment dispute is finally settled by the parties or a different decision is reached on the reference of the payment dispute to arbitration or in proceedings initiated in a court in relation to the adjudicator's decision.
>
> 6(12) The decision of the adjudicator, if binding, shall, unless otherwise agreed by the parties, be treated as binding on them for all purposes and may accordingly be relied on by any of them, by way of defence, set off or otherwise, in any legal proceedings.

Binding nature of the adjudicator's decision

10.2. It is not open to a party to contend or make an application in any enforcement proceedings that the adjudicator's decision is not binding where that decision is being challenged on the basis that the adjudicator was wrong on the facts or law. The UK Courts have not entertained such applications and the Irish Courts are likely to adopt a similar approach.

10.3. In *Macob v Morrison*[1] the court was asked to consider whether an adjudicator's error of law invalidated his decision. Morrison's argument was that the error of law did so to the extent that if one of the adjudicating parties challenged the validity of a decision, it was not binding and enforceable until the court determined otherwise. HHJ Dyson rejected this argument, stating that if Morrison were correct it would substantially undermine the effectiveness of statutory adjudication under the UK Act. Judge Dyson confirmed that

1 Macob Civil Engineering v Morrison Construction Limited [1999] EWHC Technology 254.

the decision of an adjudicator was enforceable summarily regardless of any procedural irregularity, error or breach of natural justice. The judge refused to accept that the word "decision" should be in any way qualified. The judge held that where a party challenged the validity of a decision it is still nonetheless a decision under the UK Act and is enforceable and binding until the court finally determines the challenge. Consequently, the judge declared that the amount payable under the adjudicator's decision was properly due and owing.

10.4. Judge Dyson's views were reiterated in *Carillion v Devonport*[2] where Chadwick LJ held that courts are obliged to enforce the majority of adjudicators' decisions irrespective of errors of law.

10.5. Based on the UK Courts' reasoning in *Carillion* and *Macob*, errors of law in an adjudicator's decision (provided that they do not impact on the adjudicator's jurisdiction or trigger a breach of natural justice) will not prevent the enforcement of the decision.

10.6. The UK Courts' position is also consistent with the underlying purpose of adjudication to ensure the continued cash flow of contractors and subcontractors engaged in construction projects. If it were the case that an adjudicator's decision was not binding until such time as an arbitral tribunal or court determined it binding, then clearly the recipient of a payment under an adjudicator's decision would be deprived of that payment until the outcome of the arbitration or litigation. Clearly, this was not the legislature's intention when passing the Act.

10.7. The position in relation to adjudicators' errors of fact or interpretation is treated similarly to errors of law and will not prevent the enforcement of an adjudicator's decision. However, the situation may be treated differently where the adjudicator's error or mistake amounts to more than an error in interpreting the facts or calculating the amount payable by one party to the other and instead involves the adjudicator failing to address the issues that have been put to him in the notice of intention to refer. In such circumstances, an adjudicator's decision in which he exceeds his jurisdiction in addressing the issues in dispute which have been put to him will, depending on the facts, be unenforceable on the basis that the adjudicator had no jurisdiction to make decisions on matters which are not encompassed within the notice of intention to refer.

10.8. In *Bouygues v Dahl-Jensen*,[3] Bouygues had terminated Dahl-Jensen's subcontract and arranged for the subcontract work to be completed by others. Dahl-Jensen issued an adjudication notice claiming payment for work it had carried out, but which Bouygues had not

2 Carillion Construction Limited V Devonport Royal Dockyard Limited [2005] EWCA CIV 1358.

3 Bouygues (UK) Limited v Dahl-Jensen (UK) Limited [2000] EWCA CIV 507.

valued, and also damages for breach of contract and wrongful repudiation. Bouygues issued its own adjudication notice and ultimately the parties agreed that this was to be treated as a counterclaim to Dahl-Jensen's claim in the adjudication. The adjudicator, in making an award in Dahl-Jensen's favour, made a calculation error that had the effect of releasing retention within the amount awarded to Dahl-Jensen notwithstanding that the parties had agreed that the retention was not yet due under the subcontract. Bouygues argued that the effect of the decision was outside the adjudicator's jurisdiction and it was therefore not binding on the parties. The Court of Appeal held that when an adjudicator makes an error in calculating an amount payable to a party, the effect of which is to require the payment of monies which are not due under the contract, that decision would not be void provided the adjudicator had answered the issue that had been asked of him, in which case a mistake in answering that issue did not prevent the decision being enforceable. The court held that the adjudicator had not answered a question about the release of the retention monies but rather had mistakenly released those monies in his decision. As Dahl-Jensen had gone into liquidation the court refused the application for summary judgment on the grounds that summary judgment should not be given on an adjudicator's decision when the paying party has its own claims against the insolvent party. Notwithstanding this, the court made clear that had Dahl-Jensen not been insolvent, it would have granted summary judgment in Dahl-Jensen's favour enforcing the adjudicator's decision despite the fact that the adjudicator had erred in his decision.

10.9. Although the adjudicator in *Bouygues* was offered an opportunity to correct his mistake he chose not to. If similar circumstances arose in an adjudication under the Act one would expect that the adjudicator would take the opportunity to correct the decision pursuant to Section 6(13) which entitles the adjudicator to correct a clerical or typographical error, provided that the error can be properly characterised as an error in calculation. However, if the adjudicator chooses not to correct the error because he refuses to accept that he made an error, then no matter how obvious the error, or how blatantly wrong he might be, the courts will uphold the adjudicator's decision.

10.10. In *Verry v Camden*,[4] Verry issued court enforcement proceedings seeking to enforce the adjudicator's decision in respect of the third of four adjudications arising out of a housing scheme refurbishment contract between the parties. One of the issues the court had to

4 William Verry Limited v The Mayor & Burgesses of the London Borough of Camden [2006] EWHC 761 (TCC).

consider was whether an adjudicator's decision could be defeated by a counterclaim for defects that had not been argued in the adjudication before, or considered by, the adjudicator. HHJ Ramsey considered that the legislature's underlying intention of the UK Act was that the adjudicator's decisions should be enforced, pending final determination, because of the use of the word "binding" in the legislation. Judge Ramsey noted that as Camden had not raised the issue of the defects counterclaim in the adjudication they could not be in a better position by attempting to set off in the enforcement proceedings, a "disputed, unliquidated counter-claim against the adjudicator's decision". The court held that Camden could therefore not resist payment of the amount owed by them to Verry under the adjudicator's decision on the basis of Camden's unliquidated disputed counterclaim for defects.

Set-off and cross-claims

10.11. It is now a well-established principle of UK law and practice in the context of construction adjudications that a valid adjudicator's decision is to be enforced in effect without set-off or cross-claims. If the party refusing to comply with the adjudicator's decision seeks to raise a set-off or cross-claim in the enforcement proceedings the logic is that these are matters that should have been raised in the adjudication, and the adjudicator would either have allowed or disallowed them. It is not appropriate for a losing party in the adjudication to raise, during the course of the enforcement proceedings, either the same set-offs or cross-claims which they may have already raised in the adjudication and which have been adjudicated on by the adjudicator, or raise new set-offs and cross-claims which could have been raised in the adjudication but were not.

10.12. In *Balfour Beatty v Serco*,[5] an adjudicator awarded Balfour Beatty an extension of time in respect of delays caused by the necessity to carry out an environmental impact assessment on foot of a legislative procedure, and awarded them a sum of money for loss and expense. Serco refused to pay the adjudicator's award on the basis that even taking into account the extension of time the adjudicator had decided upon, the work was still not practically completed at that time and, consequently, Serco would be entitled to LADs for the period after the extension of time that had been awarded by the adjudicator until the work was completed. Serco argued that the LADs sum would exceed any amount payable to Balfour Beatty

5 Balfour Beatty Construction v Serco Limited [2004] EWHC 3336 (TCC).

under the adjudicator's decision. In reviewing the situation, HHJ Jackson derived two principles of law, namely that:

1) Where it follows logically from an adjudicator's decision that the employer is entitled to recover a specific sum by way of liquidated and ascertained damages, then the employer may set off that sum against monies payable to the contractor pursuant to the adjudicator's decision, provided that the employer has given proper notice (insofar as is required);
2) Where the entitlement to liquidated and ascertained damages has not been determined either expressly or impliedly by the adjudicator's decision, then the question whether the employer is entitled to set off liquidated and ascertained damages against sums awarded by the adjudicator will depend upon the terms of the contract and the circumstances of the case.

10.13. Judge Jackson concluded that the key question to ask was whether the extension of time had been finally determined, or whether the question had been left open as to what the extension of time might be in due course when the contract completed. The judge held that in this case the question had been left over in the adjudicator's decision and Serco was not entitled to set off the liquidated damages. Judge Jackson held that Balfour Beatty was entitled to summary judgment for the amount of the adjudicator's decision.

10.14. While it is a well-established principle from the UK cases that a party cannot seek to set off or cross-claim against an adjudicator's decision in defence of any enforcement proceedings, there are specific circumstances in which an entitlement to set off might arise.

10.15. In *Squibb v Vertase*,[6] HHJ Coulson held that an unsuccessful party to an adjudication will generally not be entitled to avoid the result of the adjudication by relying on a right to set off any other claims. The judge held that there were exceptions to this rule, namely:

1) The right to set off may be provided by the terms of the contract but such wording must be clear and must not offend the requirement for immediate enforcement of the adjudicator's decision; or
2) If the adjudicator's decision is in the nature of a declaration as to the contractual payment machinery, rather than a one-off award, a Respondent may be able to rely on the payment provisions of the Contract to issue a withholding notice.

6 Squibb Group Limited v Vertase F.LI. Limited [2012] EWHC 1958 (TCC).

10.16. Judge Coulson held that Vertase was not entitled to set off its claim for LADs against the adjudicator's decision as it did not meet the requirements of the above-mentioned exceptions. Accordingly, the judge awarded summary judgment in favour of Squibb in the full amount claimed and as payable under the adjudicator's decision.

10.17. The exception to the rule prohibiting set-off or cross-claim against an adjudicator's decision is unlikely to arise in an Irish context unless the courts were to construe the phrase "any dispute relating to payment" in Section 6(1) as encompassing an adjudication referral by a party for an "adjudicator's decision in the nature of a declaration as to the contractual payment machinery, rather than a one-off award"; the second exception in *Vertase*. If Section 6(1) is not construed in this way and is held to mean that the payment dispute must only relate to a claim for the award of money, it would not be possible for a party to refer a dispute to adjudication seeking the adjudicator's decision in the nature of a declaration as to the contractual payment machinery, if that issue does not also require the adjudicator to reach a determination in relation to the amount of money that might be payable following the adjudicator's decision on the contractual payment machinery. It is also difficult to envisage, in an Irish context, appropriate language that might be incorporated into the terms of the construction contract entitling a party to a right of set-off against an adjudicator's decision which does not offend the requirement and entitlements of the opposing party to enforce the adjudicator's decision under the Act.

Timing of payment awarded in an adjudicator's decision

10.18. The binding nature of an adjudicator's decision is not undermined by the fact that the adjudicator may not deal with the timing of any payment that one party is required to make as a result of the adjudicator's decision. While the Act does not provide for a period within which any payment arising from an adjudicator's decision should be made, and there is no guidance on this in the Code of Practice, it is often the case that the adjudicator will specify in his decision a time frame within which payment is to be made. Section 7(1) of the Act, which deals with a party's right to suspend work for failure to comply with an adjudicator's decision, provides that where an amount due pursuant to the adjudicator's decision is not paid in full within seven days from the date of the decision, the party who is entitled to receive payment may suspend work. While on the face of it there is no correlation between the paying party's obligation to pay the adjudicator's decision within a specified time frame and the receiving party's entitlement to suspend the work for non-payment in full, it appears to be the case that the intention of the Oireachtas in respect of the timing of any payment due pursuant to an adjudicator's decision is that such payment should be made, at the latest, within seven days of the date of the decision. The binding nature of an adjudicator's decision would not be compromised by an argument that the adjudicator failed to specify a time frame within which payment should be made. An adjudicator's decision is binding until finally settled by the parties or a different decision is reached in a court or arbitration.

10.19. In *ROK v Celtic*,[7] Celtic subcontracted ROK to provide a composting facility at a site. ROK claimed an extension of time for delay arising from a flooding incident and associated loss and expense. Celtic rejected the claim and ROK referred the dispute to adjudication. The adjudicator found that the flooding event was a compensation event under the subcontract, entitling ROK to an extension of time and ROK should be paid a sum of money in respect of the compensation event. Celtic failed to pay ROK arguing that the adjudicator had not set out a time within which the sums should be paid and that Celtic was entitled to take the amount into account in subsequent interim payment certificates. In the enforcement proceedings HHJ Akenhead, re-stating that the UK Act is clear that adjudicators' decisions are binding pending final determination by a court or arbitrator and granting judgment in ROK's favour enforcing the adjudicator's decision, held that:

> Whilst … many adjudicators do call for payment within a specified period of time, that is not essential in my view if it is clear from the wording of the decision and in context that payment is required to be made. If, as here, there is the clearest contractual requirement that an adjudicator's decision is binding and is to be "implemented" the absence of a specified period for payment in the decision does not undermine the requirement that the sum is to be paid.[8]

Enforceability of an adjudicator's decision

10.20. Section 6(11) further expands on the binding nature of an adjudicator's decision, providing that "The decision of the adjudicator, if binding, shall be enforceable either by action or, by leave of the High Court, in the same manner as a judgment or order of that Court with the same effect and, where leave is given, judgment may be entered in the terms of the decision".

7 ROK Building v Celtic Composting Systems Limited [2009] EWHC 2664 (TCC).
8 At paragraph 26 of the judgment.

10.21. The guidelines for enforcement and resisting enforcement in adjudication under the UK Act and UK Scheme were summarised in *Sherwood v MacKenzie*[9] as follows:

1) A decision of an adjudicator whose validity is challenged on the basis of its factual or legal conclusions as to procedural error remains a decision that is both enforceable and should be enforced.[10]
2) A decision that is erroneous, even if the error is disclosed by the reasons, will still not normally be capable of being challenged and should, ordinarily, still be enforced.[11]
3) A decision may be challenged on the grounds that the adjudicator was not empowered by the Act to make the decision, because there was no underlying construction contract between the parties[12] or because he had gone outside his terms of reference.[13]
4) The adjudication is intended to be a speedy process in which mistakes will inevitably occur. Thus, the court should guard against characterising a mistaken answer to an issue, which is within an adjudicator's jurisdiction, as being an excess of jurisdiction. Furthermore, the court should give a fair, natural and sensible interpretation to the decision in light of the disputes that are the subject of the reference.[14]
5) An issue as to whether a construction contract ever came into existence, which is one challenging the jurisdiction of the adjudicator, so long as it is reasonably and clearly raised, must be determined by the court on the balance of probabilities with, if necessary, oral and documentary evidence.[15]

10.22. Irish Courts will likely adopt similar principles, as established in the above-mentioned UK Court decisions, and refuse enforcement in very limited circumstances where it is demonstrated that adjudicators made the decision outside their jurisdiction or in breach of natural justice.

9 Sherwood & Casson Limited v MacKenzie Engineering Limited [2000] 2 TCLR 418.
10 Macob Civil Engineering Limited v Morrison Construction Limited [1999] EWHC Technology 254.
11 Bouygues (UK) Limited v Dahl-Jensen (UK) Limited [2000] EWCA CIV 507.
12 Project Consultancy Group v The Trustees of The Gray Trust [1999] HT/99/29 (TCC).
13 Bouygues (UK) Limited v Dahl-Jensen (UK) Limited [2000] EWCA CIV 507.
14 Bouygues (UK) Limited v Dahl-Jensen (UK) Limited [2000] EWCA CIV 507.
15 Project Consultancy Group v The Trustees of The Gray Trust [1999] HT/99/29 (TCC).

Insolvency situations

10.23. The issue of insolvency of the recipient payee (under an adjudicator's decision), in the context of enforcement proceedings, has seen the UK Courts refuse, or put a stay on the execution of, summary judgment where they might have otherwise granted summary judgment enforcing the adjudicator's decision. While the UK Court of Appeal in *Bouygues* upheld the High Court decision that the award was enforceable it nonetheless refused to grant summary judgment in Dahl-Jensen's favour on the basis that it would be wrong for the sums awarded to be paid to Dahl-Jensen due to the fact that Dahl-Jensen were in liquidation.

10.24. The issue with which the court is concerned is where the payee party is insolvent and there is no prospect that they will be able to repay the monies payable to them where the court enforces the adjudicator's decision and an arbitrator or court later reaches an opposing decision. Where an arbitrator or court reaches an opposing decision (following referral to arbitration or court by the party who rejects the adjudicator's decision) it would generally result in any of the original monies paid pursuant to an adjudicator's decision being returnable to the paying party. However, in circumstances where the recipient party is no longer in business because of insolvency, it would not be possible for the monies to be repaid to the paying party. Clearly, such circumstances can give rise to injustices.

10.25. In *Wimbledon v Derek Vago*,[16] Vago engaged Wimbledon to carry out an extension and refurbishment of his London home. Disputes arose and were referred to adjudication. The adjudicator awarded Wimbledon a sum which Vago refused to pay and instead commenced arbitration proceedings. Wimbledon sought summary judgment and Vago sought to stay the summary judgment application on the basis of Wimbledon's financial impecuniosity. HHJ Coulson QC held that a stay would only apply in "special circumstances". The judge held that an inability of a party to repay where a court or arbitrator reaches an opposing decision to the adjudicator's decision could amount to special circumstances in which the court would stay or refuse a summary judgment application for enforcement. The judge went on to review the authorities dealing with a stay of an adjudicator's decision because of a receiving party's financial impecuniosity. Judge Coulson concluded that the applicable principles were:

a) adjudication (whether pursuant to the UK legislation or the consequential amendments to the standard forms of building and

16 Wimbledon Construction Company 2000 Limited v Derek Vago [2005] EWHC 186 (TCC).

engineering contracts) is designed to be a quick and inexpensive method of arriving at a temporary result in a construction dispute.

b) In consequence, the adjudicator's decision is intended to be enforced summarily and the claimant (being the successful party in the adjudication) should not generally be kept out of its money.
c) In an application to stay the execution of summary judgment arising out of an adjudicator's decision, the court must exercise its discretion taking firmly into consideration a) and b) above.
d) The probable inability of the claimant to repay the judgment sum (awarded by the adjudicator and enforced by way of summary judgment) at the end of the substantive trial, or arbitration hearing, may constitute special circumstances rendering it appropriate to grant a stay.[17]
e) If the claimant is insolvent, or there is no dispute on the evidence that the claimant is insolvent, then a stay of execution will usually be granted.[18]
f) Even if the evidence of the claimant's present financial position suggested that it is probable that it would be unable to repay the judgment sum when it fell due, that would not usually justify the grant of a stay if

    i) the claimant's financial position is the same or similar to its financial position at the time that the relevant contract was made;[19] or
    ii) the claimant's financial position is due, either wholly, or in significant part, to the defendant's failure to pay those sums which were awarded by the adjudicator.[20]

10.26. Judge Coulson, in applying these principles, concluded that the fact that Wimbledon's directors were making loans to the company demonstrated a high degree of practical faith in the company's future. Furthermore, the judge held that Wimbledon was in a similar financial position at the time of the enforcement proceedings as it was when the parties entered into the contract. In addition, the judge found that the contract with Vago represented over half of Wimbledon's turnover, and consequently it was unsurprising that Vago's refusal to pay under the adjudicator's decision led to Wimbledon's impecuniosity. In the circumstances, Judge Coulson

17 Herschel Engineering Limited v Breen Property Limited [2000] EWHC Technology 178.
18 Bouygues (UK) Limited v Dahl-Jensen (UK) Limited [2000] EWCA CIV 507; Rainford House (in Receivership) v Cadogan [2001] HT 01/014.
19 Herschel Engineering Limited v Breen Property Limited [2000] EWHC Technology 178.
20 Absolute Rentals Limited v Glencor Enterprises Limited [2000] CILL 1637.

gave summary judgment in Wimbledon's favour and refused to stay execution of that judgment.

10.27. In *Pioneer v Graham*,[21] Graham engaged Pioneer under a subcontract in respect of which two disputes arose which were referred to two separate adjudications. The adjudicator awarded Pioneer £222,394 for the first dispute but found that Pioneer had been overpaid by £29,388 in the second dispute, and that left a balance of £193,006 due to Pioneer. Graham did not pay under the adjudicator's decision and Pioneer commenced enforcement proceedings. Graham resisted the enforcement proceedings arguing that there should be a stay of execution on the principles in the *Wimbledon* case. HHJ Ramsey found that Pioneer's financial position wasn't due to Graham's failure to pay the adjudicator's award but rather Pioneer's financial problems were inherent in its business model. Judge Ramsey refused to enforce the adjudicator's decision and ordered a stay of execution, acknowledging that this was a rare case where Judge Coulson's principles set out in *Wimbledon* were satisfied. Applying the *Wimbledon* principles, Judge Ramsey found that:

1) If money was paid to Pioneer they would not be in a position to repay it if the arbitration subsequently went against them, and there was a considerable amount of evidence to support the conclusion that Pioneer were insolvent.
2) Although Pioneer's financial position was the same or similar to when it entered into the subcontract, Graham having "robustly vetted" Pioneer had been misled by Pioneer as to their financial stability and so Graham had entered into the subcontract on a false premise.

Correcting the decision

10.28. Section 6(13) of the Act provides that "The adjudicator may correct his or her decision so as to remove a clerical or typographical error arising by accident or omission but may not reconsider or re-open any aspect of the decision".

10.29. As soon as an adjudicator's decision has been completed and sent to the parties the adjudicator is, *prima facie*, *functus officio* and the decision becomes enforceable, even if the decision contains an error or a mistake. However, there are occasions when an adjudicator identifies that the decision contains an error either because it has been pointed out to him by one of the parties or because he notices it himself. The purpose of Section 6(13) is to allow for such

21 Pioneer Cladding v John Graham Construction Limited [2013] EWHC 2954 (TCC).

a scenario and empower the adjudicator to correct his decision so as to remove the clerical or typographical error, provided that the adjudicator does not seek to reconsider or re-open any aspect of the decision. This entitlement to correct a clerical or typographical error is often referred to as the "slip rule".

10.30. The adjudicator's entitlement to correct errors will depend on the nature of the error and the time within which that error is corrected. In *Bloor v Bowmer*[22] Bloor was engaged as Bowmer's subcontractor under a JCT contract incorporating amendments containing an adjudication clause. A payment dispute arose which was referred to adjudication and the adjudicator issued a decision in Bloor's favour awarding them a sum of money. Bowmer immediately realised that the adjudicator had not taken into account the payments that Bowmer had previously made on account. The error was pointed out to the adjudicator who corrected his decision within 2 hours, describing the error as an obvious slip. The overall result was that no further payment was due to Bloor. Bloor issued court proceedings seeking to enforce the adjudicator's original decision (where the obvious slip had not been corrected) claiming that the adjudicator had no power to correct errors, except perhaps clerical errors. The UK legislation at the time did not include a slip rule similar to Section 6(13) of the Irish Act.[23] Consequently, the court examined the practice in arbitration and found that arbitrators could use the slip rule to correct errors. The court found that in the absence of a specific agreement to the contrary, a term should be implied into the contract that where a dispute is referred to adjudication the adjudicator may correct an error in his decision arising from an accidental error or omission. The court held that in the circumstances, the error the adjudicator had made constituted "an obvious slip" and an error characteristic of the type of error that an implied slip rule encompassed. In an Irish context if an adjudicator were to make a similar error as the error the adjudicator made in *Bloor* then such an error could be corrected by the adjudicator pursuant to the power to do so under Section 6(13).

10.31. In *Bouygues*, the adjudicator was invited by Bouygues' solicitors to amend his decision under the slip rule, but Dahl-Jensen's solicitors opposed this arguing that the adjudicator had no jurisdiction to do so. In addition, the adjudicator confirmed in writing that he had not made a clerical slip or error, despite the fact that both Judge

22 Bloor Construction (UK) Limited v Bowmer & Kirkland (London) Limited [2000] EWHC 183 (TCC).

23 Following the enactment of the UK Local Democracy, Economic Development & Construction Act 2009 a "slip rule" has been incorporated in 108A(2) of the UK Act.

Dyson, and subsequently the Court of Appeal considered that he had made an error. However, the Court of Appeal held that the adjudicator error was made when he was acting within his jurisdiction and that while it was clearly an error it was an error that he was entitled to make. This ruling appears to suggest that, once an adjudicator issues his decision, that is effectively the end of the matter and if that decision was within his jurisdiction, it's binding and enforceable. Since the *Bloor* and *Bouygues* decisions, the UK legislature introduced the 2009 Act which enshrines the slip rule in UK adjudication law.

10.32. In *CIB v Birse*,[24] HHJ Toulmin QC held that, in cases which involved a slip or alleged slip, there were two questions. Firstly, was the adjudicator prepared to acknowledge that he had made a mistake and correct it? Secondly, was it a genuine slip that failed to give effect to his thoughts? Judge Toulmin held that it was only if the answer to both these questions was yes that the court could give effect to the amendment to rectify the slip, if the justice of the case required it. Justice Toulmin's decision was subject to the important considerations as to the time within which the correction was made and issues of prejudice.

10.33. In *ROK v Celtic*[25] the main issue before the court related to the timing of the adjudicator's decision and its enforceability. The court, however, also considered the issue of correcting errors in the decision. On receipt of the decision Celtic had asked the adjudicator to make a number of corrections. While Celtic's letter set out some obvious amendments, the preponderance of the letter related to a request that the adjudicator revisit the computation of a particular certificate. The adjudicator said that this went beyond the terms of the slip rule, and although he did correct the clerical errors, he declined to amend the substantive figures. HHJ Akenhead rejected Celtic's submission that the adjudicator failed to properly exercise his jurisdiction and held that the payment and certification position was confused and unclear and that Celtic failed to take the opportunity to present their case in a simple way. The judge declined to decide if there had been an error stating that if there had it did not affect the enforceability of the adjudicator's decision.

10.34. Judge Akenhead reached the same conclusion in *YCMS v Grabiner*[26] in which a payment dispute arose during the works as a consequence of a significant increase in the contract value due to substantial variations issued to YCMS. YCMS referred the dispute to adjudication. The adjudicator found in YCMS's favour.

24 CIB Properties v Birse Construction Limited [2004] EWHC 2365 (TCC).

25 ROK Building v Celtic Composting Systems [2009] EWHC 2664 (TCC).

26 YCMS Limited v Grabiner [2009] EWHC 127 (TCC).

Receiving the decision YCMS informed the adjudicator that he had made an arithmetical error and in fact a greater sum was due to YCMS. The adjudicator decided that both he and YCMS's calculations were wrong and amended his decision by inserting an even greater amount due to YCMS after calculating the sum by a different method. Grabiner paid neither sum under the decision and ultimately YCMS issued court enforcement proceedings in respect of the adjudicator's first decision as revised. Grabiner opposed the enforcement on several grounds including that the purported revision of the first decision was not valid. Judge Akenhead, taking *Bloor* into account, made the following observations regarding when an adjudicator may revise his decision stating that the slip rule

> can and will only relate to "patent errors". A patent error can certainly include the wrong transposition of names or the failing to give credit for sums found to have been paid or simple arithmetical errors. ... The slip rule cannot be used to enable an adjudicator who has had second thoughts and intentions to correct an award. ... The time for revising a decision by way of the slip rule will be what is reasonable in all the circumstances. In the Bloor case, the adjudicator revised his decision within several hours and before the time for issuing a decision had been given. It will be an exceptional and rare case in which the revision can be made more than a few days after the decision. The reason is that, unlike a court judgment or an arbitration award, a principle purpose of the 1996 Act is to facilitate cash flow. If an adjudicator was able to revise his decision, say, 21 or 28 days later that would necessarily slow down and interfere with the speedy enforcement of adjudicators' decisions. That would in broad terms be contrary to the policy of the Act.[27]

10.35. Judge Akenhead found that in the circumstances the adjudicator had gone further than just correcting his error and had decided to make a recalculation using different figures. He held that Grabiner were materially prejudiced by the amendment and the revised first award was not valid.

10.36. In *O'Donnell v Build Ability*[28] the main contractor, Build Ability, engaged O'Donnell as a concrete framework subcontractor. Several disputes arose and a series of adjudications ensued. In Adjudication No. 8(A) the adjudicator issued a decision in favour of O'Donnell that in respect of Interim Valuation No. 25 they were entitled to

27 At paragraph 50 of the judgment.

28 O'Donnell Developments v Build Ability [2009] EWHC 3388 (TCC).

£14,328,131.88 less retention, valid Build Ability deductions and payments to date. The day after the decision O'Donnell notified the adjudicator of two errors in the decision, namely the incorrect inclusion in the payments to date of a sum previously awarded in Adjudication No. 7 (which was after the Interim Valuation No. 25 date), and miscalculation of the retention. Build Ability objected to the adjudicator correcting these two errors on the grounds that they did not fall within the "slip rule" permission. The adjudicator considered that he had the power to correct the errors and revised his decision. In a subsequent adjudication the adjudicator awarded O'Donnell a further sum of £385,306.56. Build Ability declined to pay the sums awarded in the two adjudications and O'Donnell issued court enforcement proceedings in respect of the adjudicator's two decisions. Build Ability argued that the adjudicator's calculations were induced by figures which O'Donnell had provided to him and that the adjudicator's errors in those figures fell outside what could be defined as a "slip". O'Donnell argued that the errors fell within the slip rule because the incorrect adjudication No. 8(A) decision did not accurately reflect the true intentions of the adjudicator. HHJ agreed with O'Donnell's contentions and held that:

> if the adjudicator is asked by one party to correct a slip and he accepts that an error has been made within the slip rule, then if the adjudicator makes an error of fact or law in so doing, I consider that such an error does not take the exercise of the slip rule outside his jurisdiction. Finally, if the adjudicator is asked by one party to correct a slip which the other party agrees is a slip within the slip rule but in operating the slip rule he makes an error of law or fact, then I do not consider that the Court can interfere in that decision.[29]

Judge Ramsey held that the adjudicator was asked to correct a slip and, as he had made an error within the slip rule, he was not acting outside of his jurisdiction. The judge held that the adjudicator made an inadvertent slip, as by making a deduction which he had not intended to make this was within the slip rule. Notwithstanding that the mistake had been made as a result of the erroneous information provided by O'Donnell, this was not uncommon and did not prevent an adjudicator from concluding that what he had done was not what he had intended.

10.37. The distinction must be made between an error in the adjudicator's decision-making process and an error corrected by the slip rule.

29 At paragraph 35 of the judgment.

When requesting an adjudicator to correct an error under the slip rule, the requesting party needs to be satisfied that, without the error, the adjudicator had jurisdiction to make the decision that he communicated to the parties notwithstanding that the decision contained an error which could be corrected by the slip rule. The party that fails to make clear this distinction is at risk of waiving its rights regarding the validity of the decision itself. The thinking behind this view is that if a party requests that an adjudicator correct a decision under the slip rule, the implication is that they have accepted the validity of the decision and that the only issue that they wish to address is to correct the error within the decision, which is capable of being corrected under the slip rule.

10.38. In *Laker v Jacobs*,[30] Laker was engaged by Jacobs under a subcontract on a power plant project. Disputes arose and Laker commenced three adjudications. The adjudicator issued three decisions (Award Nos. 1, 2 and 3) on the same day. After receiving the awards, Jacobs successfully applied to the adjudicator under the slip rule for correction of Award No. 1, stating that "it fully reserved its position in relation to the adjudicator's jurisdiction and that its application was made without prejudice to that position". In (corrected) Award No. 1 the adjudicator decided that Laker was entitled to a 52-week extension of time of the contractual completion date to 11 October 2013. In Award No. 3, the adjudicator decided that Jacobs ought to have issued a taking-over certificate stating that the works were complete on 30 August. Clearly, there was an inconsistency between a taking-over certificate stating that the works were complete on 30 August and an award entitling an extension to a contractual completion date of 11 October. Two sets of proceedings subsequently arose. In the first proceedings Laker sought summary judgment seeking enforcement of the three adjudication decisions, which Jacobs opposed on the grounds that the adjudicator lacked jurisdiction and his decisions were inconsistent. In the second proceedings, Jacobs sought two declarations as to the relationship between time extension and the subcontract taking-over mechanism. Laker sought to stay those proceedings to arbitration. The court was asked to decide whether, *inter alia*, Jacobs, by applying for a correction under the slip rule, elected to affirm the decisions and was barred from challenging them. HHJ Ramsey held that Jacobs had no real prospect of successfully defending the claim and Laker was entitled to summary judgment. However, Judge Ramsey, in considering the issue as regards the slip rule and affirmation of the adjudicator's jurisdiction, held that a general reservation

30 Laker Vent Engineering Limited v Jacobs E & C Limited [2014] EWHC 1058 (TCC).

of rights is sufficient to prevent a party's participation in an adjudication being taken as an election to affirm the adjudicator's jurisdiction or a waiver of any rights to object to the adjudicator's jurisdiction. With some hesitation the court held that, following publication of a decision, a party to an adjudication can still rely on the general reservation of jurisdiction and apply under the slip rule, or make payment, without losing its right to challenge the adjudicator's decision. Judge Ramsey held that Jacobs reservation of rights when seeking correction of Award No. 1 was sufficient to prevent Jacobs from waiving its rights in respect of the adjudicator's jurisdiction and that Jacobs had not waived their jurisdictional challenge merely because they made an application under the slip rule.

Suspending work under an adjudicator's decision

10.39. Section 7(1) entitles the executing party (upon giving written notice not later than seven days before beginning the proposed suspension stating the grounds under Section 7(2)) to suspend work under the construction contract where an amount due under an adjudicator's decision is not paid in full by the end of the seven-day period. The wording of Section 7(1) is similar to that of Section 5(1) which entitles an executing party to suspend work for non-payment. The entitlement to notify suspension arises if less than full payment is received within the seven-day period.

10.40. Pursuant to Section 7(3), the executing party may not suspend work after it has received payment of the amount due. If the full amount of the award under the adjudicator's decision is not paid within seven days, then on the eighth day the successful party can issue its notice of intention to suspend the work, but the suspension cannot begin until the eighth day after the date of serving the notice of intention to suspend. Sections 7(2) and 7(3) are worded very similarly to Sections 5(2) and 5(3) insofar as they require the executing party to give notice of its intention to suspend, after the amount becomes due, and at least seven days before the actual suspension is to begin. As a result of the operation of Sections 7(1) and 7(2) together, the payer has in effect 15 days, commencing from the date of the adjudicator's decision, within which to pay the full amount of the award so as to avoid the successful party suspending the works. A payer could ignore the seven-day period within which they are required to pay under an adjudicator's decision, allow the successful party to serve notice of intention to suspend, but pay the amount due before the notice's expiry, thereby effectively extending the payment period under the decision to 15 days. This outcome is clearly unsatisfactory from the successful party's perspective.

10.41. Section 7(3)(b) prohibits the successful payee party from suspending work where the adjudicator's decision has been referred to arbitration or where proceedings are otherwise initiated regarding that decision. Consequently, the losing payer can effectively avoid any suspension of works by not paying under the adjudicator's decision and initiating arbitration (or other) proceedings at a point in time before any notice triggers the suspension. The payer could then avoid payment, and also the works being suspended until a court enforces the adjudicator's decision. This could be a significant period following the date of the adjudicator's decision and such circumstances are undesirable. It would have been preferable had the legislation been drafted so as to achieve the objective of maintaining cash flow for the payee party. As currently drafted, a payee party is forced to issue court summary proceedings seeking enforcement of the adjudicator's decision in order to finally receive payment under that decision. Any leverage the payee party may have had, through suspending the works in order to secure payment, is lost by operation of Section 7(3)(b). Summary proceedings, although designed to be speedy, can in fact be prolonged, particularly with a recalcitrant defendant, and any summary judgment granted may be subject to an appeal to a higher court which again could have the effect of ultimately delaying payment under the adjudicator's decision.

10.42. Sections 7(4) and 7(5) relate to the suspending payee party's entitlement to have disregarded any suspension period in computing the contractual time limit, or the time taken to complete the work under another construction contract. The operation of Section 7(4), which has the effect of automatically entitling the contractor/subcontractor to an extension for the time equivalent to the suspension period, is a common provision in construction contracts, particularly standard form contracts. However, the consequences of Section 7(5) are unusual insofar as it could give rise to an entitlement for a subcontractor, of a main contractor who has suspended the works, to avail of a suspension of the subcontract works for the same period of time as the main contractor's suspension and corresponding extension of time. For example, if a main contractor suspends the works for one week as a result of the non-payment of an adjudicator's decision made in a payment dispute between the employer and the main contractor, then the main contractor will be entitled to an extension of time of one week to the contract completion date under the main contract. However, all of the main contractor's subcontractors would also be entitled to a one-week extension of time to their respective subcontract completion dates. While one can readily appreciate why it is justifiable that the main contractor be entitled to the one-week extension of time on the basis that the main contractor suspends all of its works during that one-week period, the same may not be said for its subcontractors. It may be the case that a particular subcontractor had not intended to be on site during the relevant one-week suspension period due to the scheduling of their works. For example, if the one-week suspension of the main contract occurs during a period of time when an electrical subcontractor had not been scheduled in the programme to be on site because they have demobilised between the first- and second-fix works, then despite the fact that the electrical subcontractor was not intending to carry out any works during this period and there was no impact on their programme, the electrical subcontractor would nonetheless be entitled to a one-week extension of time for the completion of its subcontract works, pursuant to the operation of Section 7(5) of the Act. Clearly, while there may be an impact on the main contractor's works to the extent that the electrical subcontractor may not be able to return to site to carry out the second-fix works on the date on which they had planned to do so, it is not automatically the case that the circumstances would give rise to a one-week delay to the electrical subcontractor's works. It would have been preferable had Section 7(5) been worded in a manner which required the subcontractor to demonstrate the actual impact of the main contract one-week suspension on the subcontractor's programme. Had Section 7(5) been worded in terms of the actual impact on the related subcontractor party's programme, as opposed to an automatic entitlement to extend that programme, then that would have been a more fair, satisfactory and appropriate outcome.

11 Adjudication costs

Sections 6(15) and 6(16) of the Act

Parties' own costs

11.1. Section 6(15) provides that "Each party shall bear his or her own legal and other costs incurred in connection with the adjudication". Section 6(15) (and paragraph 38 of the Code of Practice which is similarly worded) has the effect of prohibiting the parties from seeking to recover from the other party their legal costs and expenses (or any other costs) incurred in connection with the adjudication as part of the adjudicator's decision. The position is the same under the UK Act.

11.2. Adjudication can be an expensive process despite the fact that some of the key underlying objectives of adjudication are to deal expeditiously and relatively inexpensively with disputes. The UK Courts have considered several cases where parties have sought to recover their costs in forums outside of the adjudication process and through some novel arguments.

11.3. In *Lulu v Mulalley*[1] an adjudicator's decision, including an award of "debt recovery costs", which in effect predominantly constituted the legal and other costs incurred in the adjudication, was enforced by the Deputy Judge Mr Jonathan Acton-Davis QC. The entitlement to debt recovery costs arose out of the Late Payment of Commercial Debts (Interest) Act 1998. The claim for debt recovery costs was not specifically referred to in the adjudication notice, nor in the referral notice, nor response. It was pleaded for the first time in the rejoinder and it was met by a jurisdictional objection which, on the evidence, was raised at a meeting called by the adjudicator on 28 July 2015. Deputy Judge Acton-Davis found that the debt recovery costs were essentially the costs of running the adjudication which had been instituted by the adjudication notice. He held that such costs were

1 Lulu Construction v Mulalley & Co Limited [2016] EWHC 1852 (TCC).

> clearly connected with and ancillary to the referred dispute and must properly be considered part of it. It follows ... the adjudicator was correct to say that he had jurisdiction to decide this element of the dispute: although it was not within the scope of the referral, it was something which was connected with and ancillary to that referred dispute.[2]

11.4. As a consequence, the judge held that if an adjudicator decides the Late Payment Act applies and allows the recovery of adjudication costs as a result, though whether that may be right or wrong is a matter of law, then that finding does not make his decision, including the award of the "debt recovery costs", unenforceable. It is a well-established principle from the UK cases that while an adjudicator may be wrong on a matter of fact and/or law, his decision will still be enforceable, unless there is some other separate reason for not enforcing the decision. Given that *Lulu* concerned the adjudicator's jurisdiction it is not a full endorsement that the Late Payment Act can be used to recover adjudication costs where the dispute adjudicated on relates to payment.

11.5. Whilst the Act applies only to contractual provisions in construction contracts the intention is clearly to limit the award of a parties adjudication costs to situations where it is agreed between the parties following the referral of the dispute. This would conflict with the award of adjudication costs being made under the relevant Irish legislation, the European Communities (Late Payment in Commercial Transactions) Regulations 2012. The *Lulu* judgment is very short and makes limited reference to the adjudication; therefore, it is not entirely clear whether the adjudicator considered the Late Payment Act in light of Section 108A of the UK Act. Consequently, a future UK case may establish that Deputy Judge Acton-Davis reached the wrong conclusion.

11.6. In *National Museums v AEW*,[3] National Museums engaged AEW to design, and PIHL Galliford Try to construct, the Liverpool museum. Complications arose during the construction works and National Museums brought proceedings against AEW concerning, *inter alia*, defects associated with the design and construction in the steps and seating. National Museums' claim included damages that were effectively the adjudication costs they had incurred in an earlier adjudication with the contractor, Galliford. In the adjudication between Galliford and National Museums, Galliford claimed that the steps and seating design did not fall within its scope of contractual work and

2 At paragraph 9 of the judgment.

3 National Museums and Galleries on Merseyside Board of Trustees v AEW Architects and Designers Limited [2013] EWHC 3025 (TCC).

they were not liable for defects in that design. The adjudicator agreed with Galliford's position and decided accordingly, including a decision that National Museums pay the adjudicator's fees. In the court proceedings subsequently initiated against AEW, National Museums argued that but for AEW's design defects National Museums would not have incurred costs in the adjudication against Galliford. HHJ Akenhead held that the adjudication between Galliford and National Museums would not have occurred had AEW correctly performed its contractual obligations and consequently there was a sufficient causative link between AEW's defective design and the adjudication that National Museums were compelled to instigate against Galliford. The judge held that AEW was responsible for National Museums' costs in its adjudication with Galliford and awarded damages in this regard. Judge Akenhead also addressed the issue as to whether adjudication costs were foreseeable in the sense of not being too remote and held that the causative link would only be broken if National Museums acted unreasonably or if their solicitors acted negligently in advising them that they had an arguable case in the Galliford adjudication.

11.7. The *National Museums* case potentially leaves open the option for an aggrieved party to seek to recover adjudication costs against a third party as damages if that party was negligent in their obligations leading to the aggrieved party having an adjudicator's decision made against them by another party in the contractual framework. Such an entitlement could arise in construction projects where there are myriads of contractual relationships, duties and obligations between the various stakeholders. A party on the receiving end of an unfavourable adjudicator's decision may be able to seek to recover the costs of having to defend that adjudication where the third party's negligence led the losing party to initiate the adjudication proceedings against them in the first place. For example, if an architect were to negligently issue a payment certificate or instruction resulting in an adjudication whereby a contractor (or a subcontractor) successfully recovers payment from an employer (or contractor), it may be open to the employer to issue proceedings against the architect for recovering the losses it suffered because of his negligence, in which damages could include the costs of having to defend the adjudication.

Adjudicators' fees

11.8. Section 6(16) of the Act provides that "The parties shall pay the amount of the fees, costs and expenses of the adjudicator in accordance with the decision of the adjudicator". Section 6(16) differs from the provisions of Section 108(A)(2) introduced by the UK

2009 Act[4] which amended paragraph 25 of the UK Scheme to provide that subject to any contractual provision under Section 108(2) of the UK Act, the adjudicator may determine the apportionment of payment of his fees and the parties are jointly and severally liable for any sum outstanding after such determination. These are among the several noticeable differences between Section 6(16) and paragraph 25 of the UK Scheme.

11.9. Under Section 6(16) a party's obligation to pay the adjudicator's fees, costs and expenses are premised on the supposition that the adjudicator makes a decision on those. While adjudicators under the UK Scheme need make no determination as to how or which party is to discharge his fees, under the Act an adjudicator must decide how his fees are to be paid, as otherwise the adjudicator will have failed to comply with Section 6(16). Secondly, Section 6(16) of the Act does not provide for the parties being jointly and severally liable in respect of the adjudicator's fees whereas paragraph 25 of the UK Scheme provides otherwise. Consequently, if an adjudicator appointed under the Act were to decide that one party is liable for his fees and that party fails to discharge those fees, then the adjudicator would be unable to recover those costs from the other party on the basis that they are jointly and severally liable, because the adjudicator, having made his decision, is *functus officio*, meaning he no longer has any function relating to the adjudication. Consequently, the adjudicator would be unable to make any further decision requiring the non-defaulting party to discharge his fees which in his decision he had directed the other party to pay. This could result in a serious problem for an adjudicator who has made a decision against a party in respect of his fees where that party subsequently becomes insolvent. A prudent adjudicator should ensure he addresses this issue of liability for his fees in addition to addressing the substantive issues in the adjudication.

11.10. The lack of joint and several liability for the adjudicator's fees under Section 6(16) is in contrast to the liability in respect of the adjudicator's fees that arise where the parties revoke the adjudicator's appointment. Section 6(18) makes the parties jointly and severally liable for the adjudicator's fees if they revoke the adjudicator's appointment. It is uncertain whether inconsistency between Section 6(16) and Section 6(18) as regards this joint and several liability was deliberate, and it will be interesting to see whether the matter is addressed by the courts or through an amendment to the legislation.

4 UK Local Democracy, Economic Development and Construction Act 2009.

11.11. The parties' obligations under Section 6(16) in respect of payment of the adjudicator's fees are dependent not only on the adjudicator making a decision in relation to fees but also on the adjudicator having jurisdiction to make such a decision on fees, and more importantly to make the substantive decision. Equally, if an adjudicator's decision is unenforceable for any reason other than a lack of jurisdiction then an issue will arise as to the adjudicator's entitlement to payment. For example, if an adjudicator commits a breach of natural justice, or is biased, or otherwise breaches the adjudication process, and such breaches or bias renders the adjudicator's decision unenforceable then the adjudicator is not entitled to be paid his fees in the adjudication.

11.12. In *Linnott v Halliwells*,[5] the responding party expressly rejected the adjudicator's terms of appointment and challenged his jurisdiction. The court held that by asking in the alternative for the adjudicator to deal with the dispute the responding party was found liable for the adjudicator's fees either on the basis of a contract made by conduct or as a *quantum meruit*. Judge Ramsey held that this result would be the same whether the adjudicator has had jurisdiction (as the judge found) or not. The judge held that Halliwells had asked the adjudicator to withdraw but, in the alternative, asked him to adjudicate the merits, albeit reserving their position on jurisdiction. Judge Ramsey stated that in doing so Halliwells had asked the adjudicator to proceed and carry out work and there was an acceptance by Halliwells that if the adjudicator rejected the jurisdictional argument, he would carry out work in dealing with the merits. The request from Halliwells and the fact that the adjudicator proceeded with the adjudication gave rise to a contract formed by conduct with an obligation by Halliwells to pay the adjudicator's reasonable fees and expenses. Judge Ramsey went on to say that even if he was wrong, the court held that Halliwells had been unjustly enriched by the adjudicator's decision and were therefore liable for his fees in restitution.

11.13. The *Linnett* decision is informative as regards circumstances where an adjudicator's jurisdiction is challenged, the adjudicator reaches a non-binding conclusion that he has jurisdiction, proceeds with the adjudication and issues a decision. If it is later determined by a court, in any enforcement proceedings, that the adjudicator's decision is unenforceable because the adjudicator lacked jurisdiction and his non-binding conclusion was therefore wrong, then depending on the precise wording of the jurisdictional challenge it may be that the adjudicator is still entitled to his fees in considering the

5 [2009] EWHC 319 [TCC].

challenge to his jurisdiction and his subsequent decision. If a party challenges an adjudicator's jurisdiction but in the alternative asks the adjudicator to deal with the dispute in the event that the adjudicator reaches a non-binding conclusion that he has jurisdiction, then the *Linnett* decision is instructive on the principle that the challenging party will be liable for the fees incurred by the adjudicator in reaching his non-binding conclusion of jurisdiction.

11.14. In *PC Harrington v Systech*,[6] the English Court of Appeal considered the question of whether an adjudicator could recover his fees where his decision has been held to be unenforceable for his failure to comply with the rules of natural justice. The court held that the question that arises is whether the appointment between the parties and the adjudicator was "(a) an entire contract such that the bargained-for consideration was an enforceable decision or (b) a divisible contract for the performance of a series of ancillary and anterior functions culminating in the making of a decision". If it were an entire bargained-for contract the adjudicator would not be entitled to payment of his fees if he did not produce an enforceable decision. If, however, it were a divisible contract the adjudicator would be entitled to payment for the performance of the divisible functions whether the decision was enforceable or not. The Court held that this was a question of construction of the contract and that the terms of the adjudicator's engagement had to be read together with the terms of the UK Scheme. The Scheme contained important provisions dealing with the question of remuneration in various circumstances where the adjudicator did not reach a decision. The adjudicator's entitlement to be paid did not accrue as and when services were provided but only when he had completely performed his role as adjudicator and made a decision. The Court held that the terms of the adjudicator's appointment were not inconsistent with this, and that the bargained-for performance in this case was that the adjudicator would make an enforceable decision. The purpose of the adjudicator's appointment was to produce an enforceable decision which, for the time being, would resolve the dispute. An unenforceable decision was of no value to the parties and they would have to start again on a fresh adjudication in order to achieve the enforceable decision which the adjudicator had contracted to produce. In order to circumvent the consequences of the PC Harrington decision adjudicators should consider

6 PC Harrington Contractors Limited v Systech International Limited [2012] EWCA Civ 1371.

incorporating terms into their appointments (if the parties to the adjudication are prepared to agree) covering payment of their fees and expenses in the event of a decision not being delivered or proving to be unenforceable.

11.15. The *PC Harrington* decision endorsed the proposition that the "bargained-for performance" is an enforceable decision. In other words, the adjudicator's entitlement to be paid depends on the production of an enforceable decision. If the adjudicator's decision is not enforceable, the adjudicator is not entitled to his fees. While it is not an entirely satisfactory decision it is nevertheless persuasive authority on the issue, albeit English authority.

11.16. The Court of Appeal in *PC Harrington* preserved the *Linnett* finding as the exception to the rule which applies where one party raises a jurisdiction issue and the other party nevertheless decides to proceed with the adjudication. In that situation, the parties are taken to have accepted the risk that the outcome may be a decision that is unenforceable.

11.17. Circumstances may arise following the adjudication conclusion where a party is dissatisfied with the adjudicator's fees for whatever reason, including a belief that the adjudicator's fees are excessive. In *Stubbs v Tolley*,[7] the court had to deal with a challenge that the adjudicator's fees were excessive. Mr Smart of Stubbs was appointed as adjudicator in respect of a dispute between Tolley, a contractor engaged under two building contracts with Torridge District Council. Mr Smart delivered two written decisions in July 2000 which included decisions in relation to his fees, where he claimed the sum of £1,561.50 (plus VAT). Tolley paid the adjudicator's fees but alleged that they were unreasonably excessive and issued District Court proceedings claiming to recover the fees for the alleged overpayment. Deputy District Court Judge Guest heard the claim in the Small Claims Court and found in favour of the respondent on two grounds: firstly, that an adjudicator does not as a matter of law enjoy immunity from a claim that his hours or remuneration are unreasonable, and secondly that Mr Smart's hours were in fact excessive. Mr Smart appealed the Deputy District Court Judge's findings, and, on appeal, Mr Recorder Lane QC considered the question of immunity. Mr Lane QC referred to the adjudication agreement which repeated the immunity provisions set out in Section 108(4) of the UK Act and held that the adjudicator's fees formed an integral part of the adjudication agreement and could only be challenged if the adjudicator had

7 Stubbs Rich Architects v WH Tolley & Sons Limited [2001] Gloucester County Court, Case No. BP001105.

acted in bad faith. He held that there was no suggestion or evidence the adjudicator had acted in bad faith and so the appeal succeeded, and the adjudicator was awarded his fees. While the judge's finding on the adjudicator's immunity meant that he did not actually have to deal with the allegation that the adjudicator's fees were excessive, he nonetheless held that "A court must be very slow indeed to substitute its own view of what constitutes reasonable hours". The judge rejected the allegation that Mr Smart's hours (and consequently his fees) were excessive and applied the following principles to the adjudication, which concurred with the view expressed in Mustill & Boyd on Commercial Arbitration[8] that:

> The court does not substitute its own view for that of the arbitrator. In order to make good an allegation of misconduct, very clear evidence is required, and it is not enough to show that the amount demanded is more than the court would have considered appropriate if it had been approaching the matter afresh.

11.18. In a later decision in *Fenice v Jerram Falkus*,[9] Judge Waksman rejected the proposition that the adjudicator's decision on payment of his fees was part of his decision on the substantive issue and that the fees should be paid regardless of whether they are reasonable. The judge held that the adjudicator's fees are open to the reasonableness assessment but the burden of proving reasonableness required the adjudicator to provide a level of justification for his fees such as the rate, the hours worked and a short narrative. Thereafter, the evidential burden as to the reasonableness of the adjudicator's fees shifts to the challenging party. The judge held, in the particular case, that the adjudicator's fees were reasonable and Fenice, who paid the fees on the basis they were jointly and severally liable, was entitled to recover the paid fees from Jerram.

11.19. In *Jensen v Staveley*,[10] Staveley referred a dispute between them and Salford City Council to Paul Jensen as adjudicator. A jurisdiction issue arose, and the parties asked the adjudicator to deal with his jurisdiction as a preliminary issue. The adjudicator found that he did not have jurisdiction and did not proceed with the adjudication. The adjudicator, Mr Jensen, sought to recover the fees incurred in considering the jurisdictional issue, but Staveley refused to discharge the fees. Mr Jensen pursued payment in Wigan County Court.

8 Mustill & Boyd on Commercial Arbitration, Second Edition, pages 239 to 240.

9 Fenice Investments Inc v Jerram Falkus Construction Limited [2011] EWHCR 1678.

10 Paul Jensen v Staveley Industries Plc [2001] WN 101245.

Staveley argued that the adjudicator had misinterpreted the *Atlas*[11] decision, in which the central issue was whether the parties entered into the contract before or after 1 May 1998, the operative date under the UK Act, and therefore whether the contract came within the UK Act's adjudication provisions. District Justice Donnelly held that it was not for him to consider whether the adjudicator had interpreted the *Atlas* case incorrectly or otherwise, the adjudicator had been asked to deal with the preliminary issue of jurisdiction and had done so; whether he was right or wrong was irrelevant.

11.20. The judge held that there had been no suggestion of any default or misconduct and that Mr Jensen should therefore recover his fees. The judge went on to state that under paragraph 11 of the UK Scheme (under which the adjudication had been referred) the adjudicator was entitled to such reasonable amount as he may determine by way of fees and expenses incurred. The allegation that the adjudicator made a wrongful determination could not be construed as default or misconduct by the adjudicator. The judge concluded that as the adjudicator's fees were reasonable, he was entitled to be paid them and the judge made an order to that effect.

11 Atlas Ceiling & Partition Co Limited V Crowngate Estates (Cheltenham) Limited [2002] 18 Const LJ 49.

12 The Code of Practice Governing the Conduct of Adjudications

Section 9 of the Act

12.1. Section 9 of the Act provides that "The Minister may prepare and publish a code of practice governing the conduct of adjudication under section 6".

12.2. The Code of Practice[1] is not part of the Act but rather a statutory code governing the conduct of adjudication. The Code of Practice contains 39 paragraphs dealing with, *inter alia*, the appointment of the adjudicator (either by agreement between the parties or by the Chair of the Minister's panel), the adjudicator's responsibilities, the referral, procedures, the decision and reporting on the conduct of adjudication cases.

12.3. The intention of this chapter is not to consider each of the paragraphs of the Code of Practice, as the pertinent paragraphs are discussed where relevant in the context of the issues addressed in other chapters. However, one of the issues that will be at the forefront as payment disputes are referred for adjudication will be as to the binding effect, if any, of the Code of Practice on the parties.

Binding effect of the Code of Practice

12.4. The Code of Practice is not incorporated into the Act, but rather reference is made to it in Section 6(8), which provides that the adjudicator shall comply with the Code of Practice, irrespective of whether the adjudicator has been appointed from the Minister's panel or not. While the Act requires the adjudicator to comply with the Code of Practice, it is uncertain as to the extent to which the adjudicator's jurisdiction or decision is impacted upon by any breach of the provisions of the Code of Practice by the adjudicator.

1 Construction Contracts Act, 2013, Code of Practice Governing the Conduct of Adjudications, Department of Business, Enterprise and Innovation, 25 July 2016.

12.5. Until cases on this issue come before the court there will remain uncertainty as to the binding nature, or otherwise, of the Code of Practice. If the Irish Courts adopt a pragmatic approach (as the UK Courts have done) in respect of minor procedural breaches of the Code of Practice, then it is likely that the Irish Courts will take the view that where an adjudicator commits a minor breach of the Code of Practice that should not invalidate his jurisdiction or render any decision unenforceable. However, cases of this nature will undoubtedly be determined on their own particular circumstances. What might be considered a minor breach in one case (because it has no impact on the outcome) may not be considered a minor breach in another case if the outcome is that such a breach was to be construed as having constituted, or facilitated circumstances that contributed to, a bias or breach of natural justice. It remains to be seen how the Irish Courts will approach such situations and it is inevitable that they will do so on a case-by-case basis.

12.6. While Section 6(8) provides that the adjudicator shall comply with the Code of Practice, the Act does not provide that the parties comply with the Code of Practice. Consequently, it would appear that the Code of Practice does not have the binding effect on the parties in the same way that it does on the adjudicator. The fact that the Act does not explicitly stipulate that the Code of Practice is binding on the parties could prove problematic and potentially lead to uncertainty, giving rise to allegations that a party's conduct is not consistent with the provisions of the Code of Practice. Where a complainant party alleges that the opposing party has breached some aspect of the Code of Practice (albeit that the opposing party may argue that the Code of Practice has no binding effect on them) the complainant party could seek a direction from the adjudicator in respect of the opposing party's alleged "breach" or request that the adjudicator take some action in respect of the same. If the adjudicator were to decide not to take any action, or take action which is considered insufficient or inappropriate by the complainant party, then there could arise the risk of the complainant party alleging that the adjudicator has misconducted the adjudication or has been biased against the complainant party. While such an allegation would be decided on its own merits, the risk nonetheless exists that an adjudicator's decision could potentially be unenforceable if the complainant's allegations prove successful. For example, paragraph 5 of the Code of Practice provides for the matters that should be addressed in the notice of intention to refer a payment dispute for adjudication, including each party's name, address and contact details; relevant details of the payment dispute including the disputed amount (even if it's zero), the nature of the dispute, and the site address; a copy of the relevant payment claim notice, and any response to that payment claim notice; and relevant details to identify the construction contract which must be attached if in writing, and any supporting information that may assist the adjudicator in understanding the nature of the dispute.

12.7. Paragraph 5 of the Code of Practice is similar to the provisions of paragraph 1(3) of the UK Scheme, although the precise details required to be included in the respective notices differ. However, whereas the UK Scheme is binding on the parties (where the construction contract fails to properly provide for adjudication), the Code of Practice is not binding on the parties. Consequently, a failure of a party to comply fully with paragraph 5 of the Code of Practice may have no consequences for the non-compliant party. Accordingly, if a party were to fail to include in its notice of intention the matters specified in paragraph 5, the responding party might seek to challenge the notice's validity. While a party not complying with the Code of Practice is exposing itself to unnecessary risk, it will be interesting to see how Irish Courts approach any non-compliance. Notwithstanding this, it is strongly recommended that the parties seek to comply with the Code of Practice thereby avoiding uncertainty and the potential exposure of an unenforceable decision in this regard. Where non-compliance by a party with the Code of Practice arises during the course of an adjudication, and the adjudicator proceeds with the adjudication notwithstanding same, it would be likely that the non-compliance would be raised by the compliant party in enforcement proceedings.

12.8. However, it is unlikely (albeit it will depend on the particular circumstances) that a party's non-compliance with the Code of Practice could give rise to an adjudicator not having jurisdiction thereby rendering any decision unenforceable. An adjudicator faced with a party who is not complying with the Code of Practice would be well advised to invite the offending non-compliant party to comply. If the adjudicator were to simply ignore the non-compliance and indulge the non-compliant party, then such action or omission by the adjudicator could give rise to allegations of bias or that the adjudicator was not acting impartially and/or in accordance with the principles of natural justice. If the court (in any enforcement proceedings) were to agree that such allegations were well-founded then, subject to the comments below regarding the particular circumstances surrounding the non-compliance, the court would likely refuse the summary judgment application rendering the adjudicator's decision unenforceable. The proviso to the aforementioned conclusion, that the court could potentially refuse the enforcement application, would be where the non-offending party (relying on the non-compliance of the other party in seeking to defeat an enforcement application) continued in the adjudication without raising any objection with the adjudicator at the time or reserving its position in respect of the opposing party's non-compliance with the Code of Practice. If such were the circumstances surrounding the non-offending party's reaction to the non-compliance, then the outcome in the enforcement proceedings could be quite different on the basis of the *Maymac*[2] decision.

12.9. The extent to which the adjudicator's jurisdiction is impacted upon by the failure of a party to adhere to the procedural matters set out in the Code of Practice is debatable. It will be interesting to see whether the courts view an adjudicator's tolerance of a party's non-compliance with procedural matters set out in the Code of Practice as giving rise to a jurisdictional impediment to the adjudicator.

2 Maymac Environmental Services Limited v Farraday Building Services Limited [2001] 75 Com LRHT 00/222.

13 Service of notices

Section 10 of the Act

13.1. Section 10(1) of the Act provides for the freedom of the parties to agree the manner of service of any notice (or indeed any document during the course of the adjudication). In the context of adjudication, the phrase "notices under this Act" is a reference to the notice of intention to refer the payment dispute to adjudication and the subsequent referral notice.

Agreed contractual notice provisions

13.2. While Sections 3 to 5 of the Act provide for notices to be served between the parties in respect of their payment obligations and entitlements under the Act, such notices may not be applicable in the adjudication context. It is preferable for the parties to agree contract provisions in respect of service for notices, and where they do so, and provided such provisions are accurately drafted, the parties potentially reduce potential disputes arising as to whether a particular notice has been validly served.

13.3. Section 115(6) of the UK Act provides that a reference to "notice" includes any form of written communication, implying that any agreement regarding the service of notices should be an agreement in writing. Section 10(1) has no similar provision. It is, therefore, arguable that an oral agreement as to the service of notices will be sufficient to constitute an agreement on notices for the purposes of the Act. This is not a satisfactory situation as oral agreements invariably tend to cause uncertainty as to what has been agreed, particularly where an adjudicator, or a court, is tasked with deciphering an oral agreement. Fortunately, such agreements are more likely to be in a written format and generally contained within the contract.

13.4. A contractual notice provision may be permissive or mandatory in nature. If the provision is mandatory, the courts will insist on strict compliance with it as the mere breach thereof might have a significant adverse effect for the non-compliant party. Mandatory provisions often contain the word "shall". Permissive provisions often contain the word "may", in which circumstances if a party chose not to serve a notice in a particular format or manner this would not be fatal to the notice's validity.

Effective means

13.5. In the Scottish case *Strathmore v Greig,*[1] the latter sought to rely on a phone call, resulting in a message left with a receptionist, as an effective means of service of a withholding notice. The court held that this was not an effective means as a written notice was actually required.

13.6. Notwithstanding that the Act does not include any deeming provisions in relation to service on the address of the recipient to whom the notice is to be served, it is likely that the general principles which have been established in Irish law in relation to the service of notices will be applicable. There is a significant body of case law in respect of matters such as what represents the last known address, what is meant by delivery and what constitutes effective means. As adjudication cases come before the courts on these issues the courts will likely rely on existing authority on matters relating to the proper service of notices while at the same time, in the context of adjudication, take cognisance of relevant UK case law.

13.7. In *Rohde v Markham-David,*[2] Markham engaged Rohde for building works during which issues arose in relation to delays and the quality and value of works. Problems also arose in Markham's marriage which ended during the building works and resulted in Markham moving to a new address. Markham's ex-wife also left the family home, which was subsequently rented out. In March 2001, over two years after the building works commenced Rohde decided to refer to adjudication his claim for the balance of monies they believed they were owed. The adjudication notice and various other communications were sent to Markham's old home address but returned undelivered and the adjudication proceeded. The adjudicator only received submissions from Rohde and reached a decision that Markham owed monies to Rohde. When Markham failed to pay Rohde issued court enforcement proceedings. In addition, Rohde applied for and obtained an order for service by hand to both Markham's old home and his ex-wife's new address. Markham did not respond or acknowledge service and in March 2002

1 Strathmore Building Services v Colin Scott Greig (TA Hestia Fireside Design) [2000] Scot CS 133.

2 Rohde Construction v Markham-David [2006] EWHC 814 (TCC).

the court gave default judgment against Markham. Subsequently, when Markham became aware of the judgment, he applied to the court to have it set aside. In order to set aside a default judgment the applicant party has to demonstrate that he had a real prospect of successfully defending the claim. In this regard, Markham raised several defences including the fact that he was unaware that the adjudication was proceeding. Rohde argued that all documents served in the adjudication were sent to Markham's last-known principal residence and consequently service was effective and the judgment should stand. Markham argued that Rohde could have easily contacted him through his place of business (a quarry) which remained the same throughout the building works and subsequent adjudication, and that Rohde was fully aware of the quarry as he had since done business there. Markham also contended that Rohde deliberately avoided contacting Markham at the quarry, instead choosing to serve the adjudication notice and referral on Markham's old address which Rohde fully knew would be unlikely to bring the documents to Markham's attention.

13.8. HHJ Jackson held that there was a serious issue to be tried, namely whether Rohde had available during the adjudication a ready means of contacting Markham, which Rohde chose neither to use nor to communicate to the adjudicator. Judge Jackson further stated that if, after hearing evidence, it turns out that Rohde took a deliberate decision, depriving Markham of the opportunity to make representations in the adjudication, then this may be one of those rare, exceptional cases to decline to enforce an adjudicator's decision by reason of breach of natural justice.

13.9. In *Nageh v Giddings*[3] Giddings sought to have set aside a summary judgment granted in Nageh's favour enforcing an adjudicator's decision. Giddings argued, *inter alia*, that they were unaware of the adjudication proceedings or the summary judgment. HHJ Coulson found that Nageh duly complied with the requirements for the service of proceedings and held that there was no good explanation for Giddings' non-attendance at the summary judgment hearing. Although this was sufficient to dispose of the application, Judge Coulson further considered, *obiter*, the position if he were wrong and whether the judgment should have been set aside. Judge Coulson held that if that were his decision then it would be open to Nageh to seek to enforce the adjudicator's decision again. Gidding's principal complaint in respect of the adjudication also related to the service of documentation. The judge found that there was no reason why the adjudicator should insist upon more onerous service requirements than those set

3 Nageh v Giddings [2006] EWHC 3240 (TCC).

out in the civil procedure rules and no criticism could be made of the service of documents in the adjudication. Judge Coulson held that there was no suggestion that Nageh deliberately used the wrong address, or that there was some other address known to them where the documents could have been served on Giddings. In the circumstances, the documents had been served at two different addresses. Consequently, Giddings' application to set aside the summary judgment was dismissed.

13.10. Where a party is legally represented, service may be effected on its legal representatives. However, parties should ensure that the notice is sent to a solicitor who has authority to receive the particular notice.

13.11. In *Anglian Water v Laing*,[4] Anglian engaged Laing under a UK NEC contract that incorporated a notice provision at Clause 93.1 which provided that a party could serve a notice of dissatisfaction with an adjudicator's decision within four weeks. Disputes arose under the contract that were referred to adjudication and a Clause 93.1 notice was served on Laing's solicitors who had indicated during the adjudication that they would accept service of any documentation relevant to the adjudication. Laing's solicitors immediately passed on the notice to the relevant individual within Laing. A week later Laing's solicitors advised that they needed to take instructions from their client on whether they had authority to accept service, and simultaneously sought confirmation that Anglian had served the relevant notice of dissatisfaction directly on Laing. Clause 13.2 of the contract provided that a communication became effective when it was received at the last address the recipient notified for receiving communications or, if not notified, then at the recipient's address stated in the contract data. HHJ Edwards-Stuart held that compliance with the delivery mode specified in Clause 13.2 was the only means of achieving or securing effective delivery under the contract, because the communication only takes effect when it is received at the prescribed address. The judge held that the fact that the relevant personnel at Laing actually received the notice within the required four-week period (via their solicitors) was not determinative of compliance with Clause 13.2. However, the judge further considered whether Laing's solicitors' communication advising that they could accept service of any documentation relevant to the adjudication constituted a change of address notification. The judge held that such notification under Clause 13.2 is in itself a communication contractually required to be sent to the last-notified address. There was no restriction on notifying addresses for different purposes under the contract and an email

4 Anglian Water Services v Laing O'Rourke Utilities Limited [2010] EWHC 1529 (TCC).

confirmation from Laing's solicitors that they could accept service was sufficient notification that the solicitors' address was the notified address for service of adjudication-related documentation. The judge held that whilst the notice of dissatisfaction was not a document that formed part of the adjudication process itself, it was nonetheless notice relevant to the adjudication because it prevented the adjudicator's decision from becoming final, and accordingly the communication of the notice had been effectively served in accordance with Clause 13.2.

Weekends and public holidays

13.12. Section 10(3) provides for the computation of the specified number of days applicable in respect of the delivery of a notice to take account for weekends and bank holidays. However, there is no other guidance in the Act as regards what is to be understood as the last place of residence or body corporate's registered or principal office, all of which terms are provided for in the UK Act. As no similar provision is incorporated under the Irish Act this is not helpful.

14 Enforcing the adjudicator's decision

Section 6(11) of the Act

14.1. Section 6(11) provides that the adjudicator's decision shall be enforceable either by action or by leave of the High Court in the same manner as a judgment or order of the High Court.

14.2. Statutory Instrument No. 450 of 2016,[1] which came into operation on the 22 August 2016, provides for the Rules of the Superior Courts to be amended by insertion of Order 56B.

14.3. Rule 2 of Order 56B provides that an application for the leave of the court to enforce or to enter judgment in respect of an adjudicator's decision pursuant to Section 6(11) of the Act shall be made by originating notice of motion filed in the Central Office of the High Court.

14.4. The originating notice of motion must be grounded on an affidavit sworn by or on behalf of the applicant party who is seeking to enforce the adjudicator's decision. Rule 3(2) of Order 56B provides that the grounding affidavit must:

(i) identify the construction contract to which the adjudicator's decision relates;
(ii) exhibit to the affidavit the adjudicator's decision;
(iii) set out the basis on which the court should conclude that the adjudicator's decision is binding on the respondent;
(iv) confirm that the payment dispute has not been finally settled by the parties and that a different decision has not been reached by an arbitrator or court in relation to the adjudicator's decision; and
(v) set out the facts relied upon to demonstrate that the respondent has failed to comply with the adjudicator's decision.

1 SI No. 450 of 2016 – Rules of the Superior Courts (Construction Contracts Act 2013) 2016.

14.5. Rule 3(3) provides that copies of the originating notice of motion, grounding affidavit and any exhibits must be served on the respondent not later than 14 days before the date fixed for the hearing of the motion.

14.6. Rule 3(5) provides for any respondent to deliver replying affidavits, which must be filed in the Central Office of the High Court and a copy served upon the applicant (and on every other respondent) within seven days of the service by the applicant of the originating notice of motion. This is a short time frame within which to file a replying affidavit, but it is in keeping with the objective of seeking to achieve a speedy resolution of the payment dispute. The replying affidavit should set out precisely the grounds relied upon by the respondent to resist the applicant's claim for the reliefs set out in the originating notice of motion.

14.7. The applicant may then file a further affidavit replying to any matter raised by a respondent in a replying affidavit, and that further affidavit has to be filed in the Central Office of the High Court (and a copy delivered to each respondent) within seven days after the service on the applicant of the respondent's replying affidavit.

14.8. In advance of the hearing of the motion seeking summary judgment the applicant must file an affidavit giving the names and addresses of, and the place names and dates of service on, all persons who have been served with the originating notice of motion, grounding affidavit and exhibits. If any person who ought to have been served with the proceedings has not been served, the affidavit should state that fact and the reasons why they have not been served.

14.9. Where the court upholds an adjudicator's decision, the court's judgment may be entered in the terms of the decision. In addition, the court, on the application of any of the parties or of its own motion, may make such consequential further or other orders or give such directions as the court considers will facilitate an expeditious disposal of the summary judgment enforcement proceedings.

14.10. It is critical to the success of adjudication under the Act that the courts adopt a robust and supportive approach to applications for the enforcement of an adjudicator's decision. There is nothing to suggest that the Irish courts will not adopt a similar approach as the UK Courts, particularly on the basis that the respective adjudication legislation in Ireland and the UK is similar in its composition and there are many provisions of the Act which are almost identically worded to the equivalent provisions in the UK legislation. The courts have repeatedly stressed their support for arbitration, particularly since the introduction of the Arbitration Act 2010, and it has been on very rare occasions where the courts have sought it necessary to set aside an arbitrator's award but rather have repeatedly stated that it is only in extraordinary circumstances that they would seek to do so. It is very much hoped, and expected, that the courts will demonstrate a similar robust support for adjudication as they have done for arbitration, and all the indications are that this is likely to be the case.

Part 2

A real-world perspective of adjudication

Niall Lawless

15 Knowledge

15.1 Icarus

Among many talents, Daedalus was an architect and engineer. To escape imprisonment from his own invention, the Labyrinth,[1] he crafted wings from feathers bonded together with beeswax. Daedalus cautioned his son Icarus when using the wings, first of complacency and then of hubris:

> My boy, take care
> To wing your course along the middle air;
> If low, the surges wet your flagging plumes;
> If high, the sun the melting wax consumes:[2]

If Icarus were to fly too low the spray from the sea would moisten and weigh down the wings. If Icarus were to fly too high the sun would scorch the feathers and melt the wax. Unfortunately, ignoring guidance, Icarus flew high close to the sun, and the melting wax caused him to fall into the sea, where he drowned.

Adjudication is an environment often more pressured than arbitration or litigation, especially in terms of time limits, and it is easy to be "blown off course" and err. Because of that, good adjudicators have robust personalities. The parties have a right to expect an adjudicator to be knowledgeable in the law of contract, tort and evidence; in the legal framework regulating adjudication; in adjudication practice and procedure; and in decision-making and writing. Good adjudicators will be confident in their knowledge, but not excessively so. Equally, they will not limit what they are able to do with false modesty. They will use their own knowledge in an appropriate and judicious way. They will constantly update their knowledge and embrace the knowledge of those who are wiser. Good adjudicators will avoid complacency and hubris.

1 *Apollodorus: The Library*, Volume II: Book 3.10-16/Epitome (Loeb Classical Library, No. 122), translated by J. G. Frazer (31 January 1921) ISBN-10: 0674991362.

2 Ovid, *Metamorphoses* (43 BC – 17 or 18 AD), translated by Sir Samuel Garth, John Dryden, et al. London (1717).

15.2 Construction industry and adjudication framework

Twenty-first century construction contracting is fragmented and has many different flavours.

- *Conventional or traditional*: where the contractor quotes for work based on a bill of quantities, drawings, schedules and specifications. The design of the work is a pre-cursor to, and independent from site activity.
- *Design and build*: the client submits its requirements only in sufficient detail to enable the contractors to ascertain needs and submit their proposals. The contractor undertakes detailed design and construction that overlap.
- *Cost reimbursable*: the contractor tenders against a schedule of rates used to calculate the payments due. This type of contract is usually only entered into when either the project is of a large and repetitive nature or is novel and complex and cannot be fully quantified at the time of tender.
- *Management contracting*: at an early stage, the management contractor and the design team establish a clear brief, target cost and programme. The professional team agree an estimated target price for the works. The management contractor seeks competitive tenders from construction contractors for specific packages of construction work. The client pays the construction contractors through the management contractor.

The construction services sector comprises different professional groupings: architects, civil engineers, structural engineers, electrical and mechanical engineers, fire protection engineers, quantity surveyors and project managers. These professionals are employed directly by the facility or property owner, in separate specialist firms, or in larger multi-discipline firms. They come together usually on an ad-hoc basis to form design teams, working on bespoke projects, responsible for translating the client's business needs into bills of quantities, drawings, schedules and specifications. These documents allow the contractor to understand what is required and to quote for undertaking work.

The construction materials sector comprises companies that provide the actual building materials. Primary materials encompass aggregates, cement, brick, concrete, glass, steel, etc. Secondary materials encompass doors, flooring, partitions, sanitary wares, windows, etc. There will be specialist equipment such as air conditioning, building transportation, and fire and security subsystems.

It is widely acknowledged that well-managed and successfully delivered construction projects can improve the delivery of public and private services. Unfortunately, within the construction and engineering industry disputes are common and can often be difficult to avoid. Disputes may not only damage business relationships but have both a financial cost which can be high, and an emotional cost which can be debilitating. The financial costs are loss of opportunity and productivity costs, and include internal costs for the time and resources of key personnel; as well as direct costs of third-party advisors such as claims consultants, experts, solicitors and barristers. The emotional costs are those consumed being held prisoner by negative feelings such as anger, anxiety, disappointment, fear, frustration, injustice, resentment, revenge and sleep deprivation. The cost of emotional energy expended in disputes will never be recovered.

One of the most formative documents in respect of improving construction industry performance in the UK was the Latham Report "Constructing the Team" (1994)[3] which not only identified construction industry inefficiencies, but also condemned then-existing construction industry practices as "adversarial", "fragmented", "ineffective", "lacking respect for its employees" and "incapable of delivering for its customers". One of the Latham Report's key recommendations was that adjudication should be the **"normal form"** of dispute resolution.

This led to the UK being the first country to provide for the statutory right to adjudication by introducing the Housing Grants, Construction and Regeneration Act 1996 (the "English Act"). Adjudication is now available in Australia, Ireland, Malaysia, New Zealand, Singapore and the UK for most construction activities, and for the provision of professional services ancillary thereto. The experience of those countries that have embraced adjudication is that it has made their construction industry more effective and efficient, bringing consequent economic benefit.

For adjudication to be successful, the essential ingredients are:

1. Construction industry necessity
2. Willing policy-makers and legislators
3. A supportive judiciary.

Necessity within the construction industry

In 2006, construction activity in Ireland peaked accounting for approximately 25% of GNP. However, in 2010, when adjudication legislation was being considered, construction activity was in recession and it accounted for approximately 9% of GNP. This rapid reduction in Ireland's construction industry activity highlighted "the lack of formal contract arrangements and bad payment practices" such as "problems of delayed or non-payment for work done".[4]

3 The Latham Report "Constructing the Team" (1994).

4 Summary of the Regulatory Impact Analysis on the Construction Contracts Bill 2010, published by the Department of Public Expenditure and Reform (September 2011).

The Regulatory Impact Analysis conducted in respect of what was then the Construction Contracts Bill 2010 stated:

> While there is strong anecdotal evidence of the practice of delayed or non-payment having escalated in recent times, it should be noted that the problem is not new. It is reported that many firms, mainly SMEs, are experiencing serious difficulty in obtaining payment for work done. It is therefore important that where possible, payment transactions within this sector should be facilitated to ensure prompt payment of the correct amount.

Against a backdrop of deliberate non-payment or slow payment, the Irish construction industry had a problem that needed to be remedied.

Willing policy-makers and legislators

In industrialised countries, private sector and public sector construction is a major driver or measure of economic activity. Construction output is highly responsive to the economic cycle and is often at the behest of government public policy. Because governments are major construction industry customers and paymasters, wanting to control the disbursement of money, they can have vested interest in maintaining power, sometimes used abusively, to delay or reject the implementation of adjudication.

When the Irish construction industry "cried out" for government intervention to deal with the problems of deliberate non-payment or slow payment, there were willing policy-makers and legislators. In response to the need to improve cash flow, in May 2010, Senator Feargal Quinn introduced the Construction Contracts Bill in the Seanad Éireann as a Private Members' Bill. The stated aims of the legislation were to set up on a statutory basis a requirement for minimum contract terms and review procedures to be included in construction contracts:

1. to ensure prompt cash flow improving efficiency (amount and timing of payment); and
2. to allow swift resolution of disputes by way of adjudication, allowing projects to be completed without wasting time and money in litigation.

On 19 May 2010, during the second stage debate on the Bill, Senator Fergal Quinn advised that

> The main purpose of this Bill is to provide for a mechanism whereby prior notice of an intention to withhold sums from payments otherwise due to contractors must be given. Otherwise, payments must be made in full and/or the payee may suspend the provision of works and/or services under the construction contract until payment is made in full.

During that debate, Senator Joe O'Toole stated that adjudication should be "efficient, simple to understand and time bound" and Senator David Norris clarified that adjudication

> is different from the normal method of arbitration. It is straightforward, simple and quick. It is essential to have speed in deciding the amount that should be paid by one party on an interim basis. This will have the benefit of helping with cash flow difficulties.

The Construction Contracts Act 2013 (the "Act") was enacted by the Oireachtas on 29 July 2013.

A supportive judiciary

In respect of enforcement of an adjudicator's decision, Section 6 of the Act provides

> The decision of the adjudicator, if binding, shall be enforceable either by action or, by leave of the High Court, in the same manner as a judgment or order of that Court with the same effect and, where leave is given, judgment may be entered in the terms of the decision.

An adjudicator's decision is not always respected with timely compliance. If a party does not abide by an adjudicator's decision, the other party can enforce that decision by issuing proceedings in the High Court. The Rules of the Superior Courts (Construction Contracts Act 2013) 2016 provide for the procedure to be followed in applications for leave of the High Court to enforce or enter judgment in respect an adjudicator's decision.

These rules signal judicial determination to support adjudication, and should provide for a relatively fast procedure for the enforcement of an adjudicator's decision. Adjudication decisions are temporarily binding, and any subsequent arbitration or litigation may result in a different binding outcome. As adjudication legislation provides for a short period for the adjudicator to decide a payment dispute, where the adjudicator has decided the payment of money, it should be paid as ordered.

Enforcement proceedings should not be the forum for considering the merits of a payment dispute decided by an adjudicator, even if the decision is obviously wrong or unfair. To maintain confidence and good order in the adjudication system it is essential that the High Court deals with adjudication enforcement dismissing trivial albeit inventive arguments swiftly, and in a robust manner.

15.3 Array of dispute resolution methods

Every construction and engineering project involves one or several contracts and from time to time issues arise and escalate to disputes if not managed properly and promptly. Driven by the desire to find effective, reasonable, efficient and inexpensive ways to resolve disputes many in the industry are increasingly using alternative dispute resolution (ADR). ADR developments in the last 20 years include the introduction of statutory adjudication in six countries, court and legislative support for mediation prior to or in the middle of litigation as part of case management, and the rise of arbitration in domestic contracts but more importantly in those involving an international dimension.

There is a variety of dispute resolution methods available, and one way to consider them is as on an array. The array comprises consensual outcomes, non-compulsory outcomes and compulsory outcomes (see Figure 15.3). As the parties move along the array from left to right, the opportunity for them to craft a creative agreement which is not necessarily insistent on contractual entitlement diminishes.

The majority of commercial problems are routinely agreed and amicably settled through negotiation. This is despite the fact that few construction industry professionals have had any formal negotiation skills training. For many Western ADR practitioners negotiation skills training begins with the seminal text *Getting to Yes: Negotiating Agreement Without Giving In* by Roger Fisher and William Ury.[5]

This book is based on principled negotiation focusing on four core principles: 1) separate the people from the problem; 2) focus on interests rather

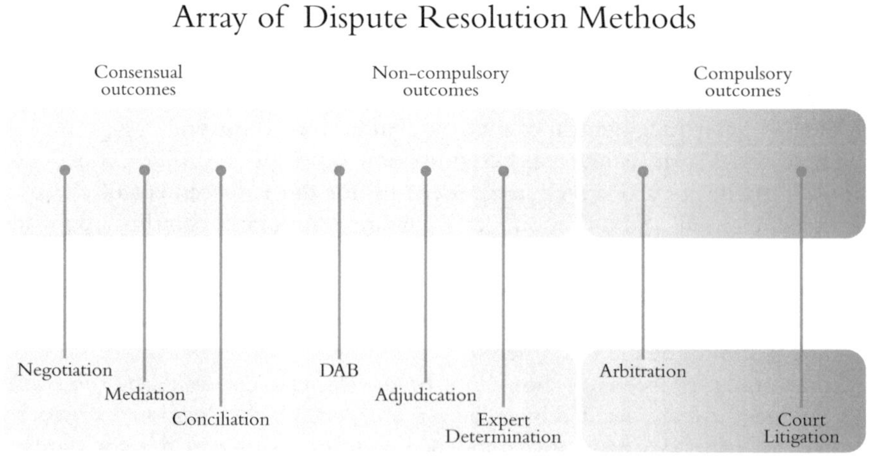

Figure 15.3 Array of dispute resolution methods.

5 Fisher, Roger & Ury, William with Patton, Bruce, *Getting to Yes: Negotiating Agreement Without Giving In*, 3rd edition, Penguin Books (3 May 2011) ISBN-10: 0143118757 ISBN-13: 978–0143118756.

than positions; 3) generate a variety of options before settling on an agreement; and 4) insist that the agreement be based on objective criteria. Fisher and Ury also identify three common obstacles to negotiation, and discuss ways to overcome those obstacles. The obstacles are what to do when the other party is more powerful, when the other party will not use principled negotiation, and when the other party uses dirty tricks.

When negotiation does not work, the parties may elect to mediate. Commercial mediation begins with the parties agreeing to mediate, and usually ends with the parties compromising their dispute. The parties are most often hardworking, common-sense business people motivated by revenue and contribution, and the desire to continue their future cooperation.

In Ireland the Mediation Act 2017 came into force on 1 January 2018. The objective of the Mediation Act is to support mediation as an attractive alternative to court proceedings, providing a statutory basis for the delivery of mediation services, and strengthening the status of mediation as a viable dispute resolution method. For example, Section 14 of the Act imposes a legal obligation on a "practising solicitor" to advise their clients, prior to issuing proceedings, to consider mediation as a means of dispute resolution.[6]

Mediation works best when there is balance and harmony between people and the participants' trust in the process. Mediation is successful because it uses a tried-and-tested process. The mediation process is confidential, private and structured. In general it has five stages: introduction, information exchange, option generation, negotiation, and conclusion. Information exchange and option generation are by far the most important.

Mediation is successful because it introduces the new dynamic of the mediator's skills. The mediator is a neutral third party who will assist the disputing parties to compromise their dispute using communication and negotiation skills, so that they can reach a solution mutually satisfactory to their interests. Essential parts of the role of the mediator are to be the guardian and guide of the mediation process, to facilitate the exchange of information, to help the parties reality-check their position, and leave no value on the table.

If the parties trust the mediator with confidential information, the mediator is in a unique position to assess if there is a zone of agreement, and guide the parties toward it. Mediation is not adversarial and works best where the parties are willing to cooperate to solve a shared problem together, and because nothing is agreed until everything is agreed in writing as a valid settlement agreement, it allows the parties to take risks when they come to deal with individual items.

Based on the premise that where a construction and engineering dispute revolves around issues of fact, the best mediation approach is evaluative carried out by an experienced construction and engineering professional, the Constriction Industry Council (CIC) in the UK has published an innovative

6 The Mediation Act 2017.

CIC Model Mediation Agreement and Procedure. The criteria for eligibility to join the CIC panel of mediators is "The panel will comprise of experienced accredited mediators who are members of CIC member organisations, and who have a minimum of 10 years' post qualification experience in their primary profession". Accordingly, membership of the panel is open to suitably qualified construction and engineering professionals in Ireland.[7] "Engineers Ireland" maintains a panel of mediators where one of the criteria for membership is to be a Fellow of Engineers Ireland or a Chartered Member of Engineers Ireland for at least ten years, or have achieved a similar status in an equivalent professional body.[8]

On 7 August 2019, the international standing of mediation as an ADR method received a significant boost when 46 countries signed the United Nations Convention on International Settlement Agreements Resulting from Mediation (the "Singapore Convention").[9] The Singapore Convention will come into effect on 12 September 2020, that is six months after the deposit of the third instrument of ratification. It will allow parties to mediated settlements to enforce the terms of those settlements in any of the convention's signatory jurisdictions, as if they were issued by a court. This will help enhance the cross-border enforceability of mediated settlement and businesses will benefit from greater certainty and assurance.

Conciliation usually takes place over a longer time than mediation, and the conciliator can make settlement recommendations that the parties may choose to take up. Conciliation and mediation are interchangeably referred to in many jurisdictions. However, on the whole, conciliation tends to be more evaluative in terms of the conciliator's more proactive and accepted role.

Conciliation in Ireland is more important than in many other countries as since the mid-1990s it has been included in almost all the standard conditions of contract, normally as a mandatory step prior to arbitration. The Public Works Contracts published by the Office of Government Procurement include a formal conciliation process. Although the parties are not required to accept any proposed or recommended settlement, the Public Works Contracts include the requirement that

> If the dispute is not resolved by agreement within 42 days after the conciliator was appointed, or a longer period proposed by the conciliator and agreed by the parties, the conciliator shall give both parties a written recommendation. The conciliator shall base the recommendation on the parties' rights and obligations under the Contract.

7 http://cic.org.uk/admin/resources/cic-model-mediation-agreement-and-procedure-first-edition-11th-june-2019.pdf.

8 www.engineersireland.ie/services/dispute-resolution/documentation.aspx.

9 www.singaporeconvention.org.

Where the parties know that the conciliator is required to base his recommendation on the parties' rights and obligations under the contract, they are less likely to share confidential information which may be key to unlocking the dispute.

Under the FIDIC Red and Yellow Book contract disputes should be referred to a Dispute Adjudication Board (DAB). The DAB is a panel of experienced, respected, impartial and independent reviewers normally convened before construction commences, and which through regular contact with problems and progress encourages the resolution of disputes at job level. Where a dispute cannot be resolved by the parties it is referred to the DAB for a decision.

Adjudication provides the parties with a quick provisionally binding decision on disputed matters. Adjudication is most often a statutory procedure by which any party to a construction contract has a right to have a dispute decided by an adjudicator. Adjudication is intended to be quicker and more cost effective than arbitration or litigation.

Expert determination can be available as a contractual agreement to jointly refer certain disputes usually of a technical or specialist nature to a third party, an expert in the relevant field for determination. Expert determination is confidential and the expert's decision is normally final and binding unless the parties provide otherwise and unless it is challenged. Rather than enter into expert determination on an ad-hoc basis, I strongly recommend using procedural rules such as the highly respected IChemE White Book "Rules for Expert Determination", fifth edition (2016).

In some ways, arbitration is more powerful than litigation. If the parties want a formal decision made by a third party which can be universally enforced, only international commercial arbitration offers this possibility. Under the 1958 New York Convention, arbitration awards will be recognised and enforced across the world, which is the essential advantage over court judgments.

As of April 2020, the New York Convention had 162 signatories[10] out of the 193 United Nations members. There are a few limited grounds on which recognition and enforcement may be refused by a contracting state. Arbitration is mostly confidential; the parties can choose the arbitrator, and help fashion a process that better meets their needs. Courts in most jurisdictions are arbitration friendly, and where required interim orders provide valuable support.

Litigation can be an efficient powerful option where both parties are within the same jurisdiction. In addition, the support of the courts can be invaluable when it is required to assist and supervise the other dispute resolution methods.

10 https://treaties.un.org/Pages/ViewDetails.aspx?src=TREATY&mtdsg_no=XXII-1&chapter=22&clang=_en.

15.4 Psychology of non-payment: why the players do not pay, and the consequences

The construction industry is particularly sensitive to cost. Seemingly oblivious to the risk profile of projects, bid profit margins are often only 2–5%. The hierarchical structure of the industry means that payment usually flows from the client to the main contractor, who then pays the project's subcontractors, who in turn pay their own subcontractors.[11] Where the employer or main contractor can simply say

> Well, I don't accept your account; therefore there is a "dispute", and the dispute must be referred to arbitration which is ordinarily long and tedious, then the subcontractor is put into considerable difficulties; he is deprived of his commercial lifeblood. … In the majority of cases, because one party or the other cannot wait any longer for the money, there is some kind of compromise, very often not based on the justice of the case but on the financial situation of one of the parties.[12]

The power imbalance between large and small businesses all too often results in "unethical relations and bullying". In the construction industry, "paying late or not at all is ruthless tactic … motivated by corporate greed".[13] In the absence of entitlement to adjudication and where the bargaining power of the parties is unequal, I have seen bespoke contracts that provide that during the tenure of the works the dispute resolution mechanism is arbitration, but a prerequisite to arbitration is mediation, to which both parties have to agree. However, when the paying party simply refuses to mediate there is no possibility of timely dispute resolution and the lack of payment problem is exacerbated.

The Latham Report identified that the UK construction industry suffered from adversarial attitudes and problems, and said that there was a need to change deep-rooted cultural attitudes.[14] The adjudication provisions of the English Act were enacted as a direct consequence of the Latham Report, and sought to deal with the endemic problems of poor payment practices and disputes in the industry. However, the culture of the industry is deeply ingrained and very difficult to change, and the industry operates opportunistically for its self-protection and the survival of individual firms.[15] There

11 House of Commons Business and Enterprise Committee Construction matters, Ninth Report of Session 2007–08, Volume I Report, together with formal minutes, printed 8 July 2008.

12 Ellis Mechanical Services Lid v Wates Construction Limited (1976) 2 BLR 57.

13 Final Report All-Party Inquiry into Late Payments in Small and Medium-Sized Enterprises Debbie Abrahams MP July 2013.

14 Constructing the Team by Sir Michael Latham Joint Review of Procurement and Contractual Arrangements in the United Kingdom Construction Industry July 1994.

15 Fellows, Richard & Liu, Anita, *Trust and Money: 20 Years of (No) Progress?* (June 2011) ISBN: 9789052693958.

remains a "deep-seated culture among main contractors of delaying, reducing or simply avoiding payment to their sub-contractors".[16]

In construction, the contractor is obliged to undertake work in accordance with the drawings and specification to the reasonable satisfaction of the contract administrator. The employer is obliged to pay the contractor for contract work which has been completed to the reasonable satisfaction of the supervising agents. On complex projects, particularly infrastructure-related ones, when unanticipated design changes and consequent unexpected costs arise, because of a lack of trust the parties often cannot address the contingencies without resorting to a legal remedy.[17]

Where change arises, the contractor has a responsibility to provide a valid accurate and timely claim with adequate supporting documentation. If the contractor over-claims advertently and excessively values its applications for payment, that may well amount to fraud. However, it will only rarely be the case that one can presume fraud where an invoice or bill is overstated. The contractor may believe that it is entitled to what it is claiming; there may be a simple and honest mistake in the formulation of the claim; and the claim may be based on a speculative but arguable point of construction of the contract or law.[18]

In the same way as they put together a tendering team, larger contractors often assemble claims preparation teams comprising commercial, engineering, legal and programming expertise. They can be sophisticated and are financially capable of procuring the services of claims consultants and lawyers. However, there is an old adage "as you sow so shall you reap", and because there are contractors who have well-deserved reputations for "claimsmanship"[19] – manufacturing claims which are without foundation – employers are often quite rightly wary of taking what is set out, at face value. Contractors should not be paid for difficulties caused by them, such as dealing with permits and permissions slowly; poor design and coordination; lack of documentation and planning; and deploying inadequate resources.

Employers and contractors withhold payment for a range of reasons; both parties can contribute, and external factors exacerbate the problem. For example:

- there is a mindset of expecting others to fund investment;
- construction was to be partly funded out of revenue, but circumstances have changed, and for reasons beyond control, money has not accrued in the way expected;

16 House of Commons Business and Enterprise Committee Construction matters, Ninth Report of Session 2007–08, Volume I Report, together with formal minutes, printed 8 July 2008.

17 Zaheer, Akbar, McEvily, Bill & Perrone, Vincenzo (1998). Does Trust Matter? Exploring the Effects of the Effects of Interorganizational and Interpersonal Trust on Performance, *Organization Science*, 9(2): 141–159.

18 SG South Limited v King's Head Cirencester & Anor [2009] EWHC 2645 (TCC).

19 Zack, J., Jr. (1993). "Claimsmanship": Current Perspective, *Journal of Construction Engineering and Management*, 119(3): 480–497. ISSN (print): 0733–9364.

- to improve cash flow and use money elsewhere;
- there is no contingency for additional costs;
- following the 2008 financial crisis the banking system was dysfunctional, failing to lend and charging excessive interest rates for overdrafts and loans;[20]
- to exercise control to obtain compliance with instructions or wishes;
- businesses are not being well managed;
- there is a shortage of management resource;
- lack of empowerment;
- supposed professionals are not willing to make decisions commensurate with their status;
- there are delays in valuation and certification;
- claims are defective or invalid;
- adequate supporting information has not been provided;
- there are disputes over entitlement.

In a project involving tunnelling, in line with international best practice the employer and the contractor shared risk on unforeseen ground conditions. The contractor met with the employer's highly qualified geologist to inspect a rock that had caused excavation problems. The contractor asks, "Well, is it a boulder?" If the answer were yes, it would create the opportunity to claim for additional costs. The geologist replies, "I cannot decide if it is a boulder; that is above my pay grade". The contractor asks, "Who will make the decision if it is a boulder?" The geologist answers, "My boss will make the decision". The contractor retorts "But your boss is a quantity surveyor!"

When asked about cash flow difficulties, contractors will emphasise that they suffer pressure at the front end of projects. For example, paying for insurance, which depending on the project can cost between 2–6% of the project sum. They can only recover this over the duration of the project.

Contractors will also stress that they suffer pressure at the back end of projects. There is evidence that some clients withhold retention money, not to protect themselves against poor-quality workmanship, but rather to use it for other purposes or just to earn interest.[21] Retention can be a cause of substantial cash flow problems. Retention is usually around 3–5%, with half being paid to the contractor upon completion of their project work, and half being paid on receipt of the final certificate or "making good defects" certificate. Retention can be more than the level of profit margin, often it is not paid automatically, and contractors have to pursue payment themselves. There can

20 Final Report All-Party Inquiry into Late Payments in Small and Medium-Sized Enterprises Debbie Abrahams MP July 2013.

21 House of Commons Business and Enterprise Committee Construction matters, Ninth Report of Session 2007–08, Volume I Report, together with formal minutes, printed 8 July 2008.

be pressure to agree retention release dates which are far in the future. Money costs money and so does time!

In the supply chain, subcontractors can be most vulnerable as they are often less well versed in both the law and claims procedures than main contractors are. One of the most disturbing examples of abusive payment practice is to be found in the January 2018 collapse of Carillion plc. Despite having signed the UK Government's Prompt Payment Code in 2013 "Carillion relied on its suppliers to provide materials, services and support across its contracts, but treated them with contempt. Late payments, the routine quibbling of invoices, and extended delays across reporting periods were company policy."

Carillion's exploitation allowed it to amass some £2 billion debt owed to its 30,000 direct subcontractors, indirect subcontractors, suppliers and other short-term creditors. Many of these businesses may have been unaware that they formed part of Carillion's supply chain until the insolvency prevented them from receiving payments owed. They are likely to get little back from the company's liquidation.[22]

Whatever the reason for lack of payment, it is easy to overlook the problem from a simple mathematical perspective. If the employer does not pay the main contractor €100,000 and the main contractor is working on a 2.5% margin, to be able to offset that shortfall the main contractor must execute €4,000,000 of work elsewhere without any problems.

15.5 Managing relationships

Most construction professionals enter the industry because they want to create things. Frequently construction professionals are paid less than other professionals are, but there is a tremendous personal satisfaction to be gained from being part of a team, designing and constructing buildings and structures. The end of a successful project is a bittersweet experience; the team has worked together for months or years. There may have been squabbles along the way, also some hard bargaining, but through compromise, no resources are being wasted on disputes. Together the project team has overcome many difficulties and obstacles, and the project is completing. There has been an outcome-focused culture of collaboration and innovation, and shared satisfaction that the team has achieved the project's quality, time and financial objectives. Some of the professional relationships that have formed will turn into lifelong friendships. However, there are demobilisation blues, familiar things are passing away, and there is sadness at parting. There can be uncertainty about what the future holds.

22 House of Commons Business. Carillion Second Joint report from the Business, Energy and Industrial Strategy and Work and Pensions Committees of Session 2017–19, HC 769, published 16 May 2018.

In the creation of this happy picture, essential ingredients help ensure site relationships work effectively. The employer's and contractor's executive management teams are the ones who should first share the project values vision, promoting a "one team" culture based on trust. Trust can be created through accumulated knowledge, the history of the relationship; through cultivating contacts and care; through fairness, honesty, predictability and reliability; through "taking each other seriously"; and through similar background or socialisation.[23] When problems arise, they create the opportunity for the development of trust.

There should be ethical contract formation and management. If there is recognition of mutual interdependence, the parties will avoid bespoke contracts designed to pass risk unfairly to the contractor; they are often a source of uncertainty. The overarching benefit of proper allocation of risks is "By extracting the risk that cannot be controlled by the contractor, the client can expect a range of tenders that concentrate on methods and cost of construction".[24]

I was involved in an engineering project worth about €600 million which offers a salutary tale. The employer met with the lowest tenderer, and as a condition of entering into the contract coerced a discount of about 5%. Some three years later during a claims team meeting comprising ten people, the project director remarked, "if they would give us the discount back, there would be no [expletive] dispute". Later I asked the project director how he managed the tender discount situation; he told me that if their preferred subcontractors and suppliers did not cut more than 5% off their prices, they did not use them.

The parties should recognise that profit is good and a legitimate driver of economic activity; it is most often the reason that the organisations want to work together. The owner will have a benefit from having the project completed, and the contractor will have a benefit, usually profit, from undertaking the works.

Often senior claim consultants advise contractors and employers that the best way to resolve disputes is to "get them off site". The irony of this is that they are recommending that disputes revert to the executive management teams with sole responsibility for the corporate culture that created them. At dispute conferences, consultants sometimes refer to Albert Einstein, and quote him as saying, "You can never solve a problem at the level on which it was created". As it happens regularly, this is clearly not true. However, Einstein's actual guidance was that "Problems cannot be solved by the same level of thinking that created them". This behoves the executive management teams responsible for the corporate culture, processes and systems, which created the dispute-breeding environment, to realise that whereas they might be called on to act as firefighters, they contributed substantially to setting the conflagration.

23 Gast, Dr Henrik (2010). Trust in Leadership. Some Observations on the Process of Trust Building in the German Bundestag and the German Cabinet, *Irish Journal of Public Policy*, 4(1): ISSN: 2009–1117.

24 Abrahamson, M. (1973). Contractual Risks in Tunneling: How They Should Be Shared, *Tunnels & Tunneling*, November: 587–598.

15.6 Right to adjudication

> What I have always understood to be required by the adjudication process was a quick, enforceable interim decision which lasted until practical completion when, if not acceptable, it would be the subject matter of arbitration or litigation.[25]

To provide for adjudication, the Act was enacted by the Oireachtas on 29 July 2013.[26] The Act introduced a new payment regime, and adjudication as a fast-track statutory dispute resolution mechanism. In July 2016, the Minister of State for Employment and Small Business published the final Code of Practice (the "Code") Governing the Conduct of Adjudications pursuant to the Act. The Act applies to **construction contracts** entered into after 25 July 2016.

By convention in UK adjudications, the parties are the referring party and the responding party. Under Malaysian adjudication law the parties are the claimant and the respondent. Section 1 of the Act says that a "construction contract" means "an agreement (whether or not in writing) between an executing party and another party, where the executing party is engaged for any one or more of the following activities...". The Code names a party who may serve a notice of intention as the "referring party", and it names the receiving party as the "responding party". The application form for the appointment of an adjudicator by the Chairperson of the Construction Contracts Adjudication panel uses the terms "applicant party" and "respondent party". Practitioners in Ireland may or may not follow UK nomenclature; but in line with the terms used in the Code, there is evidence that the terms "referring party" and "responding party" are being used in Irish adjudication. In any event, in this textbook I use the terms "referring party" and "responding party".

Adjudication is available in construction contracts (whether or not in writing) involving construction activities as defined by the Act. The definition of a "construction contract" covers all possible procurement paths and includes the provision of architectural, engineering, project management, etc. services ancillary to the construction contract. Construction activities do not include the manufacture or delivery to a construction site of equipment and materials. Adjudication is not available in contracts of employment, small value contracts, and contracts for construction dwellings occupied by one of the parties.

Whereas construction operations do not include the manufacture or delivery to a construction site of building or engineering components or equipment; materials, plant or machinery, or components for systems of heating, lighting, air-conditioning, ventilation, power supply, drainage, sanitation, water supply

25 Hansard HL, 22 April 1996, Volume No. 571, Column 989.
26 Construction Contracts Act 2013.

or fire protection, or for security or communications systems; where these items are supplied under a contract which also provides for their installation, they are covered by the definition of construction operations under the Act.

Adjudication in the UK does not cover the process engineering industry. During the period of UK parliamentary debate on adjudication, the process engineering industry made it very clear that it had not had the same dispute and payment problems as the construction industry as a whole and on that basis was excluded.[27]

However, disputes abound in the process engineering industry, and recognition of this omission frequently leads to bespoke contracts and standard forms of contract such as the Model Form of General Conditions of Contract (MF/1 Revision 4 Edition 2000), being amended and supplemented by special conditions providing for adjudication. Whereas the ethos of the Institution of Chemical Engineers (IChemE) is that it is better "to get it right" rather than "getting there fast", the IChemE Adjudication Rules (The Grey Book 4th Edition 2016) support the use of adjudication allowing for problems of a relatively minor nature to be resolved quickly, thus lessening the likelihood of a general collapse of working relationships or delays to the project programme, and a consequential escalation of the dispute to other areas of the contract.

Whereas both the Act and the English Act include "industrial plant" in their definition of construction operations, the English Act excludes operations for the "assembly, installation or demolition of plant or machinery" for the "production, transmission, processing or bulk storage (other than warehousing) of chemicals, pharmaceuticals, oil, gas, steel or food and drink". Therefore, it seems that the definition of construction operation in Ireland is much wider than the definition of construction operation in the UK, and captures process engineering works.

The Construction Contracts Bill (2010) included within its definition of construction activities "drilling for, or extraction of, oil or natural gas", as this was deleted from the definition of construction activities in the Act, it is arguable that such activities are not "industrial plant" and they are exempt from adjudication.

In Ireland, it is not possible to contract out of adjudication. The Act applies to a construction contract whether or not the parties to the construction contract purport to limit or exclude its application. A party to a construction contract has the right to refer any dispute relating to payment arising under the contract to adjudication by serving on the other party at any time a notice of intention to refer the dispute.

In arbitration it is widely recognised that allowing the parties to bifurcate their dispute into liability and quantum can be the most cost-effective and efficient way to proceed. Within a few months I was adjudicator in two

27 Hansard HC Debate, 7 May 1996, Volume No. 277, Column 51.

substantial disputes referred to adjudication where the parties did not want a decision on an amount of money, rather just a decision on principle. In one dispute the parties asked the adjudicator to decide what the conditions of contract were, and in another dispute the parties asked the adjudicator to decide the method of measurement which ought properly to be applied to thermal insulation work undertaken. The adjudicator's decision on these matters would allow the parties to move forward together. I consider the dominant and regrettable difference between adjudication under the Act and the English Act to be that parties in the UK can refer any dispute, whereas in Ireland the referral is limited to a payment dispute. This limits party autonomy, it constrains adjudication and reduces its efficacy, and can undermine the objective of the dispute being processed in the shortest time and at the lowest cost.

Arising under the contract

In a 2011 adjudication in Northern Ireland, the responding party raised a jurisdictional challenge that I did not have authority to decide what the terms and conditions between the parties were. It argued that deciding what the terms and conditions were was not a dispute arising under the alleged contract; rather it is a dispute in connection with the alleged contract. The referring party relied on *Fiona Trust v Yuri Privalov*, a case concerning arbitration. Considering a long line of cases the UK Court of Appeal said it was time for a fresh start, at least in the international commercial context. It decided that the words "arising out of" should cover "every dispute except a dispute as to whether there was ever a contract at all".

Although an arbitration case, it determined that the words "arising under" do not have a narrower meaning than the words "arising out of". The court said that the ordinary businessperson would be surprised at the nice distinctions drawn and the time taken up by argument in debating whether a particular case falls within one set of words or another very similar set of words.[28] I decided that the authority of *Fiona Trust v Yuri Privalov* although concerning arbitration did apply in adjudication, and that I could therefore decide disputes "arising out of" or arising "in connection with" the contract.

In 2014, I had to reconsider this approach when it was decided that because adjudication clauses were present or implied by reason of statutory intervention, *Fiona Trust* did not apply.[29] However, the law is a living thing and continues to evolve and in 2016 Mr Justice Akenhead while acknowledging "some uncertainty in this arguably important area of construction law" considered that adjudication cases should follow the direction of applying the *Fiona Trust* principles.[30]

28 Fiona Trust and Holding Company and Others v Yuri Privalov [2007] 4 All ER 951 (HL).
29 Hillcrest Homes Limited v Beresford and Curbishley Limited [2014] EWHC 280 (TCC).
30 J Murphy & Sons Limited v W Maher and Sons Limited [2016] EWHC 1148 (TCC).

At any time

The "at any time" provision ensures recalcitrant parties do not have an opportunity to block adjudication with artificial and unreasonable constraints. During UK parliamentary debate on adjudication, Lord Lucas stated that the words "at any time" were necessary. Otherwise

> it will be possible for a party bent on avoiding adjudication to insert a term which would allow notice to be given within an unreasonably narrow window, and we cannot allow that … I am of course aware that some have doubted the wisdom of allowing parties to refer a dispute to adjudication so long after work under the contract has ceased. However, as long as there is any possibility of disputes arising under a contract, parties will have to live with the fact that an adjudicator's decision may be sought. Indeed, there may be times, even at such a late stage, where it is desirable to have a quick and cheap procedure that can produce an effective temporary decision, particularly since this will not prevent parties from seeking a permanent decision through arbitration or the courts.

The UK Courts have noted that the "at any time" provision was carefully considered, and it was decided that adjudication legislation would not include any time limit, for the entirely rational reasons offered by Lord Lucas. Although there may be circumstances where a party is estopped or waives the right to refer a dispute to adjudication, there is nothing to prevent a party from referring a dispute to adjudication at any time, even after the expiry of the relevant limitation period. The phrase "at any time" means exactly what it says.[31] However, a company in liquidation cannot refer a dispute to adjudication. In such a case there would not be a dispute under the contract, there would be a dispute arising in the liquidation and therefore subject to the insolvency rules.[32]

It is possible to commence adjudication even if at the time of the reference, the dispute is already the subject of pending court proceedings. The UK Courts have considered arguments that it should not countenance two concurrent proceedings in respect of the same cause of action. Moreover, by starting court proceedings, that the claimant waived or repudiated the benefit of the adjudication provisions contained in the contract. Whereas a party cannot arbitrate and litigate, it can adjudicate and arbitrate, or adjudicate and litigate. Referring a dispute both to adjudication and to an arbitrator or the court are not mutually exclusive routes to dispute resolution.[33]

31 Connex South Eastern Limited v M J Building Services Group Plc [2005] EWCA Civ 193.

32 Michael J Lonsdale (Electrical) Limited v Bresco Electrical Services Limited (in Liquidation) [2018] EWHC 2043 (TCC).

33 Herschel Engineering Limited v Breen Property Limited [2000] EWHC Technology 178.

15.7 Is adjudication a profession?

Historically three great professions were assumed: the clergy, the lawyers and the physicians.[34] Modern historians and sociologists recognise "profession" in a much wider ambit. They characterise "profession" as including: a sense of commitment and vocation; an appeal to expertise; a reliance upon theoretical as well as practical knowledge and skills; a professional ethic; internal control of training, recruitment, placement and discipline; growing organisation, and flowing out of that a certain autonomy in the workplace and an evident "*esprit de corps*".[35] Professions do not have to agree, be in harmony or be uniform, and they can have different segments.

It is suggested that adjudication has not just successfully infiltrated construction dispute resolution around the world, but that with the current exception of Ireland adjudication is killing arbitration in the jurisdictions where it is available.[36] Because of this, many adjudicators were first arbitrators. Although adjudicators do not have the burden of finally determining the parties' rights as arbitrators do, adjudicators and arbitrators undertake similar work. In some ways adjudication is more demanding than arbitration, for example, such as having to respond to externally imposed time pressures. Like established professions, construction adjudication has evolved not in a vacuum, but in response to basic industry and social forces.

In principle if arbitration is a profession, then so might adjudication be. Arbitrators often refer to arbitration as a second profession, but so far, it is not commonplace for adjudicators to refer to adjudication in the same way. Adjudicators and arbitrators share similar duties and obligations to the parties; duties imposed by law, duties imposed by the parties, and ethical duties. Most adjudicators are concerned with acting in a professional way, and like arbitrators they are already professionally qualified. Whether coming from construction and engineering or the law, adjudicators and arbitrators require similar education and training. Each is qualified and accepted onto reputable appointment panels only by demonstrating, through a process of peer assessment, that they meet the required standards.

"The Guidance" by the Chartered Institute of Arbitrators (CIArb) describes a construction adjudicator as a Fellow of the CIArb who has demonstrated to an interview panel advanced knowledge and understanding of construction adjudication and its practical application, evidencing a professional approach to

34 Fox, George & Harvey, Thomas Edmund; Penney, Norman (Editor), *The Short Journals and Itinerary Journals of George Fox: In Commemoration of the Tercentenary of His Birth (1624–1924)* (Cambridge Library Collection – Religion), Cambridge University Press (31 October 2010) ISBN-10: 1,108,015,328 ISBN-13: 978–1108015325.

35 O'Day, Rosemary, *The Professions in Early Modern England, 1450–1800* (Themes in British Social History), 1st edition, Longman (21 November 2000) ISBN-10: 0582292654 ISBN-13: 978–0582292659.

36 Rana, Rashda, Is Adjudication Killing Arbitration? 75 Arbitration No. 2, pp. 223–230, May 2009.

parties and the public interest. Construction adjudicators should be competent to serve the industry in an ethical manner, and this involves an assessment of knowledge, skills, attitudes and experience. The CIArb peer review of arbitrators and adjudicators is similar, and equally demanding.

The jurisdictions that have enacted adjudication (also referred to as security of payment) law have strong historic links with the UK; five of the countries (Australia, Malaysia, New Zealand, Singapore and the UK) are part of the Commonwealth. In the UK, a clear distinction can be drawn between being designated professional and a professional who enjoys a legitimate Chartered status. In the UK, only institutions incorporated under Royal Charter by the British Monarch can award Chartered status. Whereas, since 1999, the Chartered Institute of Arbitrators confers the title "Chartered Arbitrator" as a gold standard, there is no recognised body under Royal Charter, empowered to confer the title "Chartered Adjudicator". In some occupations, membership of a professional body can be a requirement of being licensed to practise. This is generally not so in adjudication and arbitration, which is consistent with the principle of party autonomy.

Affected by cultural norms and geography, the market for providing adjudicator accreditation, certification, education and training, and having the power to nominate adjudicators is competitive. Of concern to adjudicators is whether the accreditation and training they have invested in is transportable – will it be recognised by other bodies, which have the power to nominate? Regrettably, adjudication accreditation and training is often not transportable, domestically or internationally.

Associations in different countries support adjudicators; but there is no single global organisation to ensure that standards are upheld. While control of the title "Chartered Adjudicator" is appealing, in a world of vested interest, and because of competition between professional bodies, it is unlikely that adjudicators will be able to obtain it.

15.8 Weltanschauung

In the role of adjudicator, evaluation of evidence is important. Our belief systems (how we make sense of things), and our personal prejudices affect the way that we see and understand things. It is not enough to ascertain "what do you believe?", it is important to also understand "why do you believe it?".

Truth can be defined as a finding that is true or in accordance with fact or reality. A universal truth is one where a capable, unbiased individual could verify that truth, i.e. that "it is so". For example, it is a universal truth that gravity is the force that attracts a body towards the centre of the earth, or towards any other physical body having mass. Culture or our personal perceptions do not affect a universal truth, and in adjudication, universal truths should not be controversial. Unsurprisingly, the adjudicator is seldom asked to consider truths that are immutable or universal, but is often asked to consider truths that are personal, and which result from conscious and unconscious bias.

In mediation, I describe a plausible situation as in the 2010 World Cup when England was playing against Germany. An English player, Wayne Rooney, dribbled the football to the German penalty area, and when tackled tumbled to the ground. Watching the event in real time, millions of English TV spectators scream "penalty" and millions of German TV spectators scream "dive".

The Oxford Dictionary defines the German "*Weltanschauung*" as a particular philosophy or view of life; the world view of an individual or group. This is often the key ingredient in the "why do you believe it?" "*Weltanschauung*" helps inform if it was a dive or a penalty. If millions of people can differ vehemently about something so clear in real time, how less likely are they to agree what happened in opaque circumstances, several years before?

Adjudication is a dispute resolution method commonly considered adversarial and contentious. Different truths held with equal conviction will often confront adjudicators. Therefore, adjudicators should consider whether it is in their role to decide what the truth is. If the parties have lawyers, can the adjudicator expect them to help? What opportunity is there for expert reports submitted by the parties to assist the adjudicator? Can the adjudicator realistically ask for expert opinion on his own accord?

If the adjudicator believes that his role is to decide what the truth is, the adjudicator is unlikely to enjoy the support of lawyers in his quest. Zealously partisan, lawyers want their client to prevail in disputes using any legal means. Frank summarised the role of lawyers:

> In short, the lawyer aims at victory, at winning in the fight, not at aiding the court to discover the facts. He does not want the trial court to reach a sound educated guess, if it is likely to be contrary to his client's interests. Our present trial method is thus the equivalent of throwing pepper in the eyes of a surgeon when he is performing an operation.[37]

Supposedly, British empire-building has enlightened the jurisdictions which enjoy construction adjudication. Lord Macaulay said

> It has always been held, in the most enlightened nations, that a tribunal will decide a judicial question most fairly when it has heard two able men argue, as unfairly as possible, on the two opposite sides of it; and we are inclined to think that this opinion is just. Sometimes, it is true, superior eloquence and dexterity will make the worse appear the better reason; but it is at least certain that the judge will be compelled to

37 Frank, Jerome, *Courts on Trial*, Princeton, NJ: Princeton University Press (1950), pp. 80–102.

> contemplate the case under two different aspects. It is certain that no important consideration will altogether escape notice.[38]

Experience suggests that the greater the intensity of the dispute, the more unfairly the parties will argue the merits and the truth can become more elusive and illusive. The role of the adjudicator is not to seek the truth, but to make the best decision possible in the circumstances.

15.9 Common law and civil law

Construction is increasingly international. Different countries have different legal systems and this can affect the conduct and expectations of the parties and their advisors in adjudication, and how it will be conducted. The two most common legal systems are the common law and civil law legal systems. Typically, former British colony countries use common law. Former French, Dutch, German, Portuguese and Spanish colony countries use civil law. England and Wales have a common law legal system, as does Ireland.

In adjudication some of the important differences that arise are in respect of contract formation, conduct of the proceedings, and provenance of the law.

In common law countries, there is extensive freedom of contract and fewer provisions are implied into the contract by law than in civil law countries where there is less freedom of contract and many provisions are implied by law.[39] Generally, in common law countries, as long as the underlying promise of performance is enforceable, the contract scope and structure can be wide and up to the contracting parties. In civil law countries, the structure of the contract is important, because every concept must fit into distinct legal categories that define the party's relationship to other legal principles.[40]

In common law, there are three basic essentials to the creation of a contract: (i) agreement (comprised of offer and acceptance); (ii) intention to create contractual relations; and (iii) consideration. In common law countries there can be no binding contract without consideration. In some civil law countries, consideration is not a requirement for certain types of contract to be valid.

As discussed in more detail later, common law countries use an adversarial system to resolve disputes, and civil law countries use an inquisitorial system for that. The adversarial system is a contest between two competing parties

38 Macaulay, Thomas Babington, *The Miscellaneous Writings and Speeches of Lord Macaulay*, Contributions to the Edinburgh Review, Volume II History (May 1828).

39 Nouel, Gide Loyrette, World Bank Toolkit (2006) – Approaches to Private Participation in Water Services, presentation to IFC on Some Differences between Civil Law and Common Law in a "nutshell", 2007. http://ppp.worldbank.org/public-private-partnership/legislation-regulation/framework-assessment/legal-systems/common-vs-civil-law.

40 Meldrum, Christopher, Civil Law Contracts. http://author.acc.com/legalresources/quickcounsel/clc.cfm.

where a non-interventionist decision maker will decide the dispute based on the evidence and submissions presented by the parties. The party's case will include legal precedents, which it says is authority for the decision maker to rule in its favour. In the inquisitorial system a non-passive decision maker endeavours to discover the facts, and to determine the law to apply to decide the dispute. The decision maker has the authority to examine the facts of a case independently of evidence provided by the parties. The lawyers and the decision maker will refer to written law as the basis for making a decision.

In common law countries, one of the main sources of law are cases where other judges have formulated and refined the relevant principles of law. In civil law countries, the source of law is legal code, regulations, rules and statute. Legal reasoning in common law and civil law countries is different. In common law countries, judges derive principles or rules of law from other judgments, and apply those to the case in hand. In civil law countries, judges start with the stated general principles or rules of law in legal codes, and apply those to the case in hand.[41]

Both the UK and Ireland are common law countries. The Construction Contracts Act 2013 is broadly modelled on the UK's Housing Grants, Construction and Regeneration Act 1996, the Construction Contracts (Northern Ireland) Order 1997 and the Local Democracy, Economic Development and Construction Act 2009.

During the second-stage debate of the Construction Contracts Bill in the Seanad Éireann on 19 May 2010, a significant advantage of this approach was recognised by Senator Cecilia Keaveney who said

> It is suggested that it is important to keep the statute law that will be introduced in the Republic of Ireland in the same wording and layout as that used in the United Kingdom and Northern Ireland. Some 400 cases have been decided on adjudication in the United Kingdom over the past 15 years. Clearly, there is a great wealth of case law and if the Irish Act were to use the same wording, there would be much more certainty as to how the courts here might interpret it.

As the Irish and UK legislation have similar provisions, there is therefore a significant body of persuasive, although non-binding, case law from the UK which will be helpful in understanding how the Act might be applied in practice. Because of that I refer extensively to UK adjudication case law in this text.

41 Apple, James G. & Deyling, Robert P., *A Primer on the Civil-Law System*, Federal Judicial Center (1995) ASIN: B00010XEHK.

15.10 Case law and judicial precedent

In commercial disputes, there must be a "cause of action". That is, the facts of the dispute asserted by one party must be such that the law recognises the other party as having a potential legal liability. This means matching the facts of the dispute to established legal principles. In common law systems, case law contains legal principles. When judges decide disputes, they can create and develop case law. Case law has a hierarchy depending on its genesis in the court system.

In Ireland, in terms of seniority, the courts rank as follows (from most to least senior) the Supreme Court, the Court of Appeal, the High Court, the Circuit Courts and the District Courts. The basic rule is that a court must follow the precedents from a higher court, but they are not bound to follow decisions from courts lower in the hierarchy, albeit that those decisions may be of persuasive authority. It is necessary for each lower tier to loyally accept the decisions of the higher tiers.[42] Also, the Court of Justice of the European Union has determined that its decisions in respect of the interpretation of EU law are binding on all courts in the member states.[43]

This rule provides the parties and their advisors with some certainty as to what the law is. It accords with the doctrine "*Stare Decisis*", a Latin term meaning "to stand by that which is decided". It is in the interests of consistency and predictability that disputes with the same facts should have the same decided outcome. Case law providing reasoned judicial decisions given in earlier disputes supports these objectives.

In adjudication, the parties usually cite legal principles from judgments they choose. The adjudicator must be adept at deciding if the facts in the present dispute are analogous to the facts in the cases cited, and therefore if the legal principles asserted should apply.

To achieve this, adjudicators must be able to distinguish the "*ratio decidendi*". Based on the facts of the dispute, the "*ratio decidendi*" is the rationale or reason for the decision; it is the legal principle that the judge used to decide the outcome of the case. The "*ratio decidendi*" is not always obvious, and to be able to extract and represent it in a way which compels the adjudicator's decision is a valued legal skill. For example, a party may say that their dispute is "on all fours" with a particular case, and that because the legal issues are identical (or so close as to make no difference), the adjudicator must reach the same conclusion.

The "*ratio decidendi*" is to be distinguished from other parts of the judgment known as "*obiter dictum*". *Obiter dicta* are comments or opinions of the judge made "by the way". *Obiter dicta* do not create a binding precedent. However,

42 Cassell & Co Limited v Broome and Another [1972] UKHL 3; [1972] 2 WLR 645; [1972] AC 1027.

43 Van Gend en Loos [1963] ECR 1 Costa v ENEL [1964] ECR 585; Internationale Handelsgesellschaft [1970] ECR 1125.

they offer persuasive authority, and adjudicators can consider and apply them if there are no clear or established *ratio decidendi* on a matter before them.

The British and Irish Legal Information Institute (BAILII) provides comprehensive British and Irish case law and legislation, European Union case law, Law Commission reports, and other law-related British and Irish material.[44] The Irish Legal Information Initiative (IRLII) provides a database of "leading Irish cases" classified by subject.[45] These cases are also available on the BAILII website.

15.11 Adversarial and inquisitorial systems of law

In common law countries, resolution of commercial disputes normally involves an adversarial approach. The adversarial approach involves the parties competing against each other and presenting facts, legal arguments and evidence in a way that is favourable to them. It is for each party to test or rebut the accuracy, relevance and sufficiency of the other party's contentions, before an impartial decision maker. They can do this by attacking and undermining the opposing party's position, or by advancing their own positive case. The decision maker remains neutral, ensuring that the contest is fair according to procedural guidelines. Adjudication is conducted in accordance with the rules of natural justice. The adversarial system allows the party who is best able to convince the decision maker that its perspective is the correct one, to win. However, if in the dispute parties are not equal in sophistication or resources, one side can win not because of the merits, but rather because skilful advocacy or guile distorts the outcome. There are typically strict rules of evidence that need to be complied with, for example in respect of hearsay evidence.

In civil law countries, commercial disputes are normally resolved using an inquisitorial approach. An essential premise in the inquisitorial approach is that the independent decision maker is best placed to seek and find the truth. The decision maker largely controls the conduct and content of the process. The decision maker will gather facts to assist in reaching the correct verdict, and has a duty to look for and evaluate evidence, both beneficial and deleterious. The powers of the decision maker are wide ranging, and not only include seeking evidence but also pursuing lines of inquiry it feels are important whether or not advanced by the parties.

Proponents of the inquisitorial approach say that one of its main advantages is that the adversarial approach can lead to injustice when one of the parties is not adequately skilled or represented. However, proponents of the adversarial approach say that where the role of the decision maker is both to investigate and judge, there is a danger that the decision maker will be, or be seen to be, partial to one side or the other. This is because in asking questions, the

44 www.bailii.org.

45 www.ucc.ie/academic/law/irlii/index.php.

decision maker can adduce germane evidence that influences the decision in favour of a party.

The parties usually commence adjudication using an adversarial approach. However, adjudication law can facilitate the adjudicator in adopting a hybrid approach, combining features of the adversarial and inquisitorial systems. For example, the Code of Practice Governing the Conduct of Adjudications, issued pursuant to the Act, empowers the adjudicator to take the initiative in ascertaining the facts and matters required for a decision such as the law that applies. Adjudicators can, within reason, request documents and statements from either party, they can meet with and question the parties and their representatives, and provided that the adjudicator has notified the parties of their intention, appoint experts, assessors or legal advisors.

Adjudicators should tread cautiously when exercising inquisitorial powers; they can unwittingly create snares and traps and place themselves in procedural difficulties. In an adversarial system, the parties are entitled to exploit such opportunities fully without hesitation. Adjudicators should be aware that the clumsy use of inquisitorial powers may lead to a losing party alleging that the adjudicator has breached principles of natural justice and as a consequence forfeited jurisdiction.

15.12 Burden and standard of proof

It should not surprise adjudicators or other professionals assisting the parties in adjudication that the phrase "the burden of proof" expressed by the Latin maxim "*ei qui affirmat non ei qui negat incumbit probatio*: he who asserts a matter must prove it, but he who denies it need not prove it" has its origins in the Roman legal system. Burden of proof is a sensible proposition, as the party asserting that something happened should have evidence to prove that it did.

The "burden of proof" is different to the "standard of proof". In adjudication, the standard of proof is a preponderance of evidence, so that on the "balance of probabilities" the fact is more likely to have happened than not. If the adjudicator can say that it is more probable than not, the burden of proof is discharged. However, if the probabilities are equal, the burden of proof is not discharged.[46]

In a family case, Lord Hoffmann said

> The law operates a binary system in which the only values are 0 and 1. The fact either happened or it did not. If the tribunal is left in doubt, the doubt is resolved by a rule that one party or the other carries the burden of proof. If the party who bears the burden of proof fails to discharge it, a value of 0 is returned and the fact is treated as not having happened. If he does discharge it, a value of 1 is returned and the fact is treated as having happened.[47]

46 Miller v Minister of Pensions [1947] 2 All ER 372.
47 Re D (Fact-Finding Hearing: Medical Treatment) [2014] EWHC 121 (Fam) (Mostyn J).

There is no room for ambiguity, if the disputed facts are important, the adjudicator must decide whether or not it happened, and not that it might have happened.

An adjudicator should decide a disputed issue by resort to the burden of proof only in exceptional circumstances. The adjudicator should only resort to the burden of proof if after having striven to do so, they cannot reasonably make a finding in relation to a disputed issue. It should be clear to the parties that the adjudicator has tried to decide a particular matter based on the evidence, and the adjudicator should provide reasons for not being able to do so.[48]

15.13 Evidence in adjudication

Evidence is to prove or disprove something, and is information or material helpful in forming a conclusion of judgment. In court, rules of evidence control what evidence is admissible, and thereby reduce the cost and time taken to litigate. To be admissible, evidence must be relevant. Relevance is judged by reference to the issue that the court is called upon to decide and "evidence which is irrelevant and cannot affect the outcome" is excluded.[49] "Evidence is relevant if it is logically probative or disprobative of some matter which requires proof" and "which makes the matter which requires proof more or less probable".[50]

Adjudication is an adversarial system and evidence on disputed issues of fact can include the production of contemporary documents, written and oral testimonies of witnesses of fact, written and oral testimonies of expert witnesses, inspection of the subject matter of the dispute, and the results of any experiments or tests carried out.

In adjudication the parties do not always constrain themselves to considering what evidence is relevant, and unhelpfully provide a wide range of extraneous material. Sometimes the adjudicator may receive an appeal that he request some additional documents, for example emails or letters that seem to be missing from a chain of correspondence. In the tight time scales for conducting adjudication, the adjudicator will rarely have to deal with arguments as to whether documents requested enjoy litigation privilege.

However, sometimes the adjudicator receives documents that are privileged. On more than one occasion, parties have wittingly or unwittingly, sent me "without prejudice" correspondence, for example, a copy of an offer to settle. If done wittingly, the objective could be to show that a party is behaving in a reasonable or unreasonable way. Having access to the information raises the spectre that the adjudicator is biased. Faced with such a circumstance the adjudicator should remove any offending documentation

48 Stephens v Cannon [2005] EWCA Civ 222.
49 O'Brien v Chief Constable of South Wales Police [2005] UKHL 26.
50 Director of Public Prosecutions v Kilbourne [1973] AC 729.

from the bundle, and make it clear that he will not consider it, and it is therefore irrelevant to their decision. Sensible proactive case management will protect the adjudicator from any allegation of a lack of impartiality.[51] Parties are strongly discouraged from deploying "without prejudice" communications in adjudication.[52]

Adjudicators often make decisions considering only the written testimonies of witnesses of fact. If the parties' witness statements conflict, the adjudicator can compare them to documentary evidence to expose contradictions, discrepancies or inconsistencies. The adjudicator can request to meet the party's witnesses to explore important differences that cannot be reconciled using only the documents.

Some decry the use of the evidence of expert witnesses in adjudication, but it is often very helpful in disputes requiring detailed engineering, technical, scientific or programming knowledge. A report provided by an experienced, suitably qualified expert can help the adjudicator understand complex matters, and thereby reach the best decision. Equally, some expert reports can exasperate. For example, in a dispute concerning the replacement of tens of thousands of Rosemary clay roof tiles, one party's expert said that a fair and reasonable rate was €4.00 per tile and the other party's expert said €6.00 per tile. Each expert referenced a credible price book source. In the adjudicator's decision it is not acceptable for him to simply state that he prefers the evidence of expert A to the evidence of expert B. The adjudicator should distinguish and clearly state why he prefers the evidence of one expert to the other.[53]

Adjudicators must consider any defence properly put forward, but this does not mean that the adjudicator must address each aspect of the evidence submitted by the parties. The adjudicator can decide what evidence is admissible, and helpful or unhelpful in making his decision.[54] No adverse inference is to be drawn, because a party does not offer testimonies of witnesses of fact or expert witness evidence. It is in the role of the adjudicator to decide if evidence is credible or trustworthy. Adjudicators should reach their conclusion, by applying their individual knowledge of the world to the evidence before them.[55]

Adjudication rules can provide that the adjudicator is not required to observe any rule of evidence, procedure or otherwise, of any court.[56] However, adjudicators and the professionals who advise the parties in adjudication should be more than familiar with the rules of evidence, and apply the

51 Volker Stevin Limited v Holystone Contracts Limited [2010] EWHC 2344 (TCC).
52 Ellis Building Contractors Limited v Vincent Goldstein [2011] EWHC 269.
53 Stephens v Cannon [2005] EWCA Civ 222.
54 Cynthia Jacques and Elsie Jacques Grombach (t/a C&E Jacques Partnership) v Ensign Contractors Limited [2009] EWHC 3383 (TCC).
55 Regina v Dennis John Adams [1996] 2 Cr. App. R. 467.
56 Institution of Civil Engineers Adjudication Procedure, 30 April 2012, ISBN 978-0-7277-5779-1.

underpinning principles. If you are assisting the parties in adjudication and responsible for collating evidence, the rules of court can provide helpful best practice guidance.[57]

15.14 Independence and impartiality

The United Nations Universal Declaration of Human Rights and the International Covenant on Civil and Political Rights proclaim that everyone should be entitled to a fair and public hearing by a competent, independent and impartial tribunal established by law. Judges are required to uphold the integrity and independence of the judiciary by avoiding impropriety and the appearance of impropriety in all their activities. They should decide matters in accordance with an impartial assessment of the facts without improper influences, direct or indirect, from any source.[58]

Although adjudication is generally between bodies corporate and is a private dispute resolution process, the above aims and principles are enshrined in Irish adjudication law. The Code of Practice Governing the Conduct of Adjudications requires that the adjudicator in any payment dispute under the Act shall be impartial, independent and only adjudicate where satisfied that no actual conflict of interest exists.[59]

The Code requires that prior to accepting appointment the prospective adjudicator must write to the parties to ask them to disclose any information indicating any potential conflict of interest that may arise from their appointment as adjudicator. Although it is not prescribed in the Code, it seems eminently sensible that the adjudicator should notify the parties of any conflict of interest or external factors, which they believe would give rise to a reasonable apprehension of bias; or any conflict or factors that arise during the adjudication. If a potential conflict of interest is disclosed by any party (or the adjudicator), the prospective adjudicator may accept the appointment only with the consent of the parties, and on satisfying any professional and/or ethical concerns the adjudicator may have. The adjudicator must observe the principles of procedural fairness and act impartially and independently and without bias.

Impartiality means the absence of bias or prejudice in favour of, or against parties, as well as being open-minded in considering and deciding the issues. Independence means the adjudicator's ability to act free from any control or influence, other than those established by law.

57 UK Civil Procedure Rules – Practice Direction 32 – Evidence. www.justice.gov.uk/courts/procedure-rules/civil/rules/part32/pd_part32.

58 Beijing Statement of Principles of the Independence of the Judiciary. 6th Biennial Conferences of Chief Justices of Asia and the Pacific. Beijing, 19 August 1995.

59 Construction Contracts Act 2013, Code of Practice Governing the Conduct of Adjudications (21 July 2017).

Adjudicators with integrity act impartially, independently and without bias. Integrity refers to the adjudicator's character, ethics and general reputation in the community. Adjudicators with integrity have sound character avoiding both impropriety and the appearance of impropriety in their personal and professional lives. They demonstrate fairness and independence, honesty and trustworthiness; they are even-handed and have a commitment to equality.

Before I act as adjudicator, I confirm I have no conflict of interest as with either of the parties. I confirm that I know of no reason why I cannot be properly appointed as adjudicator, and know of no matters which ought to be drawn to the attention of the parties which might appear to call into question my impartiality and/or independence. Sometimes I can go further and confirm I have no prior knowledge of Party A, Party B or the project from which the dispute has arisen. However, often the parties and the project are well known, so this further clarification is not always possible or indeed necessary.

It is a fundamental principle of adjudication that the adjudicator shall be impartial and independent of the parties. In arbitration, the IBA Guidelines[60] are widely accepted as a means of assessing impartiality and independence. An objective conflict of interest exists from the point of view of a reasonable third person having knowledge of the relevant facts and circumstances. The IBA Guidelines give practical guidance as to what can give rise to a conflict of interest requiring disclosure. It has a "Red List" (waivable and non-waivable), an "Orange List", and a "Green List" providing illustrations of situations in decreasing order of seriousness, where depending on the facts of a dispute, there may be justifiable doubts as to impartiality or independence such that there is a duty to disclose them to the parties. The courts will take account of the IBA Guidelines when determining challenges to impartiality in arbitration.[61] There is no reason why the courts will not do the same in adjudication. In adjudication, before appointment, there is very limited time for the adjudicator to go through a process of disclosure and procedurally it would be difficult to do so. Therefore, if a prospective adjudicator has doubts as to impartiality or independence, the adjudicator should immediately decline to act.

After appointment as adjudicator, conflicts can arise. For example, when appointed the adjudicator may have no information about who the responding party's representative is. If, say, a week into the adjudication, the adjudicator receives a letter confirming the responding party's representative, bias or prejudice may arise in favour of, or against that party. If the conflict is "Red List", the adjudicator should resign giving notice and an explanation to the parties. The parties are jointly and severally liable for the payment of the reasonable fees, costs and expenses incurred by the adjudicator up to the date of

60 IBA Guidelines on Conflicts of Interest, 23 October 2014, ISBN: 978-0-948711-36-7.

61 Sierra Fishing Company & Ors v Farran & Ors [2015] EWHC 140 (Comm).

resignation.[62] If the conflict is on the "Orange List", the adjudicator should write to the parties disclosing the conflict and ask the parties to waive the conflict of interest in writing within, say, 48 hours. Without confirmation of waiver, the adjudicator should resign. In not doing so, the adjudicator is exposed to later challenges of lack of impartiality.

15.15 Bias

An earlier draft of the Code of Practice Governing the Conduct of Adjudications dealt specifically with "conflict of interest" requiring that a person requested to act as adjudicator shall notify the parties of any conflict of interest or external factors which would give rise to a reasonable apprehension of bias, and that such a person shall decline the appointment unless satisfied it would be appropriate to accept.[63] Although the word "bias" is not included in the final version of the Code of Practice Governing the Conduct of Adjudications published by the Minister of State for Employment and Small Business in July 2016, it is an important concept in adjudication.

Bias is a state of mind or predisposition that would prevent an adjudicator acting with complete impartiality. Natural justice requires that the parties enjoy a fair hearing decided by an unbiased adjudicator. Bias can be actual or apparent. Whereas actual bias can be very difficult to prove, the test for apparent bias is "whether the fair-minded and informed observer, having considered the facts, would conclude that there was a real possibility" that the adjudicator was biased.[64] Bias might be thought to arise if there were personal friendship or animosity between the adjudicator and individuals involved with the dispute.[65] Adjudicators should be trustworthy and approach every adjudication with objective judgement and an open mind.[66] If apparent bias is established, the adjudicator's decision will not be enforced.

Some examples are:

- Apparent bias did not exist because of the adjudicator's past professional association with one of the party's representatives.[67]
- Apparent bias did not exist where the referring party's solicitor spoke on the telephone with a potential adjudicator to inquire about his availability to act, without disclosing this communication to the responding party.[68]

62 Construction Contracts Act 2013, Code of Practice Governing the Conduct of Adjudications (25 July 2016).

63 SI No. XXX of 2014 – Construction Contracts Act 2013 (Code of Practice) (Adjudication) Order 2014.

64 Porter v Magill [2002] 2 AC 357.

65 Locabail (UK) Limited v Bayfield Properties Limited [2000] QB 451.

66 Amec Capital Projects Limited v Whitefriars City Estates Limited [2004] EWCA Civ 1418.

67 Fileturn Limited v Royal Garden Hotel Limited [2010] EWHC 1736 (TCC).

68 Makers UK Limited v London Borough of Camden [2008] EWHC 1836 (TCC).

- Apparent bias or apparent pre-determination did not exist when the adjudicator disclosed a provisional view for the assistance of the parties. However, it is not acceptable if the adjudicator reaches a final decision before he is in possession of all relevant arguments and evidence.[69]
- Bias existed when the adjudicator drew an adverse inference from the non-participation of a particular witness at a meeting.[70]

In enforcement proceedings, the judge will consider and broadly look at the extent to which it is arguable that apparent bias is at play. If there is no real prospect of the case for apparent bias being made good, summary judgment of the adjudicator's decision will be handed down.[71]

15.16 Adjudicators' knowledge

Although appointed for their experience and knowledge, adjudicators, particularly construction professionals, need to be vigilant in how they use that expertise. There is a distinction between an adjudicator using expert knowledge to evaluate the evidence presented, and using expert knowledge to introduce new and different arguments or evidence.

The Code of Practice Governing the Conduct of Adjudications requires that the adjudicator shall observe the principles of procedural fairness, which shall include giving each party a reasonable opportunity to put their case and to respond to the other party's case.[72] The Code allows adjudicators to make use of their own specialist knowledge, providing that they shall disclose this to the parties as appropriate. It should be obvious that adjudicators must not deprive parties of a fair opportunity to deal with all relevant matters and thus cause injustice.

In broad terms, adjudicators can use their practical knowledge to adjudicate on the evidence and the submissions made by the parties. If an adjudicator uses his own experience and knowledge to "advance and apply propositions of fact or law that have not been canvassed by the parties", the adjudicator should normally set those out, and ask the parties for submissions.[73] This is in concord with the rules of natural justice that the parties have a fair opportunity to make and defend their case.

However, astute adjudicators who want to rely on propositions of fact or law not advanced by the parties should have the "opportunity test" clearly in mind. A good example of this principle is set out in *Barrs v British Wool Marketing Board*, and it is not did the decision maker's use of his own expert

69 Lanes Group PLC v Galliford Try Infrastructure Limited [2011] EWCA Civ 1617.

70 A&S Enterprises Limited v Kema Holdings Limited [2005] BLR 76.

71 Fileturn Limited v Royal Garden Hotel Limited [2010] EWHC 1736 (TCC).

72 Construction Contracts Act 2013, Code of Practice Governing the Conduct of Adjudications (25 July 2016).

73 Costain Limited v Strathclyde Builders Limited [2004] SLT 102.

knowledge lead to an injustice within the decision maker's decision, rather was there an opportunity for injustice to arise. Under the opportunity test, it is not relevant whether the adjudicator's introduction of evidence or law has in fact led to any detriment. "The test is not 'Has an unjust result been reached?' But 'Was there an opportunity afforded for injustice to be done?' If there was such an opportunity, the decision cannot stand."[74] This is an expansive concept, underpinned by the rationale that determining if an injustice was done would in most cases involve a re-examination of the merits of the case. The requirement is that the adjudicator has dealt fairly and equally with the parties in making their decision.

However, the UK Courts do not take an over-analytical approach to questions concerning natural justice arising in adjudications. Breaches of the rules of natural justice must be more than peripheral; they must be material breaches. For example, evidence advertently or inadvertently introduced by the adjudicator must lead to a material breach of the rules of natural justice. A material breach will arise when the evidence is

> either decisive or of considerable potential importance to the outcome of the resolution of the dispute and is not peripheral or irrelevant … whether the issue is decisive or of considerable potential importance or is peripheral or irrelevant obviously involves a question of degree.[75]

If the adjudicator is applying their own knowledge and experience in deciding matters that are material and not advanced by the parties, the adjudicator must invite submissions.[76] If the adjudicator goes off on a frolic of their own, making their decision upon a factual or legal basis not advanced by the parties and without giving the parties an opportunity to comment, their decision will not be enforced.[77]

There are red lines that adjudicators should not cross, but what is red is not always apparent; the palette of colour has nuance and is contingent on the specific circumstances of the case. Some red line examples of the ways that adjudicators have misuse their experience and knowledge are in:-

- Seeking advice (legal or technical) without telling the parties what that was, or inviting their comments or submissions.[78]
- Deciding what reasonable commercial rates for equipment used were, without supporting evidence.[79]

74 Barrs v British Wool Marketing Board, 1957 SC 72.

75 Cantillon Limited v Urvasco Limited [2008] BLR 250.

76 Petition of Mr and Mrs Jack Patton [2011] CSOH 40.

77 Herbosh-Kiere Marine Contractors Limited v Dover Harbour Board [2012] EWHC 84 (TCC).

78 Costain Limited v Strathclyde Builders Limited [2004] SLT 102.

79 Carrilion Utility Services v Scottish Power Systems Limited 2012 SLT 119.

- Using a method of assessment that applied a composite overall rate to overall delay, and which was not the methodology used by the parties.[80]
- Deciding an important element of the dispute using a particular contract clause which the parties had not argued or mentioned, and not referring that to the parties before issuing his decision.[81]

If one wishes to use expert knowledge to introduce new and different arguments or evidence then it should be done in a neutral way, as early as possible in the adjudication. For example, in one adjudication I wrote:

> I ask the parties to make any submissions on the relevance of *Connex South Eastern Limited v MJ Building Services Group Plc* [2005] EWCA Civ 193 to this adjudication, and in particular paragraphs 38 to 40, where Lord Justice Dyson said …

It is unlikely there will ever be circumstances where it would be appropriate or helpful for the adjudicator to send the parties a preliminary version of their decision asking for comments; this can raise the spectre of pre-judgement and apparent bias.[82] To do so invites chaos, as it facilitates the disintegration of the adjudication process and its management.

To mitigate against using experience and knowledge in an inappropriate way the adjudicator should track through the decision before publication and tag each reference to case law, contractual provisions or propositions relied upon, ensuring that the provenance of such is four-square within the parties' submissions.

15.17 Party costs

Party costs are expenses incurred in instituting or defending in litigation, as ordered that one party pay the other party. A general rule typical in common law jurisdictions is that "costs follow the event". This legal principle of costs is that a party who is substantially successful in litigation, in the absence of special circumstances, has costs awarded to it. To be recoverable party costs should be reasonable, and can include solicitors' fees and charges, expert witness fees, and disbursements.[83] The general law of costs is that except for special cause, the costs of every issue of fact or law raised upon a claim or counterclaim shall, unless otherwise ordered, follow the event.[84] This "loser

80 Herbosh-Kiere Marine Contractors Limited v Dover Harbour Board [2012] EWHC 84 (TCC).
81 ABB Limited v BAM Nuttall Limited [2013] EWHC 1983 (TCC).
82 Lanes Group PLC v Galliford Try Infrastructure Limited [2011] EWHC 1679 (TCC).
83 Tramountana Armadora SA v Atlantic Shipping Co. SA [1978] 2 All ER 870.
84 SI No. 12/2008 – Rules of the Superior Courts (Costs) 2008.

pays" principle acts as a deterrent to expense, encourages early settlement of disputes, and brings about a more efficient legal system.

However, the "costs follow the event" rule does not apply in adjudication. In Ireland, the Act provides that each party shall bear its own legal and other costs incurred in connection with the adjudication.[85] The issue of allocating party costs draws different responses. On the one hand, not being able to recover costs is a disincentive to adjudicate low value disputes. On the other hand, many smaller businesses would be discouraged from adjudication if they could be liable to pay the other side's costs.

In adjudication, there is a distinction between (a) the parties entering into a contract giving the adjudicator the power to decide the apportionment of party costs, and (b) allowing parties to draft a contract where one party, usually a referring party lower down the contractual chain, will irrespective of the adjudication outcome always be responsible for paying the adjudicator's costs and the responding party's adjudication costs – a "Tolent" clause.[86]

It seems inevitable that the issue of party costs in adjudication will come before the Irish Courts. Parties can enter into a contract which provides for adjudication, even though the scope of works does not fall within the definition of a "construction contract" under the Act. A question which will arise in respect of party costs is whether the Act ousts the doctrine of "freedom of contract", a judicial concept which holds that contracts are based on mutual agreement and free choice, and therefore contracts should not be hampered by external control such as government interference.

In the UK, whereas there is no freestanding power to allocate party costs, there are well-respected adjudication rules that the parties can incorporate into their contract that can empower the adjudicator to decide costs. For example, the TeCSA Adjudication Rules[87] provide that "If the Parties so agree in writing after any Party has given a Notice …, the Adjudicator shall have jurisdiction to award the parties' legal and expert costs in relation to the adjudication as he sees fit".

Adjudicators should anticipate that irrespective of the Act providing party costs are their own, they will be asked to decide that one party or another shall pay the other party's costs. A party claiming their costs should be mindful that if they are successful in getting the adjudicator to decide costs in their favour, the paying party may argue the adjudicator has acted without jurisdiction, that consequently the adjudicator's decision is tainted and refuse to make payment of any amounts decided, both in respect of the substantive payment dispute, and in connection with party costs. Parties should take care not to purportedly give the adjudicator jurisdiction to deal with the parties' costs as part of the decision, if this is not what they want.

85 Construction Contracts Act 2013 Ireland – Paragraph 6.(15).

86 Bridgewater Construction Limited v Tolent Construction Limited (2000) CILL 1662.

87 TeCSA Adjudication Rules 2018, Version 3.2.2.

In respect of a "Tolent" clause, as the paying party is most often the responding party, making the referring party always liable for the responding party's costs would fetter the right to refer disputes to adjudication and undermine adjudication law.[88] The main arguments against Tolent provisions in adjudication are that they provide no incentive for a party to keep its costs to a reasonable level if it knows the other side will be liable for them. A winning party could still end up out of pocket because the other side's costs could negate the amount the adjudicator awards.

In practical terms it is problematic for the adjudicator to decide the allocation of party costs. The adjudicator only has jurisdiction to make a decision during the course of the adjudication, which will most usually be a 28-day period. The parties' costs will not have fully accrued until the end of the adjudication, and therefore the costs information will at best be incomplete. In addition, the adjudicator will not have the benefit of practical submissions as to the reasonableness or otherwise of the party costs incurred, and any decision on party costs will subsequently be rudimentary.

In some adjudication, parties conduct leads to high adjudication costs, which outweigh the value of the claim. Concern has been raised that high adjudication costs revisit some of the inefficiencies and injustices of the past, and undermine adjudication as an effective dispute resolution process.[89] The Code of Practice Governing the Conduct of Adjudications requires that the adjudicator shall use reasonable endeavours to process the payment dispute between the parties in the shortest time and at the lowest cost; for example, ensuring that the procedure adopted is commensurate with the nature and value of the payment dispute.

On 1 May 2020, the Construction Industry Council (CIC) in the UK published a Low Value Disputes Model Adjudication Procedure (CIC LVD MAP).[90] The CIC LVD MAP initiative was a response to growing construction industry concerns about the increasing complexity and prohibitive costs of adjudication; its terms were informed through extensive consultation with the construction industry and other stakeholders.[91] The aim of the CIC LVD MAP is to provide a simple and cost-effective procedure to make adjudication more accessible for SMEs and others involved in lower value claims defined as for £50,000 or less, and the issues in dispute are relatively uncomplicated.

The CIC LVD MAP sets out a streamlined adjudication procedure for low value disputes, and by linking the adjudicator's fee to the amount claimed provides certainty as to how much the adjudicator will be paid for making an

88 Yuanda (UK) Co Limited v WW Gear Construction Limited [2010] EWHC 720 (TCC).
89 Primus Build Limited v Pompey Centre Limited (& Slidesilver Limited) [2009] EWHC 1487 (TCC).
90 http://cic.org.uk/services/cic-low-value-disputes-adjudication.php.
91 http://cic.org.uk/news/article.php?s=2019-06-04-cic-consultation-on-adjudication-for-low-value-disputes.

adjudicator's decision. By including an outline timetable for the procedural stages it provides a flexible, yet simple to understand, approach to the key elements of the adjudication process.

15.18 Interest

Interest is a fee paid by a borrower to a lender for assets that the former uses. Usually, it is money paid in addition to the repayment of borrowed money. Interest can be simple or compound, and interest rates can be fixed or variable. Simple interest is paid on the sum borrowed. Compound interest is paid not only on the money borrowed, but also on previously accumulated interest. Interest is an integral part of the commercial or business landscape in most jurisdictions including Ireland and the UK, but prohibited under the Sharia or Islamic law.[92]

In adjudication it is common for a referring party to ask the adjudicator to decide that an amount of interest should be paid to the date of the decision, and that an additional amount of interest should be paid per day thereafter until the sum due is paid.

The general rule in English common law is that in the absence of an agreement, there is no power to award interest as compensation for the late payment of a debt.[93] Considering the Scheme interest provision,[94] the UK Court of Appeal confirmed that for the adjudicator to have the power to award interest, the contract between the parties should include provisions allowing for the payment of interest. The adjudicator has no "freestanding power" to award interest. The adjudicator may decide questions as to interest if those questions are "matters in dispute" which have been properly referred to him, or are questions which the parties to the dispute have agreed should be within the scope of the adjudication. The parties can confer the adjudicator with the power to award interest by their conduct.[95]

In Ireland adjudicators can award interest if there is a contractual entitlement to interest, or there is an implied term providing such entitlement. The European Communities (Late Payment in Commercial Transactions) Regulations 2012 (the "Late Payment Regulations") became operative in March 2013, and replaced the European Communities (Late Payment in Commercial Transactions) Regulations 2002 (SI 388 of 2002).[96]

Under the Late Payment Regulations it is an implied term of every commercial transaction that where the purchaser does not pay for the goods or

92 Lenzin, Faris, Interest and Islamic Banking, *International Construction Law Review*, Vol. 20, Jan. 2003.

93 London, Chatham and Dover Railway Co v South Eastern Railway Co [1893] AC 429.

94 Scheme for Construction Contracts (England and Wales) Regulations 1998.

95 Carillion Construction Limited v Devonport Royal Dockyard [2005] EWCA Civ 1358.

96 SI No. 580/2012 – European Communities (Late Payment in Commercial Transactions) Regulations 2012.

services concerned, interest will arise on the expiry of the terms of payment provided for in the contract, or where the contract is silent in this regard, after 30 calendar days. This "statutory late payment interest" accrues on the amount outstanding until such time as payment is made. Statutory late payment interest is defined as the interest rate applied by the European Central Bank plus eight percentage points.

Where a term of a contract between a purchaser and a supplier purports to waive or vary entitlement to interest or the rate of interest set out, and the supplier considers that the waiver or variation is "grossly unfair", the supplier may apply to the Circuit Court for an order or to an arbitrator for an award, under the Late Payment Regulations.

Interest can compensate for late payment, and high interest can deter late payment. In the UK, the Late Payment of Commercial Debts (Interest) Act 1998 (the "Late Payment Act") gives businesses a statutory right to claim interest on overdue commercial debts. The Late Payment Act provides "Any contract terms are void to the extent that they purport to exclude the right to statutory interest in relation to the debt, unless there is a *substantial contractual remedy* for late payment of the debt".

In UK adjudications, where there are no contractual interest provisions, parties sometimes assert that interim payments in construction contracts are qualifying debts as defined under the Late Payment Act. They will say that being so, they are entitled to "statutory interest" at the applicable Bank of England base rate plus 8%. In addition, they will argue that where the contract interest rate is, say, 2.0% per annum above the base rate of the Bank of England, that does not provide a "substantial remedy" within the meaning ascribed to it in the Late Payment Act, and it would not be fair or reasonable to allow the lower interest rate of 2% cited in the contract to oust the right to statutory interest.

Parties will rely on the authority of *Yuanda (UK) Co. Limited v WW Gear Construction Limited* which provides guidance on what the courts have decided a "substantial remedy" is.[97] In this case, a contract provided for an interest rate of 0.5% over base rate on late payments, and the court decided that this was not a substantial contractual remedy as defined by the Late Payment Act. The court held that this term of the contract was void and replaced it with the statutory rate of 8% over base rate. The Honourable Mr Justice Edwards-Stuart said "Putting it crudely, it seems to me that the imposition of the statutory rate is the penalty that a contracting party pays for failing to provide in its contracts a fair remedy for late payment to suppliers".

Consider the circumstances where a construction contract between the parties in Ireland provides that the interest rate is 2%. In the adjudication, the referring party asks the adjudicator to decide that the interest rate of 2% is "grossly unfair" and to replace it with the European Central Bank interest

97 Yuanda (UK) Co Limited v WW Gear Construction Limited [2010] EWHC 720 (TCC).

rate plus 8%. Does the adjudicator have jurisdiction to make that decision? There is no definition of "grossly unfair" in the Late Payment Regulations, and the responding party says that under the Late Payment Regulations it is only an arbitrator or the Circuit Court that has the authority to decide if the interest rate of 2% is "grossly unfair". If the responding party's only grounds for contesting the adjudicator's jurisdiction to decide what is grossly unfair under the Late Payment Regulations is by virtue of the fact that the adjudicator is neither a Circuit Court judge nor an arbitrator, I would proceed to decide the matter.

It makes obvious commercial sense for an adjudicator to have the power to award and decide interest.[98] In addition, the Construction Contracts Act empowers the adjudicator to decide "any dispute relating to payment arising under the construction contract". As adjudicator I would expect to consider submissions on the relative bargaining power of the parties, the extent of negotiations in respect of the interest rate if any, the use of standard forms or terms, that the term "grossly unfair" in the Late Payment Regulations is analogous to the term "substantial contractual remedy" in the Late Payment Act, as well as other submissions. I anticipate that in due course the courts will consider whether an adjudicator has the jurisdiction to decide that the interest rate of 2% or similar is "grossly unfair" and to replace it with the European Central Bank interest rate plus 8%. A robust confirmation by the courts that an adjudicator does have jurisdiction to decide interest will support adjudication as an effective dispute resolution mechanism.

In the UK, courts have drawn a distinction between situations where there is a "bona fide dispute" and where there is "casual or feckless non-payment". It is not appropriate for a creditor without reason to delay proceedings for several years and then to expect to recover interest at the enhanced rate.[99]

15.19 Liability of the adjudicator

The doctrine of judicial immunity is rooted in the English common law.[100] The concept is that any person acting in a judicial capacity, if acting within their jurisdiction, enjoys immunity from any liability that may result from discharging their duties. The policy objects are the need for finality, to protect the independence of the judge, to maintain public confidence in the system of justice and to protect the judge from frivolous or otherwise suits.[101] These policy justifications are the umbrella that has extended immunity to arbitrators

98 Carillion Construction Limited v Devonport Royal Dockyard [2005] EWCA Civ 1358.

99 Banham Marshalls Services Unlimited v Lincolnshire County Council, and Others [2007] EWHC 402 (QB).

100 Floyd v Barker, 77 Eng. Rep. 1305 (Star Chamber 1607).

101 Block, Randolph (1980). Stump v Sparkman and the History of Judicial Immunity, *Duke Law Journal*, Volume November 1980, No. 5: 879.

and adjudicators, who like judges must also be independent, impartial and abide by the principles of natural justice.

When adjudication law was first mooted in the UK, though the government's initial view was that the parties should deal with adjudicator immunity themselves, on reconsideration it decided to specify contractual entitlement to freedom from legal liability in primary legislation.[102] The importance of immunity was summarised by Nick Rainsford, Member of Parliament,

> If adjudication is to work, it is essential for the adjudicator to enjoy immunity from litigation. Otherwise, he will not be able to act quickly and expeditiously, but will be constantly looking over his shoulder, worrying about the prospect of a writ being issued by one of the parties who is aggrieved by the way in which he is proceeding. If we want swift adjudication, the adjudicator must have immunity.[103]

In Ireland under the Act the

> adjudicator is not liable for anything done or omitted in the discharge or purported discharge of his or her functions as adjudicator unless the act or omission is in bad faith, and any employee or agent of the adjudicator is similarly protected from liability.[104]

The Act circumscribes adjudicator immunity *only* with a test of "bad faith", but it does not define what "bad faith" is. Bad faith can be considered as a contrasting mirror to good faith. Bad faith might exist where the adjudicator puts his own interests before those of the parties, or neglects or refuses to perform his duty, where the adjudicator's conduct has improper motivation; where there is dishonesty, fraudulent intent, hostility, ill will, lack of fairness or malice. Bad faith "is capable of embracing a failure to act in a commercially acceptable way and sharp practice of a kind that falls short of outright dishonesty as well as dishonesty itself".[105]

The standard of care imposed on a professional person at common law is to perform with "reasonable skill and care". If the professional person does not act with "reasonable skill and care", they are negligent.[106] However, in construing contracts it is considered inherently improbable that one party to the contract should intend to absolve the other party from the consequences

102 Hansard Housing Grants, Construction and Regeneration Bill, HL Deb, 22 April 1996, Volume No. 571, Columns 975–977.

103 Hansard Housing Grants, Construction and Regeneration Bill, HC Deb, 8 July 1996, Volume No. 281, Columns 78–88.

104 Construction Contracts Act 2013 Ireland – Paragraph 6.(14).

105 Niru Battery Manufacturing Co v Milestone Trading Limited (No 1) [2002] EWHC 1425 (Comm).

106 Bolam v Friern Hospital Management Committee [1957] 1 WLR 582; [1957] 2 All ER 118.

of the latter's own negligence, therefore the intention to do so must be made perfectly clear.[107]

In the UK, the CIC Model Adjudication Procedure[108] and the TeCSA Adjudication Rules,[109] which are well respected, clarify the immunity of the adjudicator, with the codicil that the adjudicator is not liable for anything done or not done in the discharge of his role "whether in negligence or otherwise", and any employee or agent of the adjudicator is similarly protected from liability.

Protection from third-party suit is a matter for the contracting parties, and my standard schedule of conditions for acting as adjudicator is to ask the parties to indemnify the adjudicator from any costs and liabilities incurred arising from actions taken by third parties.

15.20 Confidentiality and privacy

In adjudication, confidentiality and privacy are different. Confidentiality is the right to non-disclosure of documentation and information exchanged during adjudication. Privacy is the right for parties to have their dispute considered in camera.

There can be a wide range of reasons why parties want confidentiality. Parties often consider commercial information confidential by its very nature. Technical information can be confidential. For example, in an adjudication concerning non-destructive testing, inspection and reporting of mitigation works on an offshore wind farm, the remedial solution and method of work was confidential. Confidentiality makes adjudication attractive to parties who wish to resolve their construction disputes quickly and quietly.

An earlier draft of the Code of Practice Governing the Conduct of Adjudications dealt comprehensively with "confidentiality" providing that

> any document or information supplied for and/or disclosed in the course of the adjudication will be kept confidential. An adjudicator will only disclose the same if required to do so by law, or pursuant to an order of a court, or with the consent of all the relevant parties.[110]

However, the final Code of Practice Governing the Conduct of Adjudications provides only that "any document or information supplied for and/or disclosed in the course of the adjudication shall be kept confidential by the Adjudicator".[111]

107 Gillespie Bros & Co Limited v Roy Bowles Transport Limited CA [1973] 1 QB 400.

108 CIC Model Adjudication Procedure Revision 5, October 2011.

109 TeCSA Adjudication Rules, Version 3.2.2 which came into force on 21 March 2018.

110 SI No. XXX of 2014 – Construction Contracts Act 2013 (Code of Practice) (Adjudication) Order 2014.

111 Construction Contracts Act 2013, Code of Practice Governing the Conduct of Adjudications (25 July 2016).

The UK Scheme[112] provides that

> the adjudicator and any party to the dispute shall not disclose to any other person any information or document provided to him in connection with the adjudication which the party supplying it has indicated is to be treated as confidential, except to the extent that it is necessary for the purposes of, or in connection with, the adjudication.

Therefore, whereas the ambit of the adjudication law on confidentiality proposed in Ireland in 2014 was wider than the ambit of UK adjudication law, the ambit now is less. In Ireland in 2014, any document or information supplied for and/or disclosed in adjudication was to enjoy the cloak of confidentiality. Whereas in the UK, the party supplying the document or information must indicate it is confidential. In Ireland now, the Code of Practice places an obligation of confidentiality only on the adjudicator.

This can be contrasted with the adjudication law in Malaysia which has comprehensive provisions expressly providing that construction adjudication proceedings, and any statement, admission or document made or produced for the purposes of adjudication shall be confidential, although with some limitations and exceptions.[113]

The Public Works Contracts for Building Works in Ireland do have a "Confidentiality and Secrecy" section, which includes a requirement that for information to be confidential, it is notified or stated as confidential or secret. It is not clear whether such "Confidentiality and Secrecy" provisions will help ensure that information made or produced for the purposes of adjudication will be kept confidential.

One way parties in the UK have sought to expand entitlement to confidentiality is to agree during contract formation that the adjudication procedures, and the agreement for the appointment of the adjudicator shall be as set out in the "CIC Model Adjudication Procedure (MAP)"[114] or the "TeCSA Adjudication Rules".[115]

The CIC MAP Form of Agreement between the parties and the adjudicator provides that "The Adjudicator and the Parties shall keep the adjudication confidential, except so far as is necessary to enable a Party to implement or enforce the Adjudicator's decision".

The TeCSA Adjudication Rules provide that

112 The Scheme for Construction Contracts (England and Wales) Regulations 1998. Statutory Instrument 1998 No. 649.

113 Construction Industry Payment and Adjudication Act 2012.

114 CIC Model Adjudication Procedure Revision 5, October 2011.

115 TeCSA Adjudication Rules, Version 3.2.2 which came into force on 21 March 2018.

> Unless the Parties otherwise agree the Adjudication and all matters arising in the course thereof are and will be kept confidential by the Parties except insofar as necessary to implement or enforce any decision of the Adjudicator or as may be required for the purpose of any subsequent proceedings.

It is a matter for debate elsewhere whether the CIC MAP or the TeCSA Adjudication Rules afford a more comprehensive right to confidentiality. However, entitlement to confidentiality appears to be one area where carefully crafted contract dispute resolution clauses, adjudication rules or an adjudication procedure might benefit adjudicating parties in Ireland.

The parties' obligation to respect the confidentiality of documents or information disclosed in adjudication is a contractual commitment, and the remedy for a failure to comply is a claim for damages for breach of contract. In serious cases where prevention of disclosure is required, an injunction might be obtained.[116]

If enforcement proceedings are required because the parties fail to comply with the adjudicator's decision, the background to and other dispute details will become publically available.

Privacy is rarely unabridged. Commencing adjudication requires appointment of an adjudicator; in most circumstances, one party will communicate the nature of the dispute to the Chair of the Minister's panel of Adjudicators in Ireland, or an adjudicator nominating body in the UK. After appointment of the adjudicator, the adjudication process can remain private. Adjudicators have the power to issue directions to help safeguard this, for example, ensuring that any meetings take place in a closed setting.

In Ireland, adjudicators appointed by the Chairperson of the Construction Contracts Adjudication panel are required to provide anonymised information to the Construction Contracts Adjudication Service on each adjudication case within 21 days of completion of the case. The adjudicator summits the information using an online Form 8, the "Construction Contracts Act 2013, Adjudicator Statistical Data Return", assisted by the availability of drop-down list boxes. Form 8 does not include the names of the parties to a payment dispute, and is used for the purpose of compiling statistical information relevant to adjudications conducted in accordance with the Act.

The information provided by adjudicators is consolidated into an excellent report published annually in July by Dr Nael Bunni, Chairperson of the Ministerial panel of Adjudicators, which he submits to the Minister of State for Trade, Employment, Business, EU Digital Single Market and Data Protection.

116 Redmond, John, *Adjudication in Construction Contracts*, Blackwell Science Limited (1 November 2001) English ASIN: B00L6Z4H1Q.

16 Process

16.1 Confucius

Confucius espoused the "Doctrine of the Mean" (中庸之道), that the correct or right course of action is always some middle point between the two extremes of excess (too much) and deficiency (too little).[1]

> Tsze-lu asked about energy ... Confucius said "Therefore, the superior man cultivates a friendly harmony, without being weak.- How firm is he in his energy! He stands erect in the middle, without inclining to either side.- How firm is he in his energy!"

It is settled law that in adjudication the rules of natural justice do generally apply.[2] Adjudicators must act impartially;[3] they must be fair, making an unbiased determination not favouring one party over the other. Good adjudication requires adjudicators to control a process that seems simple, but which can however become treacherous. The parties can tussle for power; seeking to reduce the adjudicator's control of the process. They will push and pull the adjudicator to adopt a course of action advantageous to them. The adjudicator should ensure that the adjudication timetable is sufficient for them to properly appreciate the nature of the issues and understand the case of each party.[4] The adjudication timetable should be balanced so that the parties have the opportunity to "have their say" about all aspects of the dispute referred and any supporting evidence.

Adjudicators should espouse the "Doctrine of the Mean", dealing with the challenges of adjudication with balance and transparency, standing firm in the middle.

1 *Confucian Analects – The Great Learning of the Doctrine of the Mean: Chinese Classics*, new edition, translated by James Legge, Dover Publications Inc. (28 March 2003) ISBN-10: 0486227464 ISBN-13: 978–0486227467.

2 Dorchester Hotel Limited v Vivid Interiors Limited [2009] EWHC 70 (TCC).

3 Construction Contracts Act 2013.

4 CIB Properties Limited v Birse Construction [2005] BLR 173.

16.2 Dispute crystallisation and definition

In Ireland, the Act gives the right to refer a payment dispute to adjudication. In the UK, the English Act gives the right to refer any dispute or difference to adjudication. The word "difference" is taken to have a wider ambit than the concept "dispute".[5] Comparing the two words, the term "dispute" is more hard-edged than "difference".[6]

Adjudicators regularly face a jurisdictional challenge on the grounds that "No dispute has crystallised". To avoid incurring cost and time, referring parties should make sure that a dispute has crystallised and is suitable for adjudication. For a dispute to crystallise there must have been an opportunity for each of the parties to consider the position adopted by the other and to formulate reasoned arguments.[7] If a claim is ignored a dispute can also crystallise.

In Ireland, the Act provides that the construction contract between the parties will have an adequate mechanism for determining the amount of interim and final payments. The provisions apply where "not later than 5 days after the payment claim date, an executing party to a construction contract delivers a payment claim notice relating to a payment claim to the other party or another person specified under the construction contract". In such circumstances, the paying party is obliged to respond to the payment claim notice not later than 21 days after the payment claim date. Failure to accept the payment claim, or failure to respond to the payment claim should be sufficient to assert and demonstrate that a dispute suitable for adjudication has crystallised. However, contractors often fail to make payment claims or give notices in the manner set out in the Act.

Under the Act, where a payment claim is not accepted the party refuting the amount claimed should provide reasons. Where a reason for the different amount in the response is attributable to a claim for loss or damage arising from an alleged breach of any contractual or other obligation of the party which issued the payment claim notice, or any other claim, the party refuting the amount should provide details of when the loss was incurred or the damage occurred, or how the other claim arose, and the particulars of the loss, damage or claim, and the portion of the difference that is attributable to each such particular.

Accordingly, as a responding party, it is advisable to give some considerable thought as to the contents of your response to a payment claim. To do otherwise will leave open the possibility of an adjudicator, rightly or wrongly, deciding that you are not entitled to raise defences not included in your response to a payment claim in any subsequent adjudication.[8]

5 Amec Civil Engineering Limited v Secretary of Transport [2004] EWHC 2339 (TCC).
6 Allied P&L Limited v Paradigm Housing Group Limited [2009] EWHC 2890 (TCC).
7 Edmund Nuttall Limited v R. G. Carter Limited [2002] EWHC 400 (TCC).
8 M I Electrical Solutions Limited v Elements (Europe) Limited [2018] EWHC 1472 (TCC).

In the ordinary use of the English language, there is a dispute if a party has refused to pay a sum claimed, or has denied that the sum claimed is owed. The UK Court of Appeal considered, *inter alia*, what constituted a dispute for the purposes of arbitration, and the meaning of dispute adopted was that "there is a dispute once money is claimed unless and until the defendants admit that the sum is due and payable".[9] In adjudication, the word "dispute" does not have a specialised meaning. For there to be a dispute suitable for adjudication "it must be clear that a point has emerged from the process of discussion or negotiation that has ended and that there is something which needs to be decided".[10] Adjudicators should not adopt an overly legalistic analysis of what the dispute between the parties is.[11]

There does not have to be an express rejection of a claim, a dispute can arise through a period of silence. The period of silence before inferring there is a dispute depends heavily upon the specific circumstances. In fact, the period of silence can be very short:

> In normal circumstances, a gap of five days from the time that a claim or assertion is put forward will often be sufficient to give rise to an inference that it is disputed, particularly … where there had been previous claims and where liability was obviously in dispute for some two months before.[12]

Although a claim does not have to be precisely particularised, it should include fundamental information. The irreducible minimum requirements for a dispute to crystallise are that the referring party has made a claim in terms that enable the responding party to consider it, and decide whether to accept or reject it. In addition, it should be apparent from the responding party's conduct that the claim is not accepted.[13] If the claim is nebulous and ill-defined, so that it cannot be sensibly responded to, neither an express non-admission nor silence is likely to give rise to a dispute for the purposes of adjudication.[14]

Sometimes astute non-paying parties can exasperate the payment application process by asking for additional information, for information already sent, or for clarification without admitting or denying entitlement to be paid. However, a non-paying party cannot avoid a dispute by ignoring a claim or silence. Referring parties cannot curtail what would be a reasonable time to respond to a claim.[15] But they can control and manage crystallisation of

9 Halki Shipping Corporation v Sopex Oils Limited [1997] EWCA Civ 3062.
10 Beck Peppiatt Limited v Norwest Holst Construction Limited [2003] EWHC 822 (TCC).
11 Cantillon Limited v Urvasco Limited [2008] EWHC 282 (TCC).
12 Beck Interiors Limited v UK Flooring Contractors Limited [2012] EWHC 1808 (TCC).
13 Devon County Council v Celtic Composting Systems Limited [2014] EWHC 552 (TCC).
14 Amec Civil Engineering Limited v Secretary of Transport [2004] EWHC 2339 (TCC).
15 Amec Civil Engineering Limited v Secretary of Transport [2004] EWHC 2339 (TCC).

a dispute, for example by writing "If you do not make payment before [insert reasonable date], a dispute will have arisen, which we will refer to adjudication".

Faced with the jurisdictional challenge "No dispute has crystallised", the adjudicator can look at the context, what the parties were doing and saying. It is usually straightforward from the chronology of documents exchanged, and what they say, to conclude whether a dispute had crystallised at a particular date or time. It is hard to prevent a dispute from forming and the crystallisation argument is almost never successful.[16]

16.3 Evaluating your risk

Before commencing adjudication, as the referring party you should audit your adjudication risk. Risk audit is a process that helps you make sensible commercial decisions. It highlights risks, their nature and scope, and allows you to determine how to prevent or reduce the risks.

One way to evaluate adjudication risk is to use a decision tree. Widely used in engineering, a decision tree can help you decide the dispute resolution strategy most likely to secure the best financial outcome for you. Graphical representation of risk helps you to make a reasoned assessment of the adjudication outcome, and allows you to compare that outcome to the costs you will incur in adjudication, and to any settlement offer. If you have suffered a loss of cash flow because of non-payment and invested in preparing a detailed claim, there will always be a maximum recovery amount. The maximum is the total of the quantum of all the items claimed without deductions. It is natural for senior management to focus on the maximum amount possible; this is the wrong approach. Experience shows that "independent experts" often provide different credible quantum calculations, seemingly underpinned by objective standards. The maximum amount is only ever an aspirational figure, and to imbue realism I recommend the use of a decision tree to calculate an "expected cash equivalent" ("ECE"). Competent mediators will use the decision tree method, or a variation of it, to help move the parties from entrenched positions.

In an adjudication concerning the modernisation and installation of lifts in London, the referring party said that it was entitled to additional money because of 17 variations. The responding party said that it was entitled to deduct money because of defective and incomplete work. Figure 16.3 shows how a decision tree can be used to chart the lifts dispute. The referring party said one variation was for lift builders' work, another variation was for dealing with inherited faults, etc. The responding party said that it wanted credit for unfinished items.

16 CSK Electrical Contractors Limited v Kingwood Electrical Services Limited [2015] EWHC 667 (TCC).

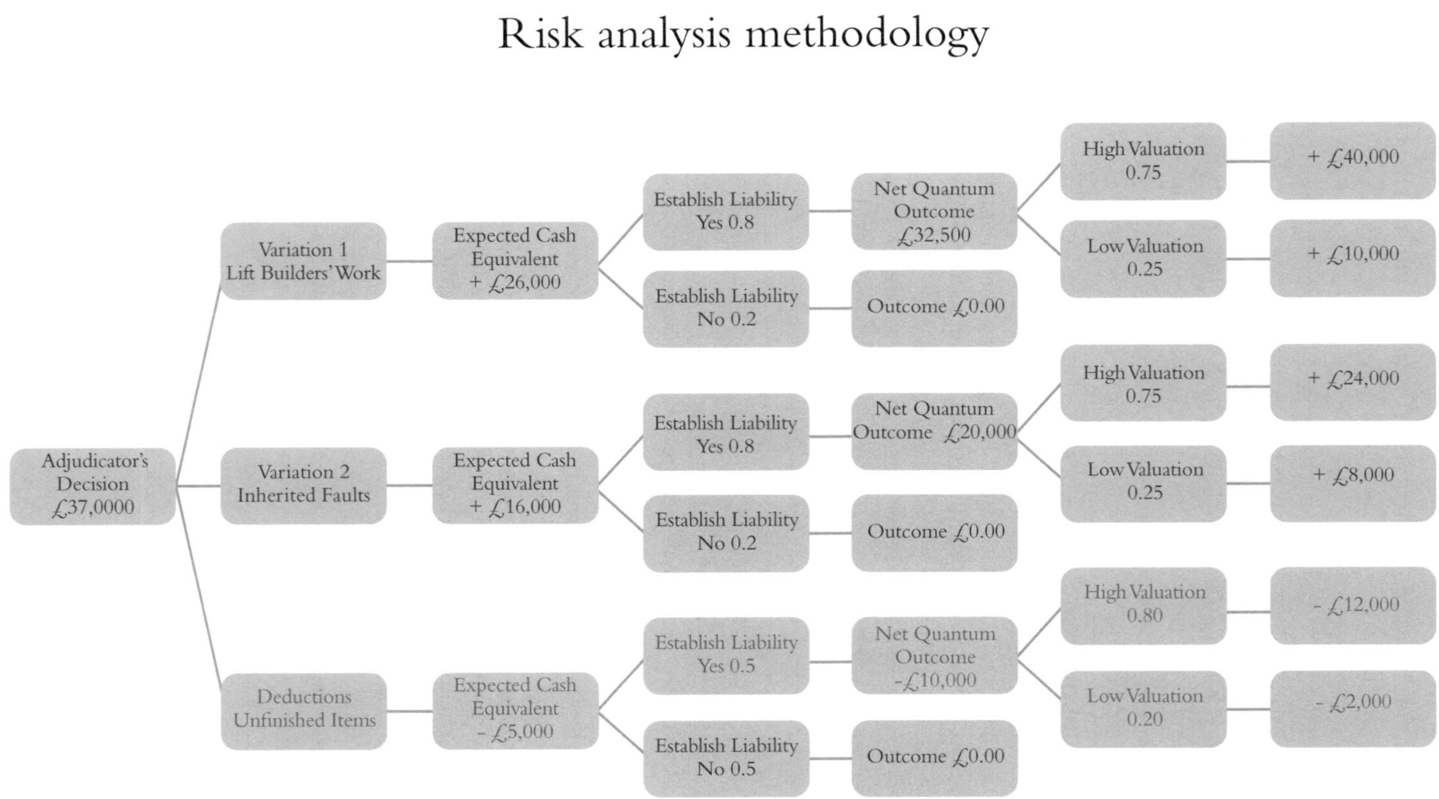

Figure 16.3 Decision tree in relation to the lifts dispute.

To grow the decision tree, each variation or claim for offset becomes a branch of the diagram, and determines the likely ECE based on statistical probability. Each branch of the decision tree represents a monetary decision event. The tree structure links the outcome of occurrences, and shows how one decision leads to the next. The use of branches indicates that each variation or deduction is mutually exclusive to another.

The efficient and normal way for an adjudicator to consider variations or deductions is first to decide liability and then to decide quantum. If the adjudicator decides that there is no liability, there is no need for him to look at quantum. However, for you, once the branches are in place it is easier to populate decision tree information starting with quantum and then moving to liability. In the lifts dispute the lift builders' work variation has a possible high valuation of £40,000 and a possible low valuation of £10,000. You assess that there is a 75% (0.75) probability that the adjudicator will decide the high valuation is correct, and there is a 25% (0.25) probability that the adjudicator will decide the low valuation is correct. Factoring the high and low valuation with their associated probability leads to a net quantum outcome of £32,500. Often the decision tree has three sub-branches representing high, medium and low valuations. Whatever number, it is essential that the probability of the branch events occurring always add to 100% (1.0).

To make the decision tree robust it is important to support each probability with reasons. Intuition is not a reason. Not preparing a full list of reasons makes the decision tree branches brittle, and is a common decision tree malady. If you identify the reasons why the adjudicator will decide that the low valuation is correct, you are identifying the scope and nature of the risks that you face. Risk is a measure of the probability and severity of adverse effects. As you may be able to do things to reduce the risks, change the risk consequences, or share the risk with another party such as a subcontractor, there might be a sequence of linked dispute decision trees. Decision trees can be drafted using paper and pencil, by using a PC-based spreadsheet such as Microsoft Excel, or proprietary low-cost decision tree software. Using a spreadsheet or software makes it easy to update the decision tree quickly when there is new information that changes your assessment of branch risks or outcomes.

Linked to the net quantum outcome is the likelihood of the adjudicator deciding that one party or the other has liability for the variation. You assess that there is an 80% (0.80) probability that the adjudicator will decide that the other party is liable for the lift builders' work variation, and a 20% (0.20) probability that you are liable. If the adjudicator decides that you are liable, you get nothing. Multiplying the probability 80% (0.8) that the other side is liable, by the net quantum outcome, £32,500, gives the lift builders' work variation ECE of £26,000. Populating the Figure 16.3 decision tree with the lift builders' work and inherited faults variations, and the unfinished items deduction gives an ECE adjudication outcome of £37,000. This contrasts starkly with the maximum amount possible, £64,000.

Each party might spend £15,000 in claims consultants' and lawyers' fees to assist it. As under the Act each party shall bear its own legal and other costs incurred in connection with the adjudication, party costs in preparing for and participating in adjudication are not recoverable.

However, the adjudicator could decide that you should pay all of his fees and expenses, or a percentage thereof. Based on the above, if the ECE adjudication outcome is £37,000, and your adjudication professional fees budget is £15,000, irrespective of whether you are risk averse or not, accepting an offer to settle for £22,500 makes sense. This would give you £500 more than the ECE, and you can invest your company's time into pursuing other more profitable opportunities.

If you are adjudicating in a jurisdiction such as Malaysia, where the party's costs in preparing for and participating in adjudication are recoverable, then not accepting an offer of £37,500 leads you to liability to pay the other parties' costs, which if not taxed, could reduce your net recovery to £37,000 – £15,000 = £22,000.

16.4 Commencing adjudication

Before referring your dispute to adjudication, you must decide that it is the best course of action for you. Have you really reached the "end of the road" with negotiation, early neutral evaluation, mediation or conciliation? Have you learned anything during those alternative dispute resolution procedures that helps inform your decision on whether to use adjudication, or what is the best time to do that? For example, when you commence adjudication, will the other party have a valid response that you owe them money, or will your referral to adjudication trigger the other party to commence adjudication against you? In addition, it makes little sense to invest energy and resources to prevail in adjudication, and then a court judgment to enforce the adjudicator's decision, if the other party will not have the money to pay you.

To commence an adjudication, you should confirm that there is an oral or written contract such that the statutory adjudication, payment or suspension provisions of the Act apply; for example, confirm that the contract is a construction contract. Depending on the circumstances, could the courts help with determining this? In Ireland an adjudicator may deal at the same time with several payment disputes arising under the same construction contract or related construction contracts. If more than one dispute needs to be decided is it better for you to ask an adjudicator to resolve them in a consolidated manner, or separately?

Adjudication should be properly planned and managed. There should be one person responsible for management of the adjudication process and team. Have you planned the adjudication like a project? Have you asked advisors such as claims consultants, experts and lawyers to give you a budget for the work that might reasonably be required? Are the fee proposals reliable and broken down against a schedule of tasks such as writing the referral, collation of relevant documents, and preparation of statements from witnesses of fact and expert witnesses? Have you estimated what internal resources and costs will be required? Is the adjudication budget proportionate to the dispute, and is it available?

At this early stage and even with limited information, undertake a quantitative risk analysis, objectively assessing as far as possible the probability that the adjudicator will or will not decide in your favour on different facets of the dispute. You might come up with a range of projected recovery, being mindful that you will not be able to recover party costs in the adjudication, or be liable for the other party's costs; you are now in a position to look at the ratio of your financial investment in adjudication over your projected recovery. Your costs are likely to increase and your recovery may fall short of expectation, with that in mind, is the investment to recovery ratio acceptable to you?

Some claims consultants and lawyers offer to assist in adjudication for a fixed fee. The fixed fee is based on an agreed scope of work, and can help ensure that costs are balanced with the amount claimed. However, if the fixed fee means that the adjudication is not properly resourced, that will be a false economy.

Have you considered the possibility of a conditional fee arrangement where advisors agree to be paid a contingency or success fee of any sum decided in adjudication, but in return charge no or reduced time-related fees? Usually you will have to pay normal time-related fees during an assessment period where advisors evaluate the strengths and weaknesses of your claim. As such arrangements usually require the advisors to have control of client staff resources, to be able to decide on the appointment of experts, and to determine when and how to proceed with adjudication, they are often incompatible with corporate governance procedures and rules. If you are an advisor offering what is effectively third-party adjudication funding, do you have the experience to be confident that your percentage success fee is adequate? Irrespective of the adjudication outcome exceeding agreed reasonable expectations or additional work being carried out by you, contractors are often recalcitrant and unlikely to agree any uplift once they have received payment of the amount decided by the adjudicator.

There are some experienced and knowledgeable adjudicators who will act as adjudication advisors preparing all of the documentation required and acting as adjudication project managers. For a variety of reasons they do not want to be visible, and they will work effectively acting as "ghost writers", usually for and in the shadow of a senior commercial manager.

Have you allocated sufficient time to prepare properly for adjudication? Commencing adjudication prematurely can result in an inefficient and ultimately more expensive process. If you are not successful, you will not recover what you claim, but may have liability for the adjudicator's fees and expenses, and in some jurisdictions the other party's costs. Remember that you can only adjudicate the dispute once.[17]

Will your professional advisors, including experts, be available during the course of the adjudication to provide assistance if required?

Success in adjudication involves both tactics and strategy. Potential disputes which can arise under a construction and engineering contract include disputed variations, loss and expenses for delay and disruption, extensions of time, and liquidated damages. It is said that adjudication is unsuitable for final account disputes and should be limited to progress payment issues only. Whereas composite and complex disputes cannot easily be accommodated within the summary procedure of adjudication, in the UK there is nothing which prohibits a "kitchen sink" final account adjudication.[18]

Parties in Ireland have a lot of flexibility as to how they might aggregate several payment disputes into one adjudication. However, part of your strategy might be to break down one large complex dispute into several smaller disputes. By limiting the issues to be tackled, the adjudication can have a clearer focus. A construction professional might decide one dispute better; a lawyer might decide another dispute better. If you like the way that an adjudicator reasons and decides one dispute, despite there being no authority for it in the Irish adjudication law, you may ask for that adjudicator to be nominated in subsequent disputes. Your reasons for this might be that the dispute involves similar principles, and as the adjudicator is familiar with the terms of the contract, it will minimise costs. You can expect the other party to protest the nomination, usually arguing that the adjudicator's previous performance was sub-standard. When parties want the nomination of a particular adjudicator, they sometimes write to the adjudicator to enquire as to availability. To avoid putting the adjudicator in an unseemly position, copy any such letters to the other party in the dispute. Then it will be straightforward for the adjudicator to write to both parties at the same time.

If you feel that the other party is disreputable and has beguiled you to work for them without any intention of proper payment, it is understandable that emotions such as anger, disappointment and frustration may be strong. Whereas such emotions can give you determination, resolve and strength, in dispute resolution they are usually your enemy.

One decision is whether to inform the other party in advance that you intend to commence adjudication or simply to confront them with the notice of intention for shock value, and to show that you are serious. No matter how challenging the circumstances are, remain calm and professional, and avoid negative emotions as far as possible. Negative emotions are an obstacle to clear thinking, and can impede making sensible choices. Forewarning the other party in advance provides a communication and dispute resolution

17 William Verry (Glazing Systems) Limited v Furlong Homes Limited [2005] EWHC 138 (TCC).
18 Quietfield Limited v Vascroft Contractors Limited [2006] EWHC 174 (TCC).

opportunity; it is possible to negotiate in parallel with adjudication, and if an agreement is reached to cease the adjudication. Whereas you may know how the other party is conducting itself, you are unlikely to know exactly what they are thinking. Remember that the adjudicator will have an impression of you from the way that you communicate and your conduct.

In some jurisdictions and situations, there is a sense that commencing adjudication will have damaging consequences for the relationship between the parties, and serious adverse consequences in respect of future work opportunities. My own view on that is to espouse the reasoning that a reference to adjudication is not a declaration of war. It is because the parties have a problem that they cannot resolve, and they need the help of an independent person to guide them by making a decision within a reasonably short time period so that their project can continue or conclude with the least possible disturbance. Such a decision can help inform the parties' conduct and thinking, allowing them to change their way of working to avoid future problems and expense.

16.5 Building the best team

Assistance required in adjudication can cover the commercial, legal, programming and technical spectrum. Before you hire external consultants, ask yourself if you can competently undertake the work yourself. If the answer is no, then you need to look for help elsewhere. How easy that is for you depends on your experience and size. Small firms may have to ask around, they can contact peers who have had similar problems, they can keyword search the internet, or ask professional organisations. Larger construction firms will have claims consultancies and law practices initiating contact regularly. Construction firms should keep commercial and personal resumes on file electronically or on paper, as you never know when a situation will arise where you require help.

In the adjudication jurisdictions claims consultancies operate at three levels:

- Tier one – full service commercial management and dispute resolution businesses with international offices offering a range of similar services.
- Tier two – made and located in the local jurisdiction, commercial management and dispute resolution businesses comprising a few key individuals, some staff and using sub-consultants (typically from a few to about 20 people).
- Tier three – commercial management and dispute resolution sole practitioners.

Commercial management covers pre- and post-contract services including assisting the preparation of contract documents, contract administration on site valuing variations and compensation events, procurement, etc.

Dispute resolution services cover claims management in respect of delay and/or increased cost: assisting in negotiation, often through a Scott schedule exchanged to distil and refine the parties' positions; helping you prepare for adjudication, arbitration and litigation; providing expert witness services covering commercial and technical matters, and planning and programming services. You can expect that tier one and tier two claims consultancies will have expertise in all or some of these areas, and particular expertise in domestic and international contract forms such as the Irish Public Works Contracts, the NEC3 and NEC4 contracts. Tier three sole practitioners should enjoy a reputation for providing one service; for example, delay analysis and planning.

Depending on you and your needs, you may feel comfortable building a relationship with a tier one, tier two or tier three firm. If you are going to do business with tier one or two companies, there can in effect be two relationships, one with the firm and another with their consultant(s). Dispute resolution services can be proactive. They can also be reactive, defending you against any documents alleging contractual entitlement to a monetised claim for extensions of time, recovery of additional costs, and preparing a counterclaim demonstrating entitlement to liquidated damages. Some claims consultancies specialise in working for contractors making claims, others in defending claims.

To procure claims consultancy services it is incumbent on you to prepare an appropriate assignment brief. This "terms of reference" document will describe the project and the services that you want. It will set a schedule for carrying out and completing the work. The claims consultancy will want to meet with you to establish a personal relationship, and to discuss the terms of reference in more detail. You should agree with the claims consultancy that it will submit a proposal for the provision of commercial or professional services before a specified time. The proposal is the claims consultancy's opportunity to demonstrate that it understands your requirements, and to detail how it can assist you to meet your goals. The proposal should include the names and biographies of the consultants it has available and whom it considers suitable.

Terms of reference will enable the claims consultancy to estimate the cost of doing the work. Often the claims consultancy will not give you a cash figure; they will say that it will take two claims writers (one technical and the other quantum) one month to undertake the work, and they will give you hourly rates. This can lead to problems: whereas as the claims consultancy envisages its operatives will work 50 (or more) hours per week, you envisage that the operatives will work 40 hours per work. At the outset, clarify the overall budget for cost and not expended hours.

You need to agree where the claims writers will work. There are two competing arguments: one is that it is best for them to work on the site where there will be easier access to documents and key personnel. The other is that an off-site (often head office) location is better because it enables the client's senior commercial staff ongoing access, and for the efficient pooling of personnel and support documentation to be used by all the team. Embedding claims writers with the client's team as a matter of practice reduces the claims consultancy's overheads and consequently should reduce hourly charge rates. It also enables the client to have confidence that the hours claimed in weekly timesheets are expended on their project.

Irrespective of the location, there should be a clear management structure in place to review progress, identify additional resources required, and remove any impediments such as lack of access to key project staff. In substantial claims or disputes, one way to achieve this is to empower a Business Management Team (the "BMT") comprising a small number of senior personnel with oversight and management responsibility. Ideally, the BMT will comprise a party senior executive, a claims consultancy director and an internal and/or external legal advisor.

Often tier one and tier two claims consultancies will have a range of people whose CVs are promoted to you. It is important that when dealing with tier one and tier two organisations that they do not dupe you into dealing with their firm; ensure that the consultants you interview are real and are the individuals that will be working for you. There can be "smoke and mirrors" where the people you meet become unavailable at short notice and you are offered a substitute. Do not accept this! But also understand that it is common business sense and commercially reasonable for the claims consultancy to qualify its proposal to say that the consultant is available now, but instructions to proceed must be received within two weeks; or that the consultant will be available within a maximum of four weeks after instructions to proceed are received. It is legitimate to enquire if the consultant is a staff member, or employed on a sub-consultancy basis. This can affect the degree of control the claims consultancy has over the consultant. It is not ethical or legitimate for the client to later approach any consultant and offer them direct employment.

Assess is your management style that the consultant will work for you under direction, or that the consultant works with you and enjoys a high degree of autonomy in achieving objectives? These are quite different things and not being clear about this can lead to a destructive mismatch. Consider how you will feel if the consultant identifies that responsibility for a problem lies with your organisation or your staff members.

Ideal candidates should possess experience appropriate to the specialisation, relevant qualifications and professional memberships. They should be able to offer references from people that you respect enough to trust. In the past, I have been able to satisfy a credibility assessment simply by reference to the endorsements on my LinkedIn profile: www.linkedin.com/in/adjudicator.

After evaluating your options and making your decision as to whom you want to employ, as a professional courtesy write to any claims consultancies that you did not choose. Unselected firms may ask for feedback as to the reasons why you made your decision. As far as possible, be frank, open and honest. If another firm offered a consultant you decided was more suitable, say so; if it is that they were too expensive, say so. If the claims consultancy is professionally managed; it will elegantly keep the door open. I have personal experience where clients later appointed my firm to assist with disputes on the same project, because the consultant/firm first appointed overpromised/underachieved and failed to meet expectations.

16.6 Adjudicator shopping

Arbitration encourages "arbitrator shopping", and it is natural that people with arbitration experience have a mindset for "adjudicator shopping". Rational parties want an adjudicator who they feel will increase their chance of success, and is likely to give them the best decision.

Parties in adjudication develop knowledge about adjudicators with whom they have been involved. Knowledge can be about how the adjudicator responded to jurisdictional challenges, whether the adjudicator controlled the process in a way they were comfortable with, whether the adjudicator was professional to work with, and based on the facts how the adjudicator previously decided a particular matter. Is the adjudicator detail-oriented? Is the adjudicator a construction professional or a lawyer? What does the adjudicator's CV say; is his background predominantly working as a consultant, a contractor or for an employer? Has the adjudicator published articles, given lectures or training, or written books indicating how he thinks? Was the billed time the adjudicator spent reasonable and resulted from their sufficient efficiency, experience and expertise?

Knowledge results from objective criteria, it is different to opinions based on personal perception. Because the adjudicator and the parties are humans, personal perception, and the mental processes that we use to form impressions of other people, is inevitable. Parties can arrive at different conclusions about an adjudicator depending on their own experiences and personality. Considering exactly the same information, one party may like and respect the adjudicator, another party may not. The consequence is that sometimes parties or their representatives have a misguided and unfair opinion of the adjudicator. If you have met the adjudicator, you will know if you have rapport.

Adjudicator shopping can lead to the adjudicator's appointment being contrived, and thereby undermining the credibility and fairness of adjudication. Experienced parties will be aware of an adjudicator's background and the details of his CV. Where possible they will suggest that the adjudicator has a combination of attributes that increases their chances of success in shopping; this can limit choice to one or a few people the party knows.

An earlier draft of the Code of Practice Governing the Conduct of Adjudications[19] did support the concept of describing the kind of adjudicator

19 SI No. XXX of 2014 – Construction Contracts Act 2013 (Code of Practice) (Adjudication) Order 2014; Code of Practice Governing the Conduct of Adjudications (25 July 2016) (Construction Contracts Act 2013).

you want, stating that a request to the Chair of the Minister's panel for the appointment of an adjudicator shall include "any information which it is considered will assist the chair in appointing an adjudicator with the appropriate expertise to deal with the payment dispute".

The final Code of Practice Governing the Conduct of Adjudications provides only that the application for the appointment of an adjudicator shall include "relevant details to identify the construction contract and any supporting information that may assist an Adjudicator in understanding the nature of the payment dispute". The Code states that the Chair of the Minister's panel may seek further information or clarification(s) from the applicant relevant to the nature of the dispute, and which the applicant should provide promptly. Applicants are advised that "No additional or other supporting information should be submitted by the applicant without a specific request for such information".

Superficially there is limited opportunity for "adjudicator shopping" in Ireland. However, this is not absolute. When providing "supporting information that may assist an Adjudicator in understanding the nature of the payment dispute" it seems to me entirely legitimate for the referring party to state that the payment dispute arises from a difference over contractual interpretation, or from a difference on how to interpret the specification for engineering works, or from a difference over evaluation of the final account, etc. This makes perfect sense as it would normally allow an assiduous adjudicator to decide if they are competent to determine the issues in dispute as required by the Code. Regrettably the abovementioned concept of an assiduous adjudicator is somewhat fanciful, as most adjudicators will say that they are competent to decide any issue, even if they have only superficial knowledge of the essence of the dispute. It is inevitable that this undermines the quality of adjudication decisions. Although in some disputes it matters less, commonly disputes are better decided by an architect, engineer, lawyer or quantity surveyor. It diminishes the adjudication process where the appointment of the adjudicator is random (for example in a taxi-rank system), and therefore the skills that the adjudicator has are not best suited to, or do not match the dispute. In reality, some adjudicators will be more competent to decide a particular dispute than others.

The referring party must refer the dispute to the adjudicator within a set time limit. In Ireland, it is within seven days of the adjudicator's appointment.[20] In the UK, it is not later than seven days from the date of the notice of adjudication.[21] If the referral is served out of time, the adjudicator does not have jurisdiction to act, unless the responding party consents to waive the irregularity of service.[22] Service within the seven days is mandatory.

20 Construction Contracts Act 2013.

21 The Scheme for Construction Contracts (England and Wales) Regulations 1998.

22 Hart Investments Limited v Fiddler & Another [2006] EWHC 2857 (TCC).

However, if the referring party does not like the nominated adjudicator and deliberately fails to serve the referral in time, can it go ahead with another reference for the same dispute or not? In the UK, although not serving the referral within seven days was considered a deliberate breach of the adjudication agreement, it may only give rise to entitlement to damages. Damages would be limited to the costs the responding party incurs in dealing with a fresh appointment, and therefore be minimal. However, as there is a right to give notice of adjudication "at any time" such a breach does not exclude the referring party from repeatedly withholding service of the referral so that the adjudication lapses, enabling it to wait for an adjudicator it liked.[23]

A similar situation in Ireland may be resolved in a different way. Firstly, although the notice of intention to refer a payment dispute can be served "at any time", the requirement is also that the referring party "shall refer the payment dispute to the Adjudicator within 7 days". This is a statutory requirement, not a contractual one. Therefore, in practice there may be a conflict between "at any time" and "within 7 days" for the courts to resolve. Secondly, the appointment procedure and policy of the Minister's panel may be that if an adjudication lapses because the referral was not served in time then the same adjudicator is appointed again. This would prevent what many would perceive as an abuse of process, in allowing the referring party *de facto* control to reject an adjudicator's appointment.

Is it legitimate to simply state your preference that particular adjudicators are not appointed? While this seems acceptable, there may be circumstances in which a stated preference could amount to a misrepresentation.[24] If you have a preference that certain adjudicators are not appointed, set out your reasons factually and honestly. Fraudulent misrepresentation irrevocably taints the appointment of an adjudicator. The elements of such a misrepresentation are "First, whether a false statement was made; secondly, whether any false statement was made fraudulently or recklessly; and thirdly, the effect of any such statement". [25]

For business promotion purposes, many adjudicators want adjudication users to retain their personal information. For example, as a minimum, name, contact details, areas of expertise, qualifications, etc. Sometimes concerns arise that inaccurate or inappropriate information is held about adjudicators. In such circumstances data protection legislation can assist.

The General Data Protection Regulation (the "GDPR")[26] became enforceable on 25 May 2018 and is designed to place legal constraints on the

23 Lanes Group plc v Galliford Try Infrastructure Limited [2011] EWHC 1035 (TCC).

24 CSK Electrical Contractors Limited v Kingwood Electrical Services Limited [2015] EWHC 667 (TCC).

25 Eurocom Limited v Siemens PLC [2014] EWHC 3710 (TCC).

26 General Data Protection Regulation (GDPR) (EU) 2016/679.

use and processing of personal data. Personal data is any information relating to an identified or identifiable natural person. Personal data may be collected only for specific legitimate purposes; must be adequate, relevant and limited to what is necessary; must be accurate and kept up to date; and may be stored only as long as is necessary and so as to ensure appropriate security, integrity and confidentiality.

A key part of the GDPR requires consent to be given by the individual whose data is held. Where organisations hold personal data, you can ask for the transfer of your personal data to you, and you are entitled to have full access to information on how your data is processed. You can make data access requests, and these must be executed "without undue delay and at the latest within one month of receipt of the request".

If you are an adjudication user retaining personal information, make sure it does not include opinions or is derogatory; faced with a subject access request this may cause you some considerable embarrassment and have legal consequences.

16.7 Appointing the adjudicator

Following a rigorous multi-stage assessment process the Public Appointments Service recommended suitably qualified persons for selection to the Department of Jobs, Enterprise and Innovation's panel of adjudicators (the "Minister's panel"). The Minister's panel comprises 30 construction, engineering and legal professionals meeting the requirements set out in Section 8 of the Act.[27] Section 8(5) of the Act mandates that only people qualified in accordance with Section 8(6) may be selected to the Minister's panel. Section 8(6) recognises the equivalence of qualifications duly obtained in other Member States of the European Union. Although a small-print complexity unlikely to affect the current make-up of the Minister's panel, post Brexit the Act does not provide for the recognition of UK construction and engineering qualifications.

Parties should start with the presumption that all the adjudicators on the Minister's panel will decide the dispute only on the evidence presented; that they have the skills to control the adjudication process; and that they are ethical, even-handed and unbiased.

The Act says that the parties may within five days, from serving the notice of intention, agree to appoint an adjudicator of their own choice or from the Minister's panel. If this confines the parties to *only* being able to agree the adjudicator after serving the notice, it seems an unnecessary curtailment of the parties' autonomy. Five days is a very short period of time, and in practice, to be effective usually the process of agreeing an adjudicator will have commenced before serving the notice of intention.

27 The Construction Contracts Act 2013.

When parties are in dispute they often find any agreement difficult, and where they are unable to jointly choose an adjudicator, either party to the construction contract may unilaterally apply to the Chairperson of the Minister's panel to seek the appointment of an adjudicator from the panel. The application must be in writing, and the applicant must send a copy of the application and the accompanying documentation to the other party/parties to the payment dispute at the same time.

The application must include:[28]

- The name, address and contact details of each party to the construction contract.
- Relevant details of the payment dispute to include the amount in dispute (even if the amount is zero), the nature of the payment dispute, and the site address.
- A copy of the notice of intention including any accompanying documents attached to that notice.
- The date as to when the notice of intention was served on the responding party/parties and how this was done.
- Relevant details to identify the construction contract and any supporting information that may assist an adjudicator in understanding the nature of the payment dispute. Where a written construction contract exists, this must be attached.

The Department of Business, Enterprise and Innovation website incorporates a section related to the Construction Contracts Adjudication Service (the "CCAS"). The CCAS has provided some template forms. Form Number 5 is the "Application to the Chairperson of the Construction Contracts Adjudication Panel for Appointment of Adjudicator" form.[29] If an applicant is applying to have an adjudicator appointed by the Chairperson to a number of payment disputes arising under the same construction contract, separate applications must be submitted for each individual dispute.

The Form Number 5 application asks the applicant to provide information on the "Applicant Party" (Part A), the "Respondent Party" (Part B), and the "Applicant" (Part C). In respect of the applicant it is requested to "Provide the name and contact details of a person, who is authorised by the applicant, to receive correspondence from the Construction Contracts Adjudication Service of the Department of Business, Enterprise and Innovation in relation to this application".

My guidance is that when a referring party makes an application for the appointment of an adjudicator, it provides the names and email addresses of

28 Code of Practice Governing the Conduct of Adjudications (25 July 2016).

29 www.dbei.gov.ie/en/Construction-Contracts-Adjudication-Service/Construction-Contracts-Adjudication-Service.html. Form 5 Application to Chairperson for Appointment of Adjudicator.

a responsible person for the referring party and for the responding party, the name and email address for the referring party's representative (if any), and the responding party's representative if known.

As an adjudicator I am reluctant to proceed with an adjudication without writing directly to the parties, as well as their representatives. It is prudent for the adjudicator to do this for a number of reasons. It is the parties that are in dispute and not their representatives. The adjudicator has no way of knowing the bona fides of a third-party applicant who is not responsible for the consequences of the adjudicator's decision. Apart from other significant implications, imagine circumstances where the adjudicator gets to the end of an adjudication and decides that the referring party is liable for their fees and expenses. However, when the adjudicator presents their fee invoice, the referring party informs them that it was unaware that an adjudication had proceeded, or similar.

The timescale for conducting adjudication is very tight, and often the adjudicator will ask for a response to a particular matter within one or two days and which will require party input, there is little time for building in delay relying on the party representatives to shuttle emails. The parties benefit from being directly connected to the authority of the adjudicator, for example it can help inform them as to whether they should comply with the adjudicator's decision. The adjudicator's control of process and even-handedness should be visible. Messages from the other party or the adjudicator should not be attenuated. It ensures that the parties have ongoing oversight of the performance of their representatives. It is the parties who are paying the adjudicator, and it is helpful if the service the adjudicator is providing is visible to them.

Some party representatives can be truculent and refuse to provide the adjudicator with the name and email address of a responsible person of the party instructing them. I respectfully ask the party representative to provide the information, or for them to provide a letter from the instructing party to the adjudicator, confirming that it does not wish to receive any correspondence in respect of the adjudication. Where neither is forthcoming, I will print out all correspondence and post it to the party's registered address with a signature required on delivery, and inform the party representative that the party instructing them will be responsible for the wasted expense and time, irrespective of the adjudication outcome.

It is usual for party representatives to correspond on behalf of the party. Unless the circumstances are exceptional, you should avoid the situation where the adjudicator is receiving correspondence from more than one person on behalf of the referring party or the responding party.

In Ireland, there are other organisations that have established adjudication panels. In 2017, I was Chair of the CIC Working Group drafting the Users' Guide to Adjudication: Ireland.[30] The Working Group comprising Damien

30 CIC Users' Guide to Adjudication: Ireland, 26 July 2017.

Keogh, Gillian Birkby, James Bridgeman, Anthony Hussey, Niav O'Higgins, Adam Williamson and I considered the viability of professional bodies or other organisations establishing their own panels and procedures to nominate adjudicators. The Working Group advised:

> There is nothing to prevent any organisation doing this and there are no external controls. If your contract specifies a particular adjudicator or the nomination of an adjudicator from a panel established by one of the professional institutions the generally accepted principle of freedom of contract suggests that you must comply with that agreement.
>
> However, in respect of adjudication under the Construction Contracts Act 2013, the position is unclear. Whereas the Code of Practice suggests that an Adjudicator can be appointed through the Contract, the Act itself does not provide for this. The Act only provides that the Adjudicator is either appointed by agreement within the five-day window pursuant to Section 6 (3) or the Adjudicator is appointed by the Chairperson. Whether or not a Contract can specify an organisation other than the Construction Contracts Adjudication Service for the purpose of making an appointment is also unclear.

This leads to a conundrum which will in due course fall to the courts to resolve. Where the parties in Ireland have chosen an adjudicator nominating organisation (an "ANO") to nominate an adjudicator, the case law from the UK is that for the adjudicator to enjoy jurisdiction they must be appointed according to the correct contractual provisions.[31] Conversely, the Act only provides that the adjudicator is either appointed by agreement within the five-day window, or the adjudicator is appointed by the Chairperson. If the adjudicator has not been properly appointed, they will not have jurisdiction; any purported decision will be a nullity, and will not be enforced by the court.

Whereas it may be legitimate for the parties to agree a contract provision for an adjudicator appointment mechanism from a list other than the Minister's panel, the ANO making the appointment should *prima facie* check that it has the authority to nominate the adjudicator. The ANO should be aware that whereas the Code provides that the Chairperson and the Department of Jobs, Enterprise and Innovation has no liability whatsoever in respect of any loss that arises from the operation of the Code, the ANO does not enjoy similar protection from suit. In the circumstances where an adjudicator's decision is not enforced because the adjudicator has been appointed improperly, the adjudicator should expect to forfeit their fees, and the ANO may face a claim for consequential loss from the parties.

If an ANO is going to accept suggestions as to the attributes, or the name of the person it will nominate, it might consider whether it should give

31 Twintec Limited v Volkerfitzpatrick Limited [2014] EWHC 10 (TCC).

notice of any suggestions to the other party.[32] As time is short, it can be difficult for the ANO to do this during the nomination process, but it could include a copy of any letter accompanying the application when it makes the adjudication appointment.

It is not permissible for the parties to agree that one of them has the right to nominate an adjudicator. This would be contrary to the policy of the adjudication legislation and inherently unsound. It would empower one party to appoint someone who although may be wholly independent, might be perceived as sympathetic to that party or its position, or who for other reasons would not be acceptable to the other party.[33]

It is unclear how the Chairperson of the Minister's panel selects adjudicators. The normal practice of some UK adjudicator nominating bodies ("ANBs") is to appoint adjudicators on a taxi-rank basis, allocating adjudicators to disputes in sequence. Some ANBs will try to match an adjudicator with the best experience and qualifications in terms of the subject matter of the dispute. Whereas there are benefits of this tailored approach, as pointed out above, one of the dilemmas that can arise is to what degree parties can have a say over the appointing bodies' choice of adjudicator. Where the Chairperson has separate applications to appoint an adjudicator to a number of payment disputes arising under the same construction contract, will the Chairperson appoint the same adjudicator or different adjudicators? Based on the same fact matrix, different adjudicators can arrive at opposing positions in their decisions, this is something which will bring adjudication into disrepute.

Where there is a series of adjudications the referring party may ask a particular adjudicator to be appointed because his "nomination would save the parties both costs and time as he is already familiar with the contract and reading-in time would be minimised". Anticipating a nomination application, responding parties can proactively object to the selection of a particular adjudicator citing a range of reasons.[34] They can also object reactively after appointment, but at this stage any discretion the appointing body might wish to exercise is spent. As adjudicator, I have received a copy of submissions about me (both good and bad) the referring party or the responding party has made to an ANB; but it is not clear whether the other party also has visibility of the submissions. Although it is not part of my role as adjudicator to provide one party or the other with copies of any submissions, in my introduction letter for UK nominations I always precisely set out a list of documents the adjudicator-appointing organisation has sent to me, and the number of pages of each. It is for the parties to be sure that they have all the information.

32 Makers UK Limited v The Mayor and Burgesses of the London Borough of Camden [2008] EWHC 1836 (TCC).

33 Sprunt Limited v Camden LBC [2011] EWHC 3199.

34 Arcadis UK Limited v May and Baker Limited (t/a Sanofi) [2013] EWHC 87 (TCC).

In section 16.8 of this chapter, "Accepting the nomination", I discuss how the Construction Contracts Adjudication Service interacts with adjudicators during the appointment process, and the information it makes available to prospective adjudicators.

It is important that the parties do not make any misrepresentations during the adjudicator nominating process; as such could nullify the appointment and deprive the adjudicator of jurisdiction.[35] It is better if parties limit unilateral contact with adjudicators, before, during and after an adjudication; the same goes for adjudicators having unilateral contact with individual parties. It can be misconstrued by the losing party, even if entirely innocent.[36] Notwithstanding this, from time to time an adjudicator will receive an email from one of the parties (not copied to the other party) stating that an adjudication is imminent, and that they wish to put your name forward to the other party for agreement, failing which it will apply to a named ANO to ask the organisation to nominate an adjudicator. They will enquire as to availability to act and ask for a copy of terms and conditions. If I have no conflict in acting as adjudicator and am available, I reply providing the information requested. I am mindful such an enquiry email does not indicate one way or the other if the party considers me suitable. It could be trying to reach out to the adjudicator and build a relationship; on the other hand, it could be looking for reasons to object to the adjudicator's appointment. In the interest of transparency, I will send a copy of the email I send to the party to the ANO.

Although I have seen it happen, it should go without saying that an ANO should not give the referring party a choice of names, and nominate the adjudicator after the referring party has made an election. The responding party is excluded from such a process, and it invites a squabble over the fairness of the appointment.

16.8 Accepting the nomination

Under the Code of Practice,[37] normally within seven days after the receipt of the application, the Chairperson of the Minister's panel appoints an adjudicator, and the Construction Contracts Adjudication Service of the Department of Jobs, Enterprise and Innovation notifies the parties in writing.

The Code requires that an adjudicator so appointed by the Chairperson to a payment dispute shall, within two days of the appointment, request the parties in writing to make a disclosure of any information indicating any potential conflict of interest that may arise from the person's appointment as adjudicator. The adjudicator shall draw the parties' attention to the provisions of paragraph 32 of the Code of Practice. The adjudicator is required at the

35 Eurocom Limited v Siemens plc [2014] EWHC 3710 (TCC).

36 Paice & Anr v MJ Harding (t/a MJ Harding Contractors) [2015] EWHC 661 (TCC).

37 Code of Practice Governing the Conduct of Adjudications (25 July 2016).

same time to provide the parties with his terms and conditions of appointment, including the basis for his fees, costs and expenses.

The adjudicator should also disclose any potential conflict of interest as far as he or she knows. In Coulson on Construction Adjudication[38] Mr Justice Coulson notes

> Should the adjudicator consider, of his own volition, and regardless of the points that may or may not have been made by the parties, whether or not he has the necessary jurisdiction? It is thought that he should. If the adjudicator does not have the necessary jurisdiction, then, *prima facie*, his decision is a nullity.

If the information disclosed indicates a potential conflict of interest, an adjudicator may only proceed with the adjudication where they are satisfied that any disclosures are frivolous or vexatious; that no professional or ethical concerns arise; and that no actual conflict of interest exists.

In adjudication if I have any concerns about impartiality, independence, or disclosure, regardless of the stage of the process, I use the International Bar Association Guidelines on Conflict of Interest in International Arbitration, applying a robust common-sense approach.

Agreeing who will act as adjudicator in advance offers advantages. It can give the parties confidence that the adjudicator will have the experience, qualifications and skills they require. It allows them to settle the adjudicator's schedule of conditions and remuneration in advance. The parties are more likely to accept as final and binding a decision from an adjudicator that they have jointly chosen. However, good, respected adjudicators are often busy, and therefore might not be available when needed. In addition, whereas the parties may have chosen the adjudicator for his technical expertise, the dispute may involve complex legal issues better dealt with by a lawyer, and vice versa.

In the UK, I have been nominated as adjudicator by ANBs such as the Chartered Institute of Arbitrators, the CIC, the Institution of Engineering and Technology, the Institution of Mechanical Engineers, the Royal Institute of British Architects and UK Adjudicators. In most adjudications, for the adjudicator the nomination process will start with an email or a telephone call.

All of the ANBs above and others will provide the prospective adjudicator with the referring party's request for the nomination of an adjudicator, and the notice of intention to refer a dispute to adjudication. Only when the prospective adjudicator has basic information about who the parties and their representatives are, the nature of the dispute and the amount in dispute, can they properly consider if they have the experience and skills to decide the

38 Coulson, Peter, *Coulson on Construction Adjudication*, 3rd edition, Oxford University Press (2011) ISBN-13: 978–0199597222 ISBN-10: 0199597227.

dispute; whether they have the time available; if there are any conflicts of interest with the parties or their representatives (often they will not know who is representing the responding party), and if are there any circumstances that would give the impression of bias.

As at April 2020, in Ireland the Construction Contracts Adjudication Service takes a different approach to UK ANBs and as a matter of policy the CCAS does *not* provide a prospective adjudicator with the Form Number 5 application and support documentation received from the applicant party for the purpose of the appointment of an adjudicator to the dispute.

The CCAS will only send the prospective adjudicator the names of the parties, which makes it practically impossible for the prospective adjudicator to comply with the Code of Practice which requires that they should only accept an appointment to a payment dispute under the Act if they are able to give the adjudication the time and attention which the parties are reasonably entitled to expect, and they believe that they are competent to determine the issues in dispute. The prospective adjudicator is placed in an invidious position of having to accept or decline the nomination effectively blind as to what the dispute is about, or the amount of time that may be required to decide it.

After receiving the adjudicator confirmation of nomination letter from CCAS, the CCAS's non-disclosure of information policy also renders it difficult for the adjudicator to efficiently take control of the adjudication from the get-go, by crafting and sending the parties a considered bespoke introduction and terms and conditions letter targeted at the dispute. The CCAS says that it is of course open to an adjudicator, if they consider it necessary for an introduction letter for the purpose of paragraph 19 of the Code of Practice Governing the Conduct of Adjudications, to request the applicant party to provide the notice of intention to them in advance of the other documentation. Whereas that it quite true, having to do so incurs cost and consumes valuable time.

Additionally, as ANBs are fallible having an experienced adjudicator look over the information provided by the applicant party before an adjudicator is appointed can help to mitigate inaccuracies and mistakes. For example, the referring party may inadvertently use different names to identify the responding party in the notice of intention and in the application for the appointment of an adjudicator, particularly where trading names are in play. The CCAS non-disclosure of information policy deprives the CCAS of the prospective adjudicator's expertise during the appointment process. The appointment email and letter from the CCAS will enclose the request "Please confirm receipt of this email". Adjudicators should reply to the CCAS quickly confirming receipt of their appointment letter, as the normal protocol of CCAS is not to write to the parties until after that has happened.

From experience I know that from time to time UK ANBs have difficulty finding a suitable and available adjudicator to act. For example, adjudicators may be unwilling to act if they know the sum in the payment dispute is very small. Even with that as context, the CCAS policy of not providing prospective adjudicators with more complete and readily available information during the adjudicator appointment process is an impediment to the smooth progress of an adjudication.

Using a service, especially for the first time can be stressful. In adjudication, it is natural for the parties to feel uneasy; it is a worrying situation for them. Irrespective of the parties' attitude they will respect and trust the adjudicator more if they are confident that they are in a skilled professional's safe hands. Because of that, taking control of the process at the outset and setting expectation in a courteous, polite and firm manner is important. When it comes to deciding on whether to accept the decision as final, how you as adjudicator made the parties feel during the adjudication will have an influence. It has an influence because how you made the parties feel during the adjudication goes toward establishing the credibility of the decision's author.

Because of that, in the UK I will ask the ANB what day they are at in the nomination process, and if there is time to wait for a few hours or until the following day before announcing the appointment. I will use that time to set up a basic adjudication administration system, and draft my letter of introduction. The introduction letter will be addressed to the parties, and if possible to both their representatives. It will describe my appointment as adjudicator, and set out what information the ANB has provided, in addition to confirming my contact details, and providing a communications protocol such as:

> The communications protocol for this adjudication is that except in the event of an emergency, all communication is to be in writing and cross copied to the other party and its representative (if any). Please would the parties ensure that all correspondence is marked in a manner that indicates that this request is being complied with. Correspondence and other documents of A4 or smaller size may be sent as an Adobe Acrobat attachment to an email or by facsimile.
>
> Where documents are sent by email, the tracking option should be set to "Request a Delivery Receipt" and only when you receive this should you assume that the other party or its representative has received the information. Where information is not suitable for transmission by email or facsimile it is to be sent by guaranteed next morning delivery.

The introduction letter will include a provisional timetable for the adjudication. For example:

> As it may be helpful to the parties, I provide an **outline** programme for this adjudication, based upon a decision to be made within 28 days, as follows:
>
> Based upon the "Referral Notice" being received by me today Monday 21 May 2018.

- The Responding Party will Respond to the Referral no later than Tuesday 5 June 2018 (Monday 28 May 2018, a holiday, is excluded from the reckoning period of time).
- If the Referring Party wishes to make any Reply it may do so in writing no later than Tuesday 12 June 2018.
- In the event that the parties or I wish to have a meeting, I provisionally set aside Friday 17 June 2018, to be confirmed or vacated as may become appropriate during the course of the adjudication.
- Accordingly, the parties should expect the adjudicator's decision not later than Tuesday 19 June 2018.

The Act requires the adjudicator to reach a decision within 28 days beginning with the day on which the referral is made or such longer period as is agreed by the parties after the payment dispute has been referred. The adjudicator may extend the period of 28 days by up to 14 days, with the consent of the party by whom the payment dispute was referred. Accordingly, I sometimes state in my introduction letter that if I decide it will be of benefit, when I receive the "Referral Notice", I will write to the parties asking for consent or agreement to extend the period for making my decision beyond 28 days.

As an adjudicator you should not arrogantly set out an adjudication timetable which exceeds 28 days, and tell the parties that if you have not heard from them to the contrary within a set period of time, you are presuming your timetable is accepted. Even if the parties acquiesce to the timetable that suits you, focused on your selfish interests and not the interests of the parties, it will diminish the respect the parties have for you. Although it is not mentioned in the Act or the Code, officially designated Irish public holidays such as Christmas Day, Boxing Day, New Year's Day, etc. should be excluded from the reckoning period of time.

Provided the adjudicator has notified the parties, the adjudicator has the power to appoint experts, assessors or legal advisors. In my introductory letter, I advise the parties that in the event that I require legal advice, which lawyer's services I propose to use and enclose a copy of the adjudicator–lawyer agreement. On occasion one of the parties (usually their legal representative) will advise that because of prior dealings there would be a conflict of interest in me using the particular lawyer. It is better to have this information available at the beginning of the adjudication when time is less critical.

The introduction letter also encloses a copy of the terms and conditions as provided for in any adjudication procedure agreed by the parties, for example the adjudicator's contract as provided for under the NEC family of contracts. Where there is no agreed adjudication procedure, I enclose a copy of my own schedule of conditions for acting as adjudicator. Sometimes both parties agree to the schedule of conditions, sometimes one party agrees to the schedule of conditions, and sometimes neither of the parties agrees to the schedule of conditions. Unless it is something that is easily resolved by consent, such as a proposed reduction in the hourly rate, the adjudicator should not haggle with either of the parties over their terms for acting. Absent agreement with the parties, the adjudicator should continue to act until the unlikely event that both parties confirm that they do not want him to continue.

16.9 Adjudication procedural rules

It is helpful to distinguish three different documents that can be interrelated and overlap: the adjudicator's agreement, construction contracts containing adjudication clauses, and procedural rules applicable to adjudication (Figure 16.9).

The adjudicator's agreement is the terms that the adjudicator enters into with the parties for providing adjudication services. This can be a bespoke agreement, or a standard form agreement used when appointing the adjudicator under, say, the JCT or NEC family of contracts. At minimum, it will record the parties' names, the adjudicator's name, a description of the dispute, and the basis for paying the adjudicator for his professional services.

Standard forms of contract can include comprehensive clauses that are consistent with the law controlling any adjudication. For example, the NEC3 and NEC4 contract option W2 requires the parties to adjudicate disputes first, limiting their discretion to proceed directly to arbitration or litigation. It clearly widens the scope of adjudication to include disputes "arising under" or "in connection with" the contract. It also fixes the status on the adjudicator's decision, in that it is final and binding if neither party notifies the other of dissatisfaction with the decision within the prescribed time.

Following the inception of adjudication in the UK, organisations such as CEDR, CIC, ICE and TeCSA have developed respected adjudication

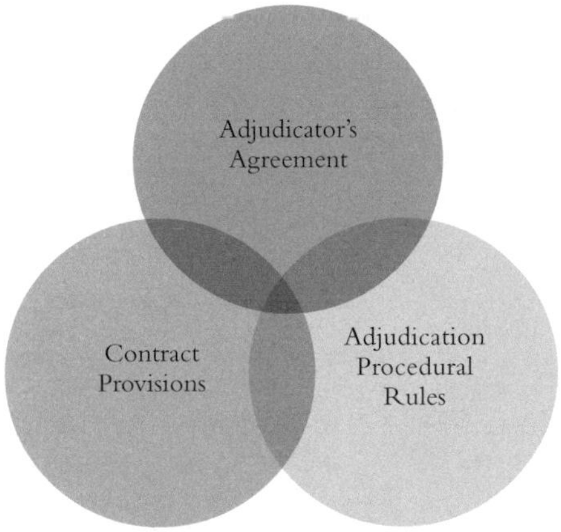

Figure 16.9 Adjudication procedural rules.

procedural rules. Rather than including adjudication provisions in the contract, some standard forms rely on third party adjudication rules. For example, the RIBA Standard Form of Agreement for the appointment of an architect provides that "The adjudication procedures and the Agreement for the appointment of an Adjudicator shall be as set out in the Model Adjudication Procedure published by the Construction Industry Council current at the date of the reference".[39]

In the early days of adjudication in the UK, the Construction Industry Board expressed concern regarding the problems that had arisen due to "bespoke" adjudication processes, and recommended the enshrinement of a single procedure in legislation.[40]

However, adjudication procedures can improve the adjudication process, promoting consistency, effectiveness and efficiency. They can confer helpful additional powers. They can cap or reduce the fees charged by adjudicators.[41] In the case of "low value disputes", defined as where the sum in dispute is £50,000 or less, they can link the adjudicator's fees to the sum claimed. This provides parties with certainty as to their maximum financial exposure in respect of fees, and in that way can make adjudication more accessible, especially to smaller firms.[42] They can target the needs of particular sectors or types of disputes.[43] They can confer a contractual right to adjudication in contracts that fall outside of the definition of construction contracts in the legislation.

The TeCSA adjudication rules include notable examples of additional powers that the parties can confer. Agreeing to use the TeCSA adjudication rules widens the scope of adjudication to include further matters the parties agree should be included, and any further matters that the adjudicator determines must be included in order that the adjudication may be effective and/or meaningful. In addition, the adjudicator may decide upon his own substantive jurisdiction, and as to the scope of the adjudication.

Adjudication's procedural rules can enhance the adjudication process, but cannot subvert or undermine adjudication law. Where adjudication rules fail to comply with the law, the courts have taken a strict approach. For example, where adjudication rules extended the time available to the adjudicator to make his decision, which was inconsistent with UK adjudication law, the

39 RIBA Standard Form of Agreement for the Appointment of an Architect (SFA/99) Updated April 2004: Article 9.2.1. Adjudication: England and Wales.

40 Improving Adjudication in the Construction Industry. A Consultation Document, January 2003, ISBN 0 7559 0653 5.

41 TeCSA Adjudication Rules, Version 3.2.2, published in 2018.

42 http://cic.org.uk/services/cic-low-value-disputes-adjudication.php.

43 Rules for adjudication. For use with the JCT Building contract for a home owner/occupier and JCT Consultancy agreement for a home owner/occupier. JCT 2011.

adjudication rules were struck out, and the provisions of the Scheme for Construction Contracts applied.[44]

Most procedural rules applicable to adjudication provide that in the absence of the parties' agreement the publishing body (CEDR, CIC, ICE or TeCSA) will appoint the adjudicator. As discussed above this may be contrary to the Act which provides that if the parties fail to choose the adjudicator, the Chair of Minister's panel will appoint the adjudicator on application. One argument is that such a restriction helps to maintain good order in the adjudication system, providing a *de facto* adjudicator quality control mechanism. In Malaysia, the Asian International Arbitration Centre exercises a similar degree of authority over the appointment of adjudicators.

Usually Ireland and the UK share the ethos that the parties' freedom to contract should be as wide as possible. If the parties join and want to agree an adjudication procedure, which they regard as fair and which meets their needs, they are more likely to accept outcomes they do not like. Concomitant to the Act, it remains to be seen if the availability and use of adjudication procedural rules will develop for use in Ireland.

44 Epping Electrical Company Limited v Briggs and Forrester (Plumbing Services) Limited [2007] EWHC 4 (TCC).

16.10 Adjudication stages and timetable

	Dispute Formation	Notice of Intention	Selection of Adjudicator	Referral	Response	Reply to Response	Decision
Referring Party	APPI 15.3; 15.4; 15.5; 15.6; 16.2; 16.3; 16.4; 16.5; 17.3; 17.6.	APPI 15.6; 15.12; 15.13; 16.11.	APPI 16.6; 16.7; 16.8.	APPI 15.12; 15.13; 16.12; 16.15; 17.7; 17.16.	APPI 16.13.	APPI 16.14.	APPI 16.18; 16.19; 16.20.
Responding Party	APPI 15.3; 15.4; 15.5; 15.6; 16.2; 16.3; 16.4; 16.5; 17.3; 17.6.		APPI 16.7; 16.8.	APPI 16.15.	APPI 15.12; 15.13; 15.13; 17.7; 17.16	APPI 16.14.	APPI 16.18; 16.19; 16.20.
Adjudicator			APPI 15.14; 15.15; 16.6; 16.7; 16.8.	APPI 15.20; 16.10; 16.15; 16.16; 17.14; 17.19.	APPI 16.13.	APPI 16.14.	APPI 15.16; 15.17; 15.18; 17.8; 17.9; 17.10; 17.11; 17.12; 17.13.

Construction Contracts Act 2013:

(6) The adjudicator shall reach a decision within 28 days beginning with the day on which the referral is made or such longer period as is agreed by the parties after the payment dispute has been referred.

(7) The adjudicator may extend the period of 28 days by up to 14 days, with the consent of the party by whom the payment dispute was referred.

Adjudication Practice and Procedure: Ireland – APPI

Figure 16.10 Adjudication stages and timetable.

16.11 Notice of intention

The first procedural step in adjudication is the notice of intention, and as set out in the Code of Practice Governing the Conduct of Adjudications it must contain the following details:

- The name, address and contact details of each party to the construction contract.
- Relevant details of the payment dispute to include the amount in dispute (even if the amount is zero), the nature of the payment dispute, and the site address.
- A copy of the relevant payment claim notice, and any response to that payment claim notice.
- Relevant details to identify the construction contract and any supporting information that may assist an adjudicator in understanding the nature of the payment dispute. Where a written construction contract exists, this must be attached.

The notice of intention in Ireland is analogous to the notice of adjudication in the UK. The notice is the foundation for the scope of jurisdiction of the adjudicator.[45] Because the adjudicator can only decide matters set out in the notice, it is an important document. The referral document must confine itself to the issues set out in the notice, and therefore the notice bounds the content of the referral. Where the exact nature of the dispute is unclear, the court might adopt the pragmatic perspective of considering the notice and the referral together with relevant correspondence.[46] Notwithstanding that, it is vital that the notice of intention is clear and comprehensive, and covers every aspect of the dispute the referring party wants decided.

Whereas the Scheme in the UK requires that the notice of adjudication shall set out briefly "the nature of the redress which is sought", the Code of Practice in Ireland does not impose a similar requirement on the notice of intention. This lacuna can get an unaware referring party into some considerable difficulty. If the referring party fails to set out clearly what it wants the adjudicator to decide it could deprive the adjudicator of jurisdiction, the corollary is also that it could confer the adjudicator with unwanted jurisdiction.

The Construction Contracts Adjudication Service provides Form Number 4 "Notice of Intention to refer a Payment Dispute to Adjudication – pursuant to section 6(2) of the Construction Contracts Act 2013" which only asks for the referring party to describe the payment dispute. Do not feel constrained to using only Form Number 4, the referring party can use the form to refer to an additional and

45 William Verry (Glazing Systems) Limited v Furlong Homes Limited [2005] EWHC 138 (TCC).

46 OFC Building Services Limited v Interior Dimensions Contracts Limited [2009] EWHC 248 (TCC).

complete document which sets out all desired elements for the adjudicator's decision, including for example allocation of the adjudicator's fees and expenses, and interest.

Whereas paragraph 5 of the Code of Conduct states that the notice of intention should include any supporting information that may assist an adjudicator in understanding the nature of the payment dispute; as at April 2020, as a matter of policy the CCAS does *not* provide a prospective adjudicator with the Form Number 4 notice of intention and any support documentation received from the referring party and sent to the responding party.

The adjudication process can work best when it is incremental, with a series of limited adjudications during the course of a project, rather than one massive dispute at the end. This is particularly so where the parties have an opportunity to modify the way that they work because of an adjudicator's decision. The title of a narrow and limited notice might be there is a "dispute in respect of the valuation of variation number 8 changes to the hydraulic system for laboratory one". The title of a wide and unlimited notice might be there is a "payment dispute in respect of the final account including variations, extension of time, loss and expense for delay and disruption and liquidated damages".

As limited information is required, it might seem that drafting the notice is relatively straightforward. However, a carelessly crafted notice can deprive the adjudicator of jurisdiction; on the other hand, it can give the adjudicator jurisdiction that you, the referring party, did not intend or want.[47] You cannot restrict the other party's ability to defend itself, by using the terms of the adjudication notice. Limiting the notice to a dispute over one valuation will not deprive the adjudicator of jurisdiction to consider a defence of overpayment on another valuation.[48]

As the dispute will be over payment of money, in the notice always leave the precise sum to be decided open by including a codicil such as "or such other sum as the adjudicator may decide". Use a similar form of words in respect of a monetised claim for time.

There is a robust connection between being successful in adjudication and the way that you define the dispute. The notice and the referral go together to do that. I advise parties that they cannot competently complete the notice of intention before they complete the referral notice.

Although business use of email to conduct day-to-day affairs is endemic, it is important to check and comply with the notice provisions in the contract. The Act confirms that the parties may agree the manner by which notices under the Act shall be delivered, and that in the absence of an agreement, a notice may be delivered by post or by any other effective means.[49] Contract notice provisions exist to ensure that important notices are served in the agreed manner therefore guaranteeing that the recipient would not be in any doubt as to their

47 Wales and West Utilities Limited v PPS Pipeline System GMBH [2014] EWHC 54 (TCC).
48 Pilon Limited v Breyer Group Plc [2010] EWHC 837 (TCC).
49 Construction Contracts Act 2013.

importance.[50] My recommendation is that sending the notice of intention by email should always be accompanied by sending a hard copy by special delivery to the intended recipient at their office address so that you can be sure that it will come to the attention of either the person concerned or the person covering any absence.

16.12 Referral

In a UK adjudication the "notice of adjudication" is equivalent to the "notice of intention" in Ireland. In a dispute, the notice of adjudication was sent on Wednesday 26 May 2010, and I was appointed as adjudicator by the Royal Institute of British Architects (the "RIBA") on Wednesday 2 June 2010 four business days later. On 10 June 2010, I received an email from the referring party:

> Sorry for the delay in submitting our Referral Notice but we received a correspondence from the Architect on Tuesday (the first correspondence from him in 5 years!). Unfortunately, it was very unhelpful as he has now changed his position and issued further documentation. This has forced me to reconsider the contents of the Referral Notice, substantially increasing the amount of paperwork within it. I hope that I should have the revised document completed by Friday and posted to you over the weekend. Trusting this does not cause undue inconvenience.

I explained that the timetable for the sending the referral notice to the adjudicator was subject to the time restrictions laid down by the Scheme for Construction Contracts,[51] and must be served within seven days of service of the notice of adjudication. In the absence of the responding party waiving the irregularity of service, the consequence of a one-day delay deprived me of jurisdiction to act as adjudicator, and any purported decision by me would not be enforced.[52] In the same dispute, due to its late service of the referral notice, the referring party went through the RIBA adjudicator nominating process three times, and consequently incurred adjudicator nomination fees three times. The responding party can expressly waive any irregularity of the service of the referral; it might also do so by its conduct.[53]

The second procedural step in adjudication is referring the payment dispute to the adjudicator (the "referral"), and as set out in the Code of Practice it must contain the following details:[54]

50 Ticket2Final OU v Wigan Athletic AFC Limited [2015] EWHC 61b (CH).

51 Scheme for Construction Contracts in Northern Ireland Regulations (Northern Ireland) 1999.

52 Hart Investments Limited v Fidler & Anor [2006] EWHC 2857 (TCC).

53 KNN Coburn LLP v GD City Holdings Limited [2013] EWHC 2879 (QB).

54 The Code of Practice Governing the Conduct of Adjudications (25 July 2016).

- The name, address and contact details of each party to the construction contract.
- Relevant details of the payment dispute to include the amount in dispute (even if the amount is zero), the nature of the payment dispute, and the site address.
- A copy of the notice of intention including any accompanying documents attached to that notice.
- The date when the notice of intention was served on the responding party/parties and how this was done.
- The contentions on which the referring party intends to rely upon to support their case.
- Relevant details to identify the construction contract and any supporting information that may assist an adjudicator in understanding the nature of the payment dispute. Where a written construction contract exists, this must be attached.

The referral will usually comprise the referral notice and supporting documentation. The referral notice should be conceived so as to be capable of standing alone identifying the material facts upon which the claim is based, and allowing for the nature of the dispute to be identified without reference to the supporting documents.

When you begin to prepare the referral notice, you should remember that it is very unlikely that the adjudicator knows anything about either the contract or your payment dispute. Even if he has some prior information, not that information but the information you provide will serve as the basis for his decision. As the referral notice is a keystone document and your principal submission in adjudication, it is important to have clear objectives in place when drafting it. If you are successful in providing adjudication support services, you will know that the core objective of the referral notice is to create a sure path to the money. The objective of the response is to make the path to the money difficult or impossible to traverse. In preparing the referral you should have at least five objectives in mind: 1) to set out completely and simply your positive case providing all the supporting evidence required; 2) to prepare for the defences that the response will include; 3) to anticipate jurisdictional challenges and consider how to deal with them; 4) to control the adjudication process; 5) to prepare all ancillary documents such as application forms, covering letters and the notice of intention.

To avoid procedural challenges, the referral notice should be properly titled, and include the information provided for in the legislation and any procedural rules. If the referral notice does not comply, that might lead to a jurisdictional challenge in the name of alleged inadequacies in the referral process. Where court support toward adjudication is robust, it will look at any alleged documentation shortcoming critically, however there may be cases where the documents included with the referral are so deficient that it affects the validity of the adjudication process.[55]

55 PT Building Services Limited v ROK Build Limited [2008] EWHC 3434 (TCC).

Avoid the unseemly situation where you send the referral notice electronically to the adjudicator and the responding party on the seventh day, but find that you have to send hard copy accompanying documents by overland mail guaranteed to arrive the next working day. Although the English Courts have distinguished the absolute necessity of the adjudicator and the responding party receiving both the referral notice and supporting information within seven days, this will needlessly expose you to the potential jurisdictional challenge that the payment dispute was referred out of time, and the adjudication is therefore a nullity.[56,57] Often the referral is sent by email or made available for the adjudicator and responding party to download using Cloud computing. Avoid using long file or document names, and be thoughtful in how you structure zipped information. Zipped directories within zipped directories are absolutely to be avoided. Carelessness in using Cloud computing can cause the adjudicator inordinate difficulty in accessing information, and consequently waste money and time.

Creativity is often required to envisage and present contractual and legal arguments that underpin your adjudication outcome. Collation of evidence to prove facts and their effects can be a delicate, intricate and time-consuming task, requiring support documentation and the deposition of witnesses of fact and expert witnesses. During the preparation of the referral collect not only statements of truth (witness statements) that you will want to include in the referral, but also the statements of truth that you will need to rebut the response (Figure 16.12).

The referral including supporting statements of truth should fit together like a well-crafted jigsaw, they should be one whole and difficult to break apart. Witnesses essential to adjudication success will have competing demands on their time, and are often not readily available. If you are providing adjudication support services, anticipate that the party instructing you might exaggerate, use hyperbole, and sequentially provide you with inaccurate information late.

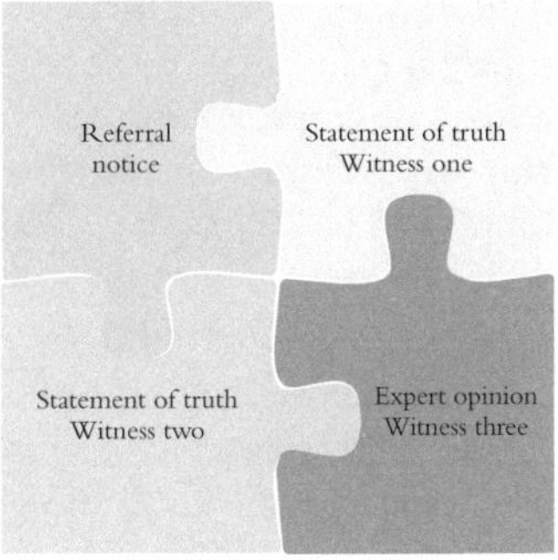

Figure 16.12 Referral jigsaw.

56 Christopher Michael Linnett v Halliwells LLP [2009] EWHC 319 (TCC).
57 KNN Colburn LLP v GD City Holdings Limited [2013] EWHC 2879 (TCC).

Notwithstanding that, there is no room for carelessness; unfortunately, it is only when a contractor or employer endures the emotional intensity of the adjudication time window that it really gets to understand why precision is required. Accuracy and attention to detail in the referral are important.

To create a sure path to the money requires experience and knowledge. Concentrate on your strengths and proactively deal with any weaknesses. Understand that the terrain on which you will do battle is hostile and uncompromising; accordingly, don't hesitate to retaliate first. If you can use the referral to destabilise the other party, so that it spends a disproportionate amount of time dealing with process rather than on content, that is a good thing. It is essential to think through and understand potential reactions. Try to deprive the other party of the intellectual headroom to focus on the problem. It is a virtue to be prepared beforehand for any contingency. To fully exploit foreknowledge, you must be able to act with blinding speed – be prepared to work overnight to accomplish this. Achieving the above requires leadership which is surefooted.

You have to have lived inside adjudication to know how it feels, and how the parties will conduct themselves to win. Well-rounded construction adjudicators will have been responsible for claims writing in adjudication or arbitration, as well as having acted as expert witness. Such experience supports the adjudicator in managing the process and making the decision.

When you transmit the referral to the adjudicator ask for the right to make a reply to the response, for example "We look forward to receiving your timetable for the adjudication, and would ask that it includes a period for the 'referring party' to reply to the 'responding party's' response". This is important as there are some adjudicators who misguidedly consider having only the referral and the response to consider to be procedurally sound.

Being concerned that the adjudicator might be difficult to control and handle, some referrals attempt to lead the adjudicator as if they are a bull with a nose ring. The referral will ask for a long series of declarations, leading the adjudicator to the inevitable finale that money and/or monetised time is due. When parties do this, they are trying to constrain the adjudicator to think in the same way that they think. This does not always sit comfortably with an adjudicator who reasons in a different way. In addition, the adjudicator is not employed to make "declarations"; the adjudicator is employed to make a decision.

What is more important than trying to lead the adjudicator is to provide the adjudicator with all the building blocks required to be able to construct a sensible decision, and which is the one that you want. Imagine that the adjudicator will use the building blocks to create linking steps, one on top of another.

The essential building blocks are: 1) facts, 2) effect, 3) contract/law and 4) decision. If for example, you provide a building block for step one, a building block for step two, a building block for step four, but no building block for step three, that can cause the adjudicator some considerable difficulty.

Most often the referral is sent by email or made available for the adjudicator and responding party to download using Cloud computing. Carefully consider if you should also send a hard copy of the referral to the responding party by registered mail. This will block any potential squabbles as to whether it was properly served, or that there was an implied term arising from the parties' conduct that the referral could be delivered electronically. In circumstances where the responding party refuses to accept the referral electronically (or to acknowledge receipt of the referral), I have shunted the adjudication timetable two or three days to allow for a hard copy to be sent and received by overland registered mail. This allows less time at the end of the process for the adjudicator to make their decision, which must normally still be made in the 28-day period.

16.13 Response

The response is your answer (as the responding party) to the allegations and/or statements set out in the referral. However, before you respond you should check and consider whether the adjudicator has jurisdiction to make a decision, whether there are objections that you can make, and if you want to do that?

The adjudicator usually sets the time within which the response is required. In some circumstances, allowing only seven days for the response may not be materially unfair.[58] My standard is to give the responding party 14 days, half of the time normally available in adjudications. This is not universal and other adjudicators give much less. In that event, if you need more time ask the adjudicator for it. For example,

> We request that you allow the responding party until 5.00 p.m. on Tuesday 24 July 2018 to respond. Given that the referring party has had some considerable time to formulate and draft the referral, we respectfully suggest that allowing the responding party only 11 days out of the 28-day period to respond to the referral does not accord with the principles of natural justice.

Within the overall adjudication time constraints, the rule of natural justice "*audi alterem partes*" gives the parties the right to be heard.[59] Faced with such a request and being alert to the potential of a jurisdictional challenge, most adjudicators will think carefully and sensibly about adjusting the time allowed. If the timetable is unfair, the courts may not enforce the decision. Provided you are courteous and professional in your approach, do not worry about upsetting the adjudicator. If you challenge the adjudicator in a respectful and legitimate way, the adjudicator will be more alert to everything you write.

It is important to me, consistent with any agreed procedure and the law, that the parties not only have the right to be heard and to have their evidence and arguments considered by me as their adjudicator, but also so far as is possible that they are satisfied with the adjudication process. Therefore, if the responding party asks for more time than 14 days, I will set out a revised timetable and ask the

58 Gibson (Banbridge) Limited v Fermanagh District Council [2013] NIQB 16

59 CJP Builders Limited v William Verry Limited [2008] EWHC 2025 (TCC)

referring party if it will consent to a timetable beyond 28 days. Because of its own good standing, most often, the referring party will consent. Sometimes, it consents because it believes a refusal will diminish it in the adjudicator's perception. Where the referring party does not consent, I simply confirm the original timetable. It is relatively easy for an organised adjudicator to accommodate a one- or two-day timetable adjustment, without extending the adjudication beyond 28 days.

The response is your opportunity to refute all of the allegations that the referring party has made. It should rebut the factual and legal claims advanced; it should explain the basis for the rebuttal by clarifying what the facts are, and referring their effect to the contract and the law; it should set out full details of any counterclaims.

With good practice of construction management, much of the information required to prepare a response should be readily available. The information required will be the same information that you have relied upon to reject the referring party's claims, and therefore should be on file. Where there are disputed facts, it is normal that the response will include statements of truth. Where the statements of truth target specific contemporaneous documents such as emails and letters that will increase their cogency.

In the UK it is decided law that in the response, subject to considerations of natural justice, you are entitled to raise any defence open to you to defend itself against the claim, regardless of whether you raised the defence prior to the adjudication. The adjudicator will consider such defences on their merits.[60] Generally speaking, no matter how narrowly a dispute referred to adjudication might be, at law the dispute is deemed to include all defences available to the responding party.[61] Therefore, in accordance with the rules of natural justice the adjudicator should consider the merits of any legitimate set-off and/or counterclaim. Although in adjudication a counterclaim can only be used as a shield and not a sword, where its value far exceeds the amount claimed in the notice and referral it can cause some considerable unease, and even panic a referring party. An astute responding party can use this to its advantage to negotiate a settlement. If the counterclaim is substantial and valid consider expeditiously commencing an adjudication based on that.

The response should include all the evidence that you rely on. Rather than increasing the amount of support documentation, where possible refer to a document provided by the other party, accurately giving its location in the referral. In particular, highlight additional documents pejorative to the referring party that you have included which the referring party omitted. This is important, because although the adjudicator knows not to make any decision until he has heard both sides, it would be surprising if the referral does not give the impression that there is a case to be answered. Where possible raise the spectre that the referring party is trying to hoodwink the adjudicator.

60 Quartzelec Limited v Honeywell Control Systems Limited [2008] EWHC 3315 (TCC).

61 Cantillon Limited v Urvasco Limited [2008] EWHC 282 (TCC).

Develop a relationship with the adjudicator. Tell your story. Give an outline of your business and say why the construction project was important to you. Describe any novel technology in a way that is easy to understand. If a party cannot do this, it can indicate that they themselves do not understand it. Use jargon-free simple terms; engineers love it, and lawyers may need it.

Introduce your response with an overarching proviso such as "Unless expressly admitted or not admitted herein, the allegations set out in the referral are denied".

If you want to accept a claim previously rejected, do that without fuss. Offer that the concession is in the interests of clarity, cost and speed, but where possible lay responsibility for this with the other party. For example, perhaps previous payment applications were contradictory, or difficult to understand, or lacked substantiation, and now in the adjudication additional support information is provided. Be careful that accepting one claim does not establish principles, which can render you liable in other areas.

Take control of the adjudicator's process for deciding the dispute by using a reference framework. For example, split variations into categories such as:

- Variation accepted in principle, but quantum claimed not substantiated.
- Variation accepted on a commercial basis, but quantum claimed not substantiated.
- Variation not accepted because the works were not additional.
- Variation not accepted because the alleged additional works were not authorised.
- Variation not accepted because the agreed contract procedure was ignored.

Such an approach will give you another opportunity to evaluate your risk.

Often the contractor will have more advanced dispute management and technical skills compared to the employer. In adjudication, this can give the contractor an advantage. In larger disputes, some of the most persuasive responses considered by me have included an expert's report dealing with technical matters or quantum. During adjudication, experts can update their reports. However, the commissioning and in large measure the writing of any response expert reports takes place sometime before the adjudication commences. Revealing the existence of the expert report in the response can be powerful, and can be considered as a reverse ambush. It can put the referring party completely on the wrong foot, as it then finds itself in the position of trying to find another expert to assess the responding party's expert report and to comment on it. The referring party anticipates its expert will challenge the conclusions of the responding party's expert. Achieving this in seven days is problematic. Another advantage of including an expert report in the response is that the responding party can emphasise that the general approach taken by it is driven by an overriding concern to be fair and reasonable. The conclusions that it arrived at in respect of claims advanced before the adjudication are not just its conclusions, but are conclusions informed by an independent expert.

Never act when you are angry. If you find that you are spending a disproportionate amount of time dealing with process rather than on content, "go to the balcony", take a break and come back strong but clear-headed, proceed based on facts but not on emotions.

16.14 Reply to response, rejoinder and surrejoinder

Managing party submissions can be a delicate process, and adjudicators should be mindful of jurisdictional challenges that they have breached the rules of natural justice by adopting an unfair adjudication procedure. To accord with the common law rules of natural justice the parties must have "an effective opportunity to make representations before a decision is made".[62]

For the purposes of the adjudication, the adjudicator may invite written submissions and evidence from both parties, and issue other directions relating to the conduct of the adjudication.[63] It is clear that the minimum procedural stages are that the responding party has the opportunity to respond to the referral. Whereas the referring party does not have an automatic right to reply to the response,[64] in setting the adjudication timetable it is standard adjudication practice that competent adjudicators anticipate the response raising matters which the referring party will want to address.

The adjudicator should give the referring party the opportunity to deal with evidence or issues arising in the response. In his directions, if the adjudicator does not offer you (the referring party), the opportunity to reply, to protect your position write to the adjudicator saying that in accordance with normal judicial procedure you will want to reply to the response.[65]

However, an adjudication phenomenon is that parties will go to great lengths to try to dominate the submissions process, and irrespective of relevance or value, strive hard to have the "last word". In the adversarial adjudication process, it is natural that parties want to present the adjudicator with persuasive evidence. Therefore, if the adjudicator directs a right to reply to the response, it is common for the responding party to assert immediately that it has the right to serve a rejoinder (its answer to the referring party's reply). The adjudicator is entitled and may need to limit the number of rounds of submissions.[66] Whereas some adjudicators try to contain the exchange of submissions, by forbidding them in advance, I do not.

My position is that it is easier and more practical to guide angry waters, rather than stand in their way, or try to swim against them. Faced with the responding party's warning that it will serve a rejoinder, I accept this by issuing additional directions. I write saying that if the responding party wishes to serve

62 Amec Capital Projects Limited v Whitefriars City Estates Limited [2004] EWCA Civ 1418.

63 Code of Practice Governing the Conduct of Adjudications (25 July 2016).

64 CIC Users' Guide to Adjudication: Ireland (26 July 2017).

65 The Rules of the Superior Courts (Order 23).

66 GPS Marine Contractors Limited v Ringway Infrastructure Services Limited [2010] EWHC 283 (TCC).

a rejoinder to the reply; it may do so within [two days – insert date]. I also direct that the referring party may serve a surrejoinder (its answer to the responding party's rejoinder) within [two days – insert date] after the rejoinder.

The reply to the response, the rejoinder and the surrejoinder should not be used to advance surprise evidence or raise new issues. If they do, the adjudicator should give the opposing party the opportunity to deal with it.

There is normally only 28 days within which to make a decision. Some adjudicators feel that allowing parties additional time for further rounds of submissions will leave them not enough time to actually make the decision. In reality the adjudicator and the parties are not competing for time, they are sharing it. When the adjudicator gets the response, the adjudicator can be working productively starting to document and weigh up the merits of both parties' positions. The reply to the response should enrich the available information, as might any rejoinder and surrejoinder.

Adjudication costs will increase and can be high because of the number of rounds of submissions. It is a concern if the costs and fees outweigh the amount of the payment dispute. Where this happens, it undermines adjudication as a cost-effective and time-efficient way to deal with disputes.[67]

The adjudicator must promptly notify the parties of any matter that will slow down or increase the cost of making his decision, and the adjudicator has an obligation to use reasonable endeavours to process the payment dispute between the parties in the shortest time and at the lowest cost.[68] In practice, the rejoinder and surrejoinder can contain little that is new. As the adjudicator's time costs for dealing with the rejoinder and surrejoinder will be comparatively small, I do not raise their submission as a cost or time concern.

Irrespective of what the adjudicator directs, some parties will submit as many documents as they want, and continue to write to the adjudicator even after their rejoinder and surrejoinder. Rather than a party feeling duty-bound to have the last word, it should use active judgement and sense to evaluate its position in a discerning way. Often there is one important issue, and only if you decide that you are losing on that should you quickly and succinctly make a further submission. If necessary, you might use colourful language to focus the adjudicator's attention on a specific assertion. For example, I assisted a referring party in a JCT Standard Building Contract dispute where we wrote in a concluding letter that the "Responding Party's attempt to argue that a 'sum due' is a 'further sum' is sophistry, it is smoke and mirrors, and 'Referring Party' asks the adjudicator not be duped by that".

From a process perspective it can be very powerful to allow the other party to have the last word, subject to the adjudicator being the guardian of procedural fairness.

67 Primus Build Limited v Pompey Centre Limited (& Slidesilver Limited) [2009] EWHC 1487 (TCC).

68 Code of Practice Governing the Conduct of Adjudications (25 July 2016).

In the UK, the adjudicator must consider any relevant information submitted to him by any of the parties to the dispute.[69] Adjudicators in Ireland will be required to do the same. The adjudicator's overriding obligation is to complete his decision within the time limit, and he does not have to consider late submissions in any detail. It is enough to look quickly at any document, to ascertain its general nature and to consider whether it contains anything of real significance.[70] Notwithstanding this, although it is not mandatory, when the adjudicator sets out the procedural chronology, it is best practice for him to list all of the parties' submissions, and confirm that he has considered them in making his decision.

16.15 Jurisdiction

Jurisdiction is an adjudicator's authority to make a decision. Applicable adjudication legislation and the contract between the parties give the adjudicator authority to act. The notice of intention confines the limits of the adjudicator's jurisdiction. If the adjudicator does not have jurisdiction, or acts in a way to lose jurisdiction, then a competent court will not enforce any purported adjudicator's decision.

In adjudication, it is common for the responding party to raise jurisdictional challenges. An obvious reason is that believing something is inherently wrong, it wishes to avoid the expense and nuisance involved in dealing with adjudication. It wants to stop the adjudication, and therefore payment of any decided amount. Jurisdictional challenges are sometimes part of a strategy for dealing with an adverse decision. Casting doubt about jurisdiction introduces an element of risk into enforcement proceedings, and makes the enforcement of a decision potentially more consuming of resource. Responding parties can use this predicament to negotiate a settlement agreement, which although not in the terms the adjudicator decided, is nonetheless based on the adjudicator's decision. Many referring parties will accept less now, than having to go through enforcement proceedings, no matter how streamlined that process is. On the other hand, many responding parties will look at an adverse decision and conclude that although they could subsequently arbitrate or go through court proceedings incurring additional expense and energy, the final binding outcome might not be more advantageous to them.

To reserve its position, the responding party should raise any jurisdictional challenge at the earliest opportunity. By participating in the adjudication and not raising any jurisdictional challenge, or not expressing any objection clearly enough, the responding party will usually forfeit its right to challenge jurisdiction. The jurisdiction challenge can be in general terms or it can be specific. Whereas a general challenge might be sufficient, it is undesirable because

69 AMEC Group Limited v Thames Water Utilities Limited [2010] EWHC 419 (TCC).
70 Scheme for Construction Contracts (England and Wales) Regulations 1998.

without specific information the adjudicator and the other party cannot investigate the objection, and decide how best to proceed.[71] If the responding party raises a specific challenge and it is wrong about that, the adjudicator will have jurisdiction. A general reservation may not be effective if at the time it was provided, the responding party knew or should have known of specific grounds for a jurisdictional objection but failed to articulate them, or the court concludes that the general reservation was worded in that way simply to try and ensure that all options (including ones not yet even thought of) could be kept open.[72] The responding party cannot later rely on other good grounds that existed, which it knew about or could reasonably have known about.[73] Ensure jurisdictional challenges are preserved throughout the process.

It is a regrettable feature of adjudication that the adjudicator and the parties spend a great deal of time and money dealing with jurisdiction.[74] The referring party should consider if there is an advantage to them to offer a stay in the adjudication proceedings, should the responding party serve within the next two working days proceedings in a competent court to seek a binding declaration as to the adjudicator's jurisdiction. Whereas such an application by consent might be considered, the UK Courts have shown a reluctance to interfere in adjudication by way of injunctive relief that would stop an adjudication from proceeding. Quibbles or challenges to an adjudicator's jurisdiction should, in a conventional case, be taken upon enforcement.[75]

In the UK faced with a challenge to jurisdiction, the guidance from the courts is that

> it is clearly prudent, indeed desirable, for an adjudicator faced with a jurisdictional challenge, which is not a frivolous one, to investigate his own jurisdiction and to reach his own non-binding conclusion to that challenge. An adjudicator would find it hard to comply with the statutory duty of impartiality if he or she ignored such a challenge.[76]

For adjudicators starting to gain experience, and who are not lawyers, dealing with jurisdictional challenges can be problematic. Often they will require

71 GPS Marine Contractors Limited v Ringway Infrastructure Services Limited [2010] EWHC 283 (TCC).

72 Ove Arup and Partners International Limited v Coleman Bennett International Consultancy plc [2019] EWHC 413 (TCC).

73 Allied P&L Limited v Paradigm Housing Group Limited [2009] EWHC 2890 (TCC).

74 Christopher Michael Linnett v Halliwells LLP [2009] EWHC 319 (TCC).

75 Billingford Holdings Limited and BFL Trade Limited v SMC Building Solutions Limited and Another [2019] EWHC 711 (TCC).

76 Christiani & Nielsen Limited v The Lowry Centre Development Company Limited [2000] HT 001/59 (TCC), 26 June 2000.

research and thought, which will be time consuming. Where the adjudicator's remuneration is hourly based, the expense of jurisdictional challenges will be of the parties. If the adjudicator is rewarded on an *ad valorem* basis, there is inherent unfairness that considerable time can be spent doing work without being paid. However, as many jurisdictional challenges arise repeatedly, with experience, the adjudicator's time involved in dealing with them will diminish.

To help investigate jurisdiction, it is sensible to ask each party to make submissions, and to have the opportunity to reply to any submissions made by the other party; this often makes it easier for the adjudicator to do his job properly. However, it is not necessary to invite submissions. As the decision of an adjudicator as to his jurisdiction is non-binding, it has no legal effect and cannot affect the rights of the parties. Because of that, the rules of natural justice do not apply when an adjudicator is considering jurisdiction.[77] In practical terms, it is common sense to apply the rules of natural justice.

Whereas in the UK, the adjudicator should investigate a challenge to jurisdiction, such direction is not universal. For example, in Malaysia, faced with a challenge to jurisdiction the adjudicator can elect to proceed and complete the adjudication proceedings without further examination.[78] However, an unenforceable decision is of no value to the parties.[79] To help maintain confidence and good order in the adjudication system, irrespective of whether the adjudicator is paid for their time, or has the jurisdiction, my view is that the adjudicator should investigate any challenge to their jurisdiction. To avoid incurring unnecessary expense, the best time to do that is as soon as possible.

Usually the referring party will want the adjudicator to withstand any jurisdictional challenge, and the responding party will want the adjudicator to succumb. For a moment consider the transition block Figure 16.15 **"Confidence that you have jurisdiction"**, black or 10 is that after considering the challenge or objection you, as adjudicator, have confidence that you have jurisdiction, and white or 0 is that you have confidence that you do not have jurisdiction. Faced with a challenge, at what score on the continuum would you decide to continue to act as adjudicator?

Post *Systech International Limited v PC Harrington Contractors Limited* [2012], when I posed this question to a large group of experienced adjudicators in the UK, the trigger point for continuing or stepping down was in the range from 7 to 2. Some adjudicators are robust in their view that if there is any real prospect that they enjoy jurisdiction, they should proceed to a decision and leave it to the courts to make the determination.

Many adjudicators believe that the trigger point for proceeding is on the balance of probabilities, and there is some support for adjudicators applying

77 AMEC Capital Projects Limited v Whitefriars City Estates Limited [2004] EWCA Civ 1418.

78 Construction Industry Payment & Adjudication Act 2012 (CIPAA).

79 Systech International Limited v PC Harrington Contractors Limited [2012] EWCA Civ 1371.

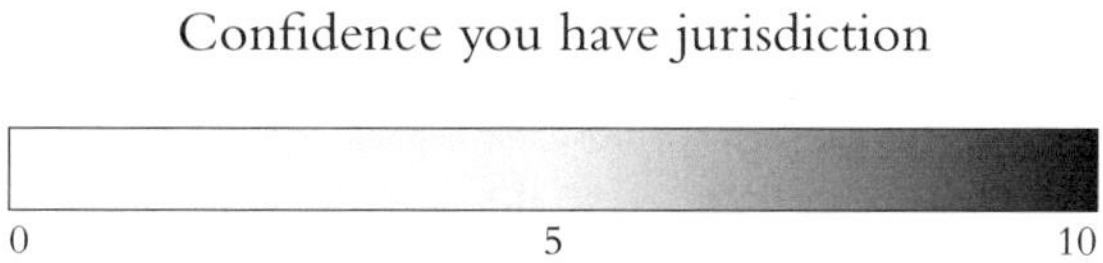

Figure 16.15 Confidence you have jurisdiction as adjudicator.

the balance of probability test when considering jurisdiction. For example, in *Aedifice Partnership Limited v Mr Ashwin Shah*, Mr Justice Akenhead said

> It is of some real relevance that the adjudicator did not actually decide as such the jurisdictional issue. In the Decision, he proceeds expressly on the basis that he does not have jurisdiction to decide that he has jurisdiction. All that he does … is to inquire into his jurisdiction to satisfy himself that he probably does have jurisdiction; that is a legitimate exercise for him to have performed; if he had formed the view that probably he did not have jurisdiction, I assume that he would simply have declined to issue a decision and told the parties accordingly.[80]

One of the lauded benefits of adjudication is that the resolution of disputes on a contemporaneous basis is more cost effective than trying to resolve disputes some considerable time after they arise. Adjudicators have to act impartially, and they should avoid incurring unnecessary expense. With that in mind, and recognising that adjudicators' decisions are widely accepted, or used to negotiate a settlement, should adjudicators not adopt a more robust approach in deciding if they should write the decision? Might it not be that arriving at a non-binding conclusion based on the test of "is there an arguable case that I have jurisdiction?" is more likely to avoid incurring unnecessary expense than arriving at a non-binding conclusion on jurisdiction based on the test of balance of probability.

Is it bad to postulate that adjudicators being courageous rather than being timid in dealing with jurisdictional challenges is a factor in encouraging the parties to a construction contract to refer their disputes to adjudication? The diminishment of any professional approbation, should a competent court subsequently decide that the adjudicator's decision is unenforceable, is enough to keep almost every adjudicator alert, straight and on their toes.

Do not write to the parties and say that having considered their arguments, your non-binding conclusion is that the jurisdictional challenge does have merit, and you resign. This is logically incongruous; if the challenge is valid, you were never adjudicator and although you cannot proceed, you cannot resign either.

80 Aedifice Partnership Limited v Mr Ashwin Shah [2010] EWHC 2106 (TCC).

16.16 Meeting with the parties

The Code of Practice Governing the Conduct of Adjudications empowers the adjudicator to decide the adjudication procedure. For example, the adjudicator can take the initiative in ascertaining the facts and the law necessary to reach a decision, decide whether it is necessary or helpful in their decision-making to meet jointly with the parties and their representatives, if any, to enable further investigation; and subject to obtaining any necessary consent from a third party decide whether to make site visits and inspections, accompanied by the parties or not.[81]

Whereas most adjudication is on a documents-only basis, there are times when the adjudicator or the parties want to have a meeting. Although the Code refers to both meeting jointly with the parties and holding an oral hearing, in adjudication, I use the term "meeting" rather than "hearing". In legal terminology, a "hearing" is a legal proceeding where a disputed fact or issue of law is tried and evidence is presented to help determine an issue. The term "meeting" connotes a less formal, less adversarial proceeding, with more relaxed standards of evidence and process.

A meeting can be helpful where the quality of submissions is inadequate, there is conflicting evidence from experts or witnesses of fact, to inspect site-based physical evidence, or for other reasons. For example, in a €400,000 dispute over management costs, each party presented a spreadsheet, which they said was accurate and supported by a summary of the information held on its computer system. So that the parties could demonstrate the provenance of their spreadsheets, I asked for a meeting so that I could sample the base data held in their computer systems. In advance of the meeting, each party agreed to access its financial systems remotely from the meeting room. Because one of the parties was better able to demonstrate how its computer system collated base data, and used records to create information, I decided that party's spreadsheet calculation of management costs was more reliable.

In adjudication when a party believes its position to be the correct one, it also believes that the other party's position is the wrong one. Adjudicators should not be surprised if parties are angry because they believe that the other party's position is unreasonable. Whereas formal, logical and rational submissions can mask negative emotions, contemporaneous documents and statements of witnesses of fact can be a better pointer. So sometimes parties want to meet for confrontation and emotional reasons. In mediation, mediators have a range of tools that they can use to deal with difficult emotions, for example intervention in private meetings with the parties. The role of adjudicator is different to the role of mediator, and the conduct options for dealing with emotion in meetings are also different. The mediator serves the

81 Construction Contracts Act 2013, Code of Practice Governing the Conduct of Adjudications (25 July 2016).

parties facilitating compromise using an agreed procedure, and adopting an interests-based approach. As the adjudicator is in charge, the parties do not have to agree with the procedure, and the adjudicator will decide the dispute adopting a rights-based approach. If both parties want to have a meeting, I will accommodate their request, even if the meeting costs will be large compared to the amount in dispute. If only one party wants to have a meeting, I will accommodate its request, only if the meeting will assist me.

To allow for a meeting, I provisionally set aside a day in the fourth week of a 28-day adjudication timetable, to be confirmed or vacated as may become appropriate during the course of the adjudication. The purpose of any meeting is to gather information to help the adjudicator make decisions. Dependable, relevant information helps the adjudicator answer important questions. For example, in one adjudication the parties had opposing views as to what they had agreed at a site meeting. I used my inquisitorial powers to request for signed witness statements concluding with the sentence "I believe that the facts stated in this witness statement are true". I said that I would be testing the reliability of the witness statements by requesting further documentary evidence, and meeting with the witnesses.

To organise a meeting, initially I ask the parties to make a cost-effective venue suggestion. The options are in the offices of the parties or the parties' representatives, or a convenient place not linked to the parties. Factors in determining the meeting venue include the whereabouts of everyone involved and the location and size of offices. After receiving suggestions and if the parties seem cooperative, I normally ask the referring party to arrange the meeting place. I ask the referring party's representative to telephone the responding party's representative to agree which premises will be used. I will advise that two rooms are required; one for the main meeting and another room to which either party may retire. I also request that while arranging the meeting venue, the parties take the opportunity to discuss any settlement possibilities. Whereas in mediation the mediator should be mindful of the emotional and psychological impact the meeting location has on parties, in adjudication the meeting location is not so important. Sometimes the parties have feelings about relative levels of power and safety, but they should not generally be concerned if the meeting takes place at the office of the other party. Indeed advantages can accrue, for example if the meeting refers to additional adverse documents, if you (the adjudicator) are not in your own office, you cannot produce them. If there is impasse, as to where to meet, I choose an independent venue and the associated costs become expenses, to be covered by the parties.

The meeting should have a specific purpose. It is for the adjudicator to decide what questions the adjudicator wishes to ask, or which issues the adjudicator wishes to address, and to set the agenda for the meeting. In so far as is possible and reasonable, the meeting agenda should set questions in a way which is neutral, and does not give any impression that the adjudicator has reached conclusions or pre-judged issues. To help ensure the most productive use of time, it is respectful and useful to ask the parties if there are things they would like the meeting to incorporate, and for them to make any suggestions as to the conduct for the meeting. For example, do they wish to have the opportunity to ask questions, or to make a written submission following the meeting? In the circumstances above, where I asked for two witness statements, I anticipate that the witnesses will be available to answer questions. It is important to ask neutral questions and not leading ones. A neutral question allows the witness to decide their answer without direction or pressure, sometimes started with who, what, when, where. A leading question suggests which answer is expected or desired, and deprives the witness of the opportunity to articulate what happened in their own terms. For example, "can you tell me what happened at the meeting?" is a neutral question, as opposed to "isn't it true that at the meeting you agreed … ?" which is a leading question.

In respect of who represents the parties at a meeting, there are two options. One is that the adjudicator asks specific people to attend. Another is that the adjudicator having sent the parties the questions he wants answered, allows each party to decide who will attend the meeting on its behalf. If the parties decide who will represent them, the adjudicator does not fetter them. The adjudicator should not attempt to do so, but in terms of proportionality, the adjudicator might suggest that the number of attendees is limited to a specific number. It is good practice to ask the parties to confirm in advance who will attend the meeting. Where the adjudicator has asked specific people to attend the meeting and one party's witness is available at the meeting but the other party's witness is not, the adjudicator can consider the evidential value of the witness statements as different. The adjudicator must be careful not to construe or criticise the absence of a witness in a way which would give rise to the appearance of bias and which would be a breach of the rules of natural justice.[82]

The adjudicator and the parties should have properly prepared for the meeting.

As adjudicator, at the start of the meeting, using positive language rather than negative, I set ground rules, or my expectations. I expect the parties to allow each other to speak without interruption and to listen. Consistent with exercising control, this also applies to me. I do not place any constraints on the vocabulary used, but I expect the parties to treat each other in a way that the other party considers respectful. If I anticipate heated exchanges, I normalise it recognising that it sometimes happens. I get a clear agreement that the parties will direct their answers to me, and not the other party. This can be an invaluable control mechanism, allowing me to stop any emotional escalation. Whereas I may acknowledge emotion, and reframe and summarise what the parties say, it is not in the role of the adjudicator to neutralise emotional discord through empathy. I confirm the agenda, allocation of time and the use of

82 A&S Enterprises Limited v Kema Holdings Limited [2004] EWHC 3365 (QB).

breaks as required. I will ask the parties if they have any questions. A structured meeting conducted in a business-like manner is less likely to go awry.

As at least two rooms will be available, the parties can separate during scheduled and unscheduled breaks. I will not remain with either of the parties when they are alone, as adjudicators do not bring anything to the meeting that is private, or which might indicate to one party or another any preliminary thinking.

At the beginning of the meeting, reading from my notes, I present the background and a summary of the dispute, I ask the parties to confirm that my understanding is accurate and complete. I send that information to the parties after the meeting and it becomes part of my decision. It is possible that during the meeting the parties agree on elements of the dispute, and the adjudicator therefore does not have to invest time deciding them and providing reasons. To ensure that the provenance of such agreements is clear, I also send the parties a schedule of anything that is agreed, and a list of agreed actions with associated times.

Conduct the meeting in an even-handed manner. If I have challenging questions for one party, I will also have challenging questions for the other party. When parties channel their questions and answers through the adjudicator, this reduces any power imbalance.

At the end of the meeting I will ask the parties if there is anything that I can do to assist them to compromise the dispute without their reference to the adjudicator's decision. For example, where the adjudicator is also a mediator, and the ethos of both parties is cooperation and partnership, the adjudicator could offer a hybrid process: adjudication/mediation. The parties can agree that the adjudicator withholds his decision for a period, say seven or 14 days, to allow the adjudicator to act as mediator. If mediation is unsuccessful, the adjudicator forthwith sends his decision to the parties. The CEDR Solve Rules for Construction Adjudication[83] offers a procedure for adjudication incorporating mediation.

16.17 Bullying and intimidatory tactics toward adjudicators

Some claims consultants, lawyers and parties choose "the method" of bullying and routinely use intimidatory tactics in adjudication. It seems that their conduct is not an adjudication-specific aberration, but rather in their nature. It is usually part of a longer-term "reputation-building" strategy to exercise control over the independence of the adjudicator. My experience of adjudication bullies is that they have always been men, aggressive and lacking subtlety. It behooves us, as a community, to try and deal with adjudication bullies in an appropriate, professional way.

Adjudication bullies use tactics such as making spurious challenges as to jurisdiction, deliberately seeking to confuse the adjudicator by the use of

83 CEDR Rules for Construction Adjudication Incorporating Provisions for Mediation 2016.

technical or esoteric legal arguments, threatening to take legal action against the adjudicator or to report them to their professional institution.[84] They seek extensions of time alleging that the procedure or timetable is unfair or biased toward the other party, and a breach of natural justice. They use bellicose language. They unreasonably refuse to pay the adjudicator's fees and expenses.

When confronted by an adjudication bully, I conduct myself with a Brazilian Jujitsu (BJJ) frame of mind. BJJ promotes the concept that a smaller person can successfully defend against a bigger, stronger assailant by timely using proper technique, leverage, and most notably skilled movements. Newton's third law of motion is that for every action there is an equal and opposite reaction. In terms of timing of a movement, the adjudicator should not feel compelled to act by pushing back at an adjudication bully directly or immediately. As adjudicator you can ask appropriate questions or for clarifications, then step back and allow the bully to move forward, providing information. If you push back at the right time in a skilful manner, then you can step to the side, allowing the adjudication bully to advance, often overreaching and stumbling, figuratively speaking.

Adjudication bullies may lack empathy but they will be confident in their abilities and knowledge; they often have a superior sense of self. They feel in control when they intimidate, and feel admired when they act out. The adjudication bully will try to bait negative emotional responses such as anger, fear and frustration, to provoke you to an intemperate response; work hard and smart to keep these emotions under control. Do not allow the adjudication bully to control tempo. Do not give the adjudication bully face. Adjudicators should remain calm and collected, dealing with all the challenges of adjudication including bullies, with balance and transparency, standing firm in the middle. As they are empowered to control the process, adjudicators should not be weak.

In two parallel adjudications, the responding party's solicitor argued that I did not have jurisdiction based on estoppel, Henderson abuse, time bar, and there being no contractual or statutory right to refer the disputes to adjudication. When I sent my decisions, he issued writs of summons in the High Court of Justice in Northern Ireland Queens Bench Division. The writs said that I did not have jurisdiction to reach my decisions, and as they were made in breach of natural justice and/or abuse of process, should be set aside and declared to be of no effect. Because of this, the responding party said it did not have to pay the sums decided, or my fees and expenses.

The successful referring party made two summary judgment applications, pursuant to Order 14 of the Rules of the Court of Judicature (NI) 1980, for the enforcement of my adjudication decisions. Having managed to delay enforcement proceedings for two months, the responding party

84 "Guidance for Adjudicators" Construction Umbrella Bodies' Adjudication Task Group July 2002.

settled the first enforcement dispute on the court hearing day. Despite a court order confirming that the terms of settlement were the exact sum I decided, the responding party continued to assert that I acted without jurisdiction and therefore was not entitled to payment. The responding party only felt compelled to pay the outstanding fees in respect of the first adjudication when I issued a statutory demand under the Insolvency (Northern Ireland) Order 1989.

After delaying enforcement proceedings for three months, the responding party settled the second enforcement dispute on the court hearing day. The court order did not disclose the terms of settlement recording only

> This action being called on for trial this day, and Counsel having informed the Court that a settlement has been come to between the parties on terms endorsed on Counsels' brief. THE JUDGE ORDERED that all further proceedings in this action be stayed upon the terms of the said settlement agreed between the parties.

I had no visibility of the settlement terms, but despite settlement the responding party continued to assert that I acted without jurisdiction and was therefore not entitled to payment. Forcing me to initiate legal action, causing expense and wasting time, was a form of bullying. Such behaviour gives the adjudication bully the space to allege bias in any subsequent adjudication, inhibiting the adjudicator's ability to act.

The matter was finally settled in the County Court for the Division of Armagh and South Down Small Claims Court where District Judge Duncan remarked on the responding party's conduct

> Thus, despite the institution of four High Court actions in which the jurisdiction to reach the decisions he did was one of the primary issues, it falls to the Small Claims Court to determine the validity and enforceability of the Applicant's decision on the second adjudication.

After I recovered my fees in full, I informed the solicitor that there was *prima facie* evidence that his conduct had breached the Solicitors Regulation Authority (the "SRA") mandatory principles 2 and 6, and I would be reporting his conduct. His response was to accuse me of defamation, demanding an immediate apology and retraction. Then going on to say that he would be complaining to the ANB, and the Professional Bodies of which I am a member. He concluded telling me that I should pass his letter to my professional indemnity insurers forthwith.

The SRA's mandatory principle 2 is that a solicitor must act with integrity; and principle 6 is that a solicitor must behave in a way that maintains the trust the public places in him and in the provision of legal services. The SRA responded promptly, but advising that it protects the public by focusing on the most serious issues it receives, and that the conduct matters raised by me did not meet their threshold for further investigation. The SRA's assessment surprised me.

I was chair of the professional conduct investigations panel of the Chartered Institute for IT (formerly the British Computer Society) for six years, and a member of the professional conduct disciplinary panel of the Institution of Civil Engineers for six years. From that experience, I can say that other professional bodies look at any allegations of misconduct against a construction or IT professional with more rigour.

Bullying and intimidatory tactics toward adjudicators are premeditated conducts inherently different in character to "flaming and venting" in mediation. Bullying is intended to intimidate, and is more than being assertive or being aggressive. The highly regarded barrister Mr Anthony Bingham posed the question "is bullying behaviour or circumstance, which creates a feeling of actual intimidation in another person ('subjective test') or behaviour, which an informed observer would conclude is likely to do so ('objective test')?".

Some lawyers consider bullying to be just part of the combative environment of the solicitor, it becomes a sort of daily norm, just part of the job, ordinary, and so nothing to fuss about. I admire passionate advocacy and zealous promotion of clients' interests. But these are to be distinguished from bullying and intimidatory tactics toward adjudicators, which are detestable.

16.18 Dealing with the decision

It is now clear that the construction industry regards adjudication not simply as a staging post towards the final resolution of the dispute in arbitration or litigation but as having in itself considerable weight and impact that in practice goes beyond the legal requirement that the decision has for the time being to be observed.[85]

The aim of adjudication is to swiftly deal with disputes so that the parties can devote their efforts to moving an ongoing project forward. Alternatively, if the project were nearing completion, or completed, they might use the adjudicator's decision to reach an agreement on the underlying dispute. "Adjudication is not just a method of debt-collecting but a means of ensuring that projects are carried out and completed by the parties working together without a dispute clouding their relationship."[86]

85 Balfour Beatty Construction Limited v The Mayor and Burgesses of the London Borough of Lambeth [2002] BLR 288.

86 Herschel Engineering Limited v Breen Properties Limited (No2) [2000] Adj.L.R. 07/28.

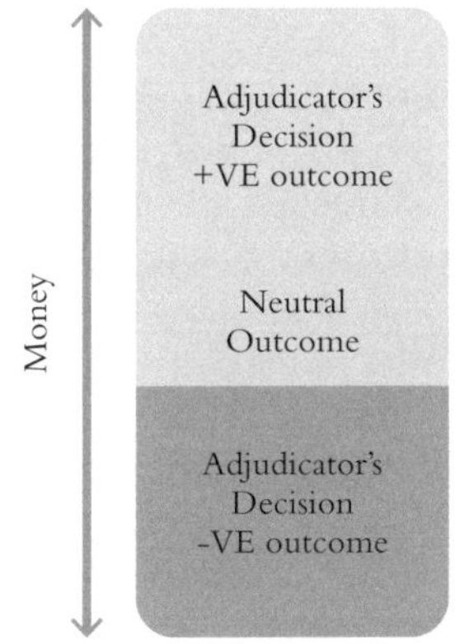

Figure 16.18 Dealing with the decision.

Money will be the measure of almost all adjudication decisions in Ireland. In financial terms only, the adjudication outcome is positive when the referring party gets more money than it invested, and the outcome is negative when it gets less. If the outcome meets or is beyond the referring party's reasonable expectations, avoid schadenfreude. Even if the outcome is positive, it can still fall below the referring party's expectations, and it may experience feelings akin to "buyer's remorse". If the outcome is negative, the referring party can feel angry, confused, disappointed, and have injudicious thoughts. Plan in advance how to deal with an adverse decision.

"When you always do what you have always done – you always get what you have always gotten" is an apposite quotation credited to the classical Greek philosopher Socrates.

How you as the referring party deal with the adjudicator's decision will depend on a mix of factors: your circumstances (financial and otherwise), company ethos, your relationship with the other party, the nature of the person in charge, and how you feel about the decision.

One of the things I have used with success is an adjudication audit, an independent objective review of documentation prepared by the referring party before it sends its notice of intention and the referral. Construction industry professionals will be familiar with value engineering which is a creative, organised process that analyses how to achieve functionality at the lowest cost. Faced with an adverse decision, consider implementing a decision workshop using some of the principles and techniques that make value engineering successful.

In a substantial dispute, whereas the decision workshop may not alleviate the cost of an adverse decision, it can allow you to derive important benefit from it. The decision workshop can help you to prepare for the future, and answer the questions:

- Based on our previous risk analysis, what are the differences in liability and quantum?
- Was the decision fair, reasonable and understandable?
- How could we have achieved a better outcome?
- Can we identify errors or mistakes that the adjudicator has made?
- Do we validate the information the adjudicator used?
- Did we make it easy for the adjudicator to decide in our favour?
- Do we really understand the other parties' perspective?
- Do we stand by the assumptions and strategy we used?
- What are the key criteria supporting the decision?
- What can we do to avoid future disputes?
- Will changing processes and systems, internally or externally, help to achieve a better decision next time?
- Can we improve our argument for a future battle?
- What could we have done differently?
- Were the costs of participating in the adjudication in line with expectations?
- How did our professional team do?
- What did we learn?
- Do we agree to accept the decision as finally determining the dispute?
- Should we proceed to arbitration or litigation (how to prepare best for that)?

Similar to value engineering, the decision workshop stages are information gathering, alternative generation, evaluation of alternatives and taking action. People who have been involved with an adverse decision can be emotionally involved, and therefore less detached or objective. One of the core value engineering principles is to involve an independent person or team to bring a fresh outside perspective. In any event, where the dispute involves significant sums, a multi-discipline team with relevant experience and expertise should conduct the decision workshop. Do not lose sight of the key objective, which is not to assign blame for the past, but plan and take action for the future!

Faced with an adverse decision do not write to the adjudicator asking for clarifications, elaborations or additional reasons; do not present the adjudicator with a missive repleading your case, or trying to make good deficiencies. Once an adjudicator has made their decision and sent it to the parties, they are *functus officio*; that is, the adjudicator's jurisdiction has expired because it has accomplished the purpose for which it was created. If you write to the adjudicator as set out above, you should expect a curt reply along the lines of "I cannot assist you further".

Meet with the other party without pre-judgement. Maybe they feel they got more than they expected. Maybe they are also unhappy with the outcome. Explain factually, where and why you disagree with the adjudicator's decision. There is collateral in uncertainty if there are jurisdictional challenges. There is collateral in uncertainty if there is the prospect of future arbitration or litigation. Can you use that collateral to reduce the amount to be paid, or to schedule the payment over time?

Senior management has first responsibility for setting aside bruised feelings and ego. If the dispute has sown discord, can you bring personnel together in a business or social setting to repair damage to teamwork and relationships?

In construction and engineering, mistakes are inevitable. When you manage a team of people you will face the understandable question, "who is accountable for this?" I have always maintained that when we have determined how to fix the mistake in cooperation with the customer, and implemented corrective action, then that is the time to look at responsibility. When the initial panic has receded and there is a plan for the future, there is less call for "heads to roll". Avoiding negative emotions, and taking a positive proactive stance to deal with the problem, is a rich and wonderful relationship-building/development opportunity. Try to take a similar approach when faced with an adverse adjudication decision, and get the other party involved in finding ways to avoid future disputes.

16.19 Enforcement

My experience is that there are honourable business people who will comply forthwith with adjudicators' decisions. In addition, government, quasi-government agencies and public bodies will also comply, as the adjudicator's decision empowers them with a mandate to make payment.

However, if parties can ignore or refuse to implement an adverse decision adjudication is undermined. One of the essential ingredients in the success of adjudication is courts dealing with challenges to enforcement of adjudicators' decisions quickly and robustly. An adjudicator's decision can be challenged on the basis that the adjudicator acted without or in excess of his jurisdiction; or in serious breach of the rules of natural justice.

The courts should give no encouragement to parties "simply scrabbling around to find some argument, however tenuous, to resist payment".[87] The adjudicator's role is to provide speedy interim solutions to disputes, and that purpose is "obstructed unless the courts protect adjudications from disputes of fact and law on appeals which can better be resolved subsequently".[88] One bizarre example of "scrabbling around" was for a party to argue that an adjudicator's decision I made in China lacked legitimacy because it did not comply with personal data obligations under the General Data Protection Regulation (GDPR) (EU) 2016/679.

Usually the adjudicator's decision will direct that a sum of money be paid by a certain date. Rather than just wait and see if that happens, on receipt of the adjudicator's decision it is best practice to write to the paying party, recording receipt of the decision and setting out the paying party's indebtedness. Raise a combined invoice that accords with sums decided by the adjudicator. For example: "Adjudication Decision of Mr Niall Lawless dated

87 Carillion Construction Limited v Devonport Royal Dockyard Limited [2006] 1 BLR 15.

88 Lead Technical Services Limited v CMS Medical Limited [2007] EWCA Civ 316.

4 June 2018. In accordance with the adjudication decision of Mr Niall Lawless dated 4 June 2018, 'Referring Party' invoices the 'Responding Party' [insert 'Sum Decided'] plus VAT." Ensure that the *precise date* the invoice is due for payment is highlighted, this will normally be a date decided by the adjudicator. If you send a consolidated invoice for the payment of sums decided plus interest due, to be paid by the date the adjudicator decided, then the paying party cannot argue that it does not have the documentation it requires. It also clearly knows what you expect from it.

After adjudication where the adjudicator has decided that a particular invoice is not to be paid, I have seen parties issue a credit invoice to offset the original invoice. I can understand that from an accounting perspective doing so is attractive, because it entitles you to reclaim the VAT included. Nevertheless, it is best to wait until the adjudicator's decision is complied with, before issuing any credit notices.

Whereas most parties take an adverse adjudication decision in their stride, comply with its terms and move on, some parties do not. This can be a time of mixed emotions, whereas there is some relief that the adjudicator has directed a payment be made, you can be apprehensive that payment might not happen. Your situation is different if there has or has not been jurisdiction or natural justice challenges. The paying party may say that the adjudicator was wrong, but in the absence of challenges it is difficult for the paying party to contest that it should not comply with the adjudicator's decision forthwith; albeit the paying party has arbitration or litigation options it can invoke later.

What happens next often depends on the parties' relationship. If it is ongoing and you want to preserve that, it may be that a meeting will serve the parties best. At that meeting, the parties can succinctly express their views about what the adjudicator got right, and what the adjudicator got wrong. Outcomes from such a meeting are that the paying party forgoes arbitration or litigation, but pays less than the adjudicator decided; the paying party is given more time to comply with the adjudicator's decision; or there is an agreement that the paying party should make instalments. Another outcome could be that the parties agree to alter the way that they work to avoid a repetition of the dispute.

If you have not received payment by the date the adjudicator decided, then you should send another letter giving clear warning of your intention to commence court proceedings without further notice for the summary enforcement of the adjudicator's decision. In my experience the firms, who help parties during adjudication, often do not provide additional services associated with enforcement of the adjudicator's decision. However, they should be able to recommend a specialist who does, and provide them with all the information that they require. Unfortunately, this will lead to an additional instruction and to additional costs.

In Ireland, the High Court is the court in which the enforcement of an adjudicator's decision and ordinarily any other business connected with adjudication are undertaken. The Act provides that the

> decision of the adjudicator, if binding, shall be enforceable either by action or, by leave of the High Court, in the same manner as a judgement or order of that Court with the same effect and, where leave is given, judgment may be entered in the terms of the decision.

An application to enforce or to enter judgment in respect of the adjudicator's decision shall be made by originating notice of motion. The Rules of the Superior Courts[89] set out the form requirements and the process to be followed. It is envisaged that the High Court will deal with adjudication enforcement proceedings on an expedited basis.

In the UK, although parties cannot recover costs associated with the adjudication, the majority of the costs associated with successful enforcement of the adjudicator's decision are recoverable. If the paying party has offered an apparently comprehensible defence to any enforcement proceedings, the judge has the discretion to award the costs of preparing for and commencing enforcement proceedings. These may be recoverable, even if the paying party complies with the adjudicator's decision as soon as proceedings are issued.[90]

As a professional courtesy confirm (or otherwise) to the adjudicator that his decision was complied with, or provide a copy of court's judgment in respect of enforcement proceedings. This is helpful to the adjudicator if the adjudicator needs to take action in respect of being paid fees and expenses. Adjudicators are unlikely to enjoy the same unfettered access to the High Court as the parties in respect of any claim for unpaid fees as directed in the adjudicator's decision.

16.20 Preparing for the future (arbitration or litigation)

The Act provides that

> The decision of the adjudicator shall be binding until the payment dispute is finally settled by the parties or a different decision is reached on the reference of the payment dispute to arbitration or in proceedings initiated in a court in relation to the adjudicator's decision.

Therefore, on receipt of an adverse decision your procedural options are to accept it, come to an alternative agreement, or move to binding dispute resolution through arbitration or litigation.

89 SI No. 450/2016 – Rules of the Superior Courts (Construction Contracts Act 2013) 2016.

90 Amber Construction Services Limited v London Interspace HG Limited [2007] EWHC 3042 (TCC).

Access to arbitration or litigation is important. The

> possibility that repayment may be required is not theoretical since the experience of the courts has shown that, given the limitations of adjudication and indeed the ordinary susceptibilities of human nature, some decisions are ones which might well not stand in the ultimate arbitration or litigation.[91]

Under the Act, there is nothing to fetter a losing party commencing arbitration or litigation at any time after an adverse decision. However, under the contract it may have to give notice of intention to do that. For example, under the NEC3 and NEC4 contract, the decision is final and binding if neither party notifies the other of dissatisfaction with the decision within the time allowed, and states its intention to arbitrate or litigate. In addition, the contract between the parties may place restrictions on issuing proceedings, requiring a party to wait until after practical completion. To protect your entitlement to arbitration or litigation ensure that you comply with any relevant clauses in the contract.

In the context of development of Irish adjudication law, it is important to emphasise that in accordance with Act the adjudicator's decision binds the parties until the outcome of arbitration or litigation. An earlier draft of the adjudication law, the "Construction Contracts Bill 2010", contained a cause which would have rendered adjudication in Ireland a pastiche,

> The decision of the adjudicator shall not be binding if the payment dispute is referred to arbitration or proceedings are otherwise initiated in relation to the decision unless the parties agree to accept the decision as finally determining the payment dispute.

This controversial and offensive provision would have enabled a reluctant "payer" to go through the motions of adjudication knowing that it could refer the adjudicator's decision to arbitration or litigation regardless of the outcome, rendering it ineffective.

In making the decision to arbitrate or litigate, it is helpful to understand the status of the adjudicator's decision in subsequent proceedings. Eminent and learned commentary is that in arbitration or litigation, the adjudicator's decision does not have any status and the aggrieved party does not have to demonstrate the adjudicator was wrong in reaching their decision.[92] In subsequent proceedings the winning party in adjudication will have to make good its assertions and discharge the burden of proof "where the law places it, and

91 Herschel Engineering Limited v Breen Properties Limited (No2) [2000] Adj.L.R. 07/28.

92 Coulson, Peter, *Coulson on Construction Adjudication*, 3rd edition, Oxford University Press (2011) ISBN-13: 978-0199597222; ISBN-10: 0199597227.

is unaffected by the terms of the adjudicator's decision". The adjudicator's decision does not change or reverse the onus of proof.[93] However, the UK Court of Appeal expressing *obiter* comments has disturbed the clarity of what seemed to be a settled legal principle, saying that it had real difficulty with the proposition that the adjudicator's decision "has no effect whatsoever on the onus of proof in subsequent proceedings".[94] It may well be that successful parties in adjudication will be emboldened to argue that the adjudicator's decision shifts the evidential burden in subsequent litigation.

If the referring party is dissatisfied with the adjudicator's decision, it can commence arbitration or litigation, within the limitation period applicable to the underlying cause of action. If the responding party is dissatisfied with the adjudicator's decision, it has six years from the date of the decision to issue proceedings.[95]

There is nothing in the Irish adjudication law to prevent the parties calling the adjudicator as a witness in arbitration or litigation concerning the adjudication subject matter. In the UK, the CIC Model Adjudication Procedure contains an agreement that no party will call the adjudicator as a witness. The TeCSA Adjudication Rules provide that the adjudicator shall not be joined as a party to, nor shall be summoned or otherwise required to give evidence, or provide his notes in any subsequent arbitration or such litigation. My standard terms and conditions for acting as adjudicator go further. The parties agree that they shall not call the adjudicator or any employee or agent of my company as a witness or as an expert in any subsequent hearing, litigation or arbitration involving the parties arising from or related to the dispute, nor shall they subpoena any notes or other materials generated by the adjudication process. Moreover, the parties agree to indemnify the adjudicator, my company and its agents from any costs and liabilities incurred arising from the service of any subpoenas from any third party.

93 City Inn Limited v Shepherd Construction Limited 17 July 2001 Scotland Outer House, Court of Session CA101/00.

94 Walker Construction (UK) Limited v Quayside Homes Limited & Anor [2014] EWCA Civ 93.

95 Aspect Contracts (Asbestos) Limited v Higgins Construction Plc [2013] EWHC 1322 (TCC).

17 Skills

17.1 Athena

Athena, the Goddess of War, took no pleasure from battle, preferring instead to settle conflict through diplomacy and mediation. In Eumenides, Athena reconciled different concepts of justice promoting a legal system based on fairness, the presentation of opposing evidence, and trial. She promotes fairness and impartiality "But to speak ill of others who are free of blame is far from Justice, and Right will have none of it".[1]

Athena was also known as a goddess of skill, and the ability to spin and weave is credited as her invention.[2]

Arachne, a Lydian maiden, admired and famous for being a very skilful weaver and spinner, claimed she had learned the art all by herself, and that there was no one else who could weave as delicately as she could. Hearing of her boast, Athena challenged Arachne to a weaving contest in order to prove herself. Arachne's work was not only beautiful when it was done, but also beautiful in the doing. Although Athena is favourably impressed with Arachne's flawless weaving, she cannot tolerate the maiden's disrespect, and transforms her into a spider so that she might weave forever.

In myth, Arachne's fate is preserved as a warning to presumptuous mortals not to compete with, or compare themselves with the divinities.

This book is part of our adjudication journey, and no matter what role we choose to play, be it assisting the parties in adjudication or acting as adjudicator, we must be mindful that we are not judges and we should not compare ourselves to them.

17.2 Culture and role awareness

Culture affects the ways that construction and engineering disputes are resolved. In international disputes erudite parties are aware of this, and will

1 Aeschylus, *The Oresteia*, translated by Hugh Lloyd-Jones (1 December 1993) ISBN 13: 9780520083288.

2 Murray, Alexander S., *Manual of Mythology*, David McKay (1895) ASIN: B000M0JF6I.

want to engage ADR professionals with cross-border experience. However, culture is one of the most difficult words in the English language. "Culture is a derivative of individual experience, something learned or created by individuals themselves or passed on to them socially by contemporaries or ancestors".[3]

In adjudication, culture is at play in wide-ranging ways. There is a cultural difference between peoples. Culture is part of why in Malaysia the Asian International Arbitration Centre is the single adjudicator nominating organisation, whereas in the UK there are many adjudicator nominating bodies. There is a cultural difference between companies. That is why some companies are averse to resorting to legal remedies, while others embrace them. Construction professionals and lawyers working on the same team have differing cultural backgrounds and views too. This is important to consider even in a routine involvement with adjudication.

Engineers and lawyers have different belief systems. Engineers should work with integrity and objectivity having regard for the welfare of the people, of the organisations and the environment with which they interact.[4] Lawyers should work in a way that maintains the trust the public places in them and in the provision of access to legal services.[5] Construction professionals such as engineers and lawyers are aware that they belong to professions with expectations and guidelines, but there are cultural differences between them, and they define problems in different ways. There is an argument that the engineer's ability of creating things that work well, and which do not cause damage within their environment is a culture that lawyers could learn from.[6]

Usually for engineers the core focus is on technical problem-solving. The first step is to define the problem, and to draw a boundary around it. Sometimes a creative solution requires abstract thinking, but usually the engineer applies appropriate scientific methods in a systematic way. Engineers deal with certainty and facts, and are concerned with creating tangible solutions that work.

The core role of the lawyer is to act as advocate, to advise what it is possible to achieve, and to evaluate risk. Lawyers apply legal knowledge to advance the interests of their client. The first step in legal problem-solving is to identify the issue. Although a creative solution requires abstract thinking, usually the lawyer will identify relevant law, apply it to the facts of the problem and reach a conclusion. Lawyers deal with ambiguity, and are concerned with creating intangible solutions that win.

3 Avruch, Kevin, *Culture & Conflict Resolution*, United States Institute of Peace Press (12 January 1998) ISBN-10: 1878379828; ISBN-13: 978-1878379825.

4 The Institution of Mechanical Engineers Code of Conduct, 9 April 2008.

5 Solicitors Regulatory Authority Code of Conduct, 2011.

6 Howarth, David, *Law as Engineering: Thinking About What Lawyers Do*, Edward Elgar Publishing Limited (28 Feb 2014) ISBN-10: 178254013X, ISBN-13: 978-1782540137.

Whereas engineers and lawyers use similar skills such as analytical ability to solve problems, generally their way of defining problems is conceptually different. In addition, engineers will use a technical vocabulary and lawyers a legal vocabulary. One consequence of the difference is that through rigorous training and fierce experience lawyers have a deep understanding of the process and reasoning that other lawyers use to arrive at a conclusion, and so many are generally more comfortable when the adjudicator is also a lawyer. They believe that the decision of an adjudicator lawyer will be more predictable and reliable, and therefore they are more in control of risk. The corollary of this is that the adjudicator lawyer also well understands the way that the party's lawyers think, and the methods they use to succeed.

In adjudication, it is inevitable that the domains of the construction professional and the lawyer come into contact. If they are not to collide, the ability to work effectively in diverse and multicultural environments is an important skill. This is a global competency, defined as "having the knowledge, ability, and predisposition to work effectively with people who define problems differently than you do". Global competency has two important dimensions: engaging people from different cultures, and engaging understanding and ways of thinking about work that differs from your own.[7]

In practical terms, in any sizeable adjudication, construction professionals should involve lawyers as early as possible. An early definition of the problem in a way that meets both the legal and technical requirements, and a cooperative dualistic approach to resolving the dispute, or winning, is a wise investment. It will help avoid dissent and rancour in the adjudication team. A team is balanced and more effective when both the construction professional and the lawyer can be articulate in speaking and writing, to convince the other that their approach has merit. Such interaction leads to an outcome which is the fusion of the best that both have to offer. Sadly, most construction professionals lack the confidence, the skills and the knowledge to participate on an equal footing.

In day-to-day adjudication, cultural diversity is not so much about the differences between countries and ethnic groups; it is more about the corporate behavioural differences between companies and professional groups.

17.3 Competent contracting and record-keeping

Competent contractors have the right combination of experience, knowledge and skills required to carry out the work safely. What is right will depend on the complexity and scale of the project. Well-organised, complete and coordinated production information is a prerequisite for the proper management of construction. To effectively control and monitor a project, capable

7 Downey, Gary Lee et al. (2006). The Globally Competent Engineer: Working Effectively with People Who Define Problems Differently, *Journal of Engineering Education*, April: 107–122.

contractors maintain good, accurate and complete records. They have a formal notice system, a change control system, and maintain a fit-for-purpose project documentation and record-keeping system. Record-keeping will enable the contractor to comply with the contractual reporting obligations, as well as to establish and/or defend claims. Building information modelling ("BIM") awareness and its use as a means of specifying construction industry information management and building information modelling is growing. One anecdote of the use of BIM in construction disputes was of a contractor taking 2D CAD drawings from consultants, and developing its own BIM model to identify clashes and consequently the possibility for making claims later.

Documentation can provide contemporaneous records of what actually happened on a project and the parties' positions regarding past events. "A party to a dispute, particularly if there is 'adjudication', will learn three lessons (often too late): the importance of records, the importance of records and the importance of records." It is for the party making a claim to satisfy the adjudicator to the civil standard of proof, on the balance of probabilities, that it enjoys entitlement. The adjudicator will not know about the project, which must be reconstructed for him with all its complexities and nuances. If not backed by meticulously established records, lawyers are ingenious in finding grounds, often quite real, on which to cast doubt on evidence.[8]

In an adjudication involving the design, procurement, supply, installation and commissioning of photovoltaic systems onto about 3,000 roofs there was a dispute about how many damaged tiles the contractor had to replace. The contractor's works were accelerated; it employed additional resources and more subcontractors. It extended weekday business hours and worked weekends. The contractor argued that neither it nor the employer could keep up with the administrative burden of authorising or recording tile replacements, and a requirement to strictly follow the contractual procedure would have rendered any programme impossible to maintain. The contractor paid not enough attention to record-keeping, and it only had evidence to support claims on about half the properties. The contractor said that the records it had were indicative of the tile replacement work for other properties where it was without records, and that it was reasonable to use the records it had to establish the total quantity of tile replacement. It asked the adjudicator to calculate an average number of tiles/slates replaced per property based on the records it had, and then to pro-rata an average cost across the remaining properties to estimate its total entitlement. In the circumstances, without contemporary records I decided that this methodology was without merit, and the contractor's claim in respect of replacing tiles was not successful. Contemporary records are "original or primary documents, or copies thereof,

8 Abrahamson, Max, *Engineering Law and the I.C.E. Contract*, 4th edition, Routledge (1979) ISBN 0 419 16080 9.

produced or prepared at or about the time giving rise to a claim, whether by or for the contractor or the employer". Witness statements can record the recollections of those who were involved with the works, but are not a substitute for contemporary records.[9]

Compensable delay and disruption can be the "Holy Grail" in claims preparation, particularly in infrastructure projects, where the contractor will want to claim or recover for head office and site overheads. However, such claims can be complex to prove, and the degree of success of such claims is often related to the quality of records available.

The "Delay and Disruption Protocol" published by the Society of Construction Law provides guidance on dealing with extensions of time and compensation for delay and disruption. The protocol recommends that the parties reach a clear agreement on what records are to be kept, and emphasises that records kept in a suitable format should considerably reduce the cost of analysing delays. In my view, it is essential for the parties to agree which commercially available software will be used to produce the programme.[10]

Whereas the protocol is not law, it reinforces that extensions of time should be dealt with at or soon after the time of the delaying event, so that the parties know where they stand and can adopt mitigation measures to limit the impact of the delay event.

Whereas contracts provide that the contractor shall keep records and make these available for inspection, they can be silent on whether the contract administrator should receive, or is allowed to take copies of the information.[11]

It has been said that in the UK the protocol first published in 2002 is not in general use in contracts in the construction industry, nor has it been approved in any reported case.[12] However, there is evidence that its acceptance in common law jurisdictions is growing, and setting the standard for what constitutes a judicially acceptable methodology for delay analysis. When delay is analysed in conjunction with a large volume of contemporaneous records such as reports, spreadsheets and correspondence, it will be preferred by the courts.[13]

There would be fewer disputes when the parties agree on the form and preparation of the construction programme, and the procedure for updating it. During contract formation, they might further agree what technique they will use in the event they have to analyse delay, and what formula they will use for the purpose of quantifying the head office overheads during any

9 Attorney General for the Falkland Islands v Gordon Forbes Construction (Falklands) Limited (2003) 6 BLR 280.

10 Society of Construction Law Delay and Disruption Protocol (2nd edition, February 2017) ISBN 978-0-9543831-2-1.

11 NEC3 and NEC4 Engineering and Construction Contracts.

12 Adyard Abu Dhabi v Sd Marine Services [2011] EWHC 848 (Comm).

13 Alstom Limited v Yokogawa Australia Pty Limited and Anor (No 7) [2012] SASC.

period of delay. If there is a dispute, followed by a dispute resolution procedure, they will have saved money and time.

Psychologically it is likely that parties will defer to adjudication decisions arising out of a process in which they had adequately and jointly participated, regardless of the extent to which the outcome is unfavourable.[14] In addition, adjudication decisions made early in the contract also support the parties adopting alternative methods of working, and mitigation measures.

The conundrum is this: with good record-keeping there is less likely to be a dispute; with poor record-keeping, there is more likely to be a dispute, and it is less likely that a claim for loss and expense for delay and disruption can be adequately substantiated.

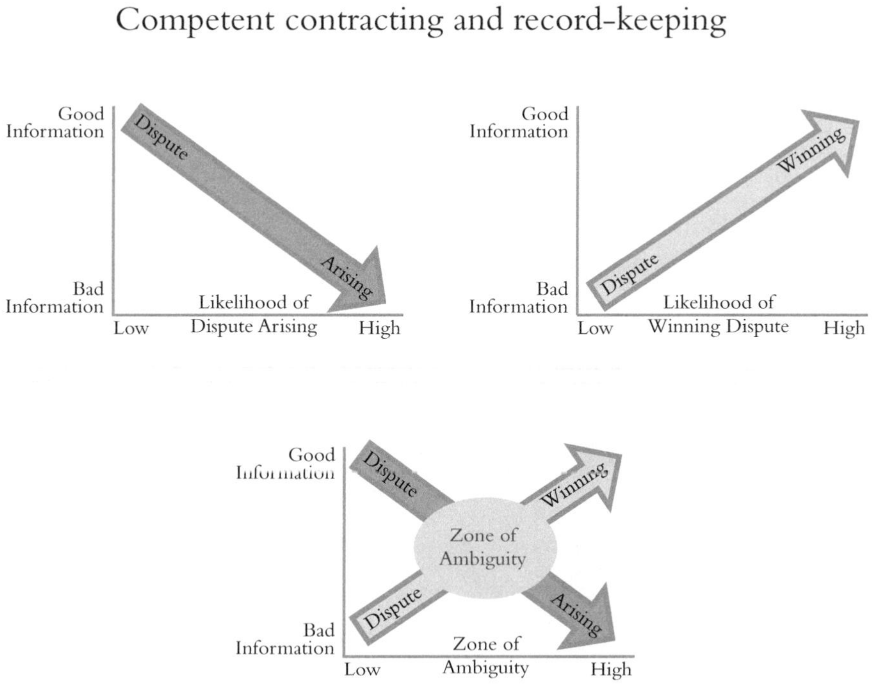

Figure 17.3 Information and the zone of ambiguity.

14 Ajibade Ayodeji Aibinu, Avoiding and Mitigating Delay and Disruption Claims Conflict: Role of Pre-contract Negotiation, *Journal of Legal Affairs and Dispute Resolution in Engineering and Construction*, ASCE, February 2009.

17.4 Managing a large volume of complex information

Information overload is where the adjudicator has difficulty understanding issues, and making decisions because they have been presented with too much information. Adjudicators are human with limited cognitive processing capacity, and when information overload occurs, it is likely that a reduction in decision quality will result.[15] Sometimes the problem is not simply too much information, but also that the information is complex too.

Adjudicators should not be spending time filtering, reducing or transforming data into accurate information. Nevertheless, sampling in adjudication regularly shows that data has not been accurately transformed, and that the opposing party cannot be relied upon to point this out, or even to point out that a column of numbers have been added incorrectly. Because of this, it is good adjudication practice to copy and paste important figures into a spreadsheet for checking.

Data is often absent, but seldom contrived. Information can be wrong because of an inadvertent mistake; but in some cases, it is deliberately wrong to influence the adjudicator. However, the parties should be able to tell the

Managing a large volume of complex information

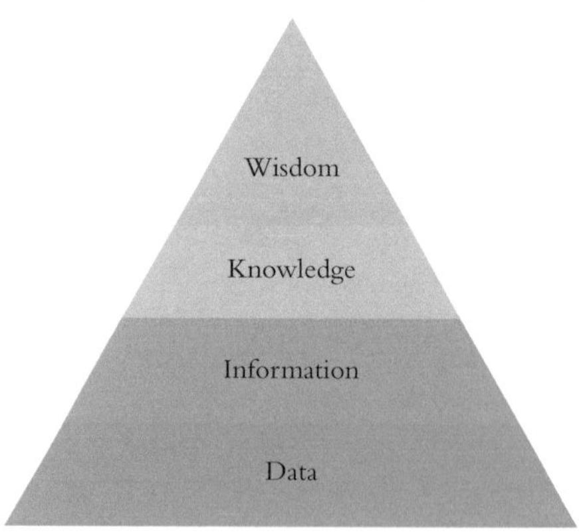

Figure 17.4 The Data, Information, Knowledge and Wisdom Pyramid.

The Data, Information, Knowledge and Wisdom Pyramid posits that data can lead to information, which can lead to knowledge and in turn lead to wisdom.[16] However, Ackoff wrote that "each of these includes the categories that fall below it", and estimated that "on average about forty percent of the human mind consists of data, thirty percent information, twenty percent knowledge, ten percent understanding, and virtually no wisdom".

15 Gross, Bertram, *The Managing of Organisations*, 1st printing edition, Free Press/Collier (1964) ASIN: B000IQ72CK.

16 Ackoff, R. L. (1989). From Data to Wisdom. *Journal of Applied Systems Analysis*, 15: 3–9.

adjudicator on what information they agree, and on what they do not agree. Where the adjudicator does not concur with the parties' common assessment, and the discrepancy is important in making a decision, the adjudicator should bring that difference to the parties' attention for clarification.

In adjudication, the parties will use their own knowledge to tell the adjudicator what any piece of information means, and thereby guide him to the decision that they seek. It is in the role of the adjudicator to use wisdom to reconcile the conflicts of knowledge. If the adjudicator has knowledge that he wishes to use, he must use it in a transparent way that meets the requirements of natural justice.

The parties can agree in advance to limit the extent of documentation constituting a referral.[17] However, although managing a large amount of information can be disconcerting, adjudicators need to remain in control and find practical ways to minimise the stress of any overload.

Often where there is a large amount of complex information, the parties will provide the adjudicator with Cloud accessed zipped directories (historically multiple CDs). It is often problematic to make any meaningful assessment of the information presented through browsing Cloud accessed zipped directories. Where engineering drawings are included these can be difficult to read even using a large monitor, and almost impossible to manipulate. For example, if it is necessary to check dimensions to decide which party's quantities to accept, this is easiest done with a paper copy, a scale rule and a highlighter. When presented with information on Cloud accessed zipped directories I usually ask for a paper copy of the documentation; it assists document navigation and understanding of the evidence.

Adjudicators should allocate and manage their time efficiently; they should plan the adjudication and execute it using a plan, and review progress. Schedule breaks, set limits, and keep electronic and physical spaces clutter-free. Recognise that there will be times in the day when you are biologically at your most productive.

Although parties compromise their disputes in about 10% of adjudications,[18] that should not impede the adjudicator allocating his time to the adjudication as early as possible. Spending time early in the adjudication will allow the adjudicator to sense where both information and procedural difficulties might arise. Adjudicators should recognise that things will happen in their personal and professional life that can cause difficulty with concluding the decision before the deadline.

17 London & Amsterdam Properties Limited v Waterman Partnership Limited [2003] EWHC 3059 (TCC).

18 Trushell, I., Milligan J. L. & Cattanach L., Glasgow Caledonian University Adjudication Reporting Centre Report 12 – October 2012 (Research analysis of the progress of adjudication based on returned questionnaires from adjudicator nominating bodies (ANBs) and from a sample of adjudicators) ISBN-9781905866632

Adjudicators should develop tools, processes and systems to manage a large amount of complex information. Adjudicators should consider if they can escape from or share the problem of a large amount of complex information by assigning tasks to the parties; for example, by asking for information to be presented in a different format, re-organised or summarised. Adjudicators can ask for the assistance of others including appoint experts, assessors or legal advisors, provided the parties have been notified of their identity and their terms of reference. Adjudicators can request any reasonable supporting or supplementing documents.[19] Bearing in mind confidentiality and information safety and security issues, adjudicators should use appropriate technology wisely for communications, and processing of data or information.

Adjudicators can structure the way that they will make decisions about individual, but similar parts of the information. For example, when faced with a large number of similar claims they could first determine what rates they would apply in the event that they decide liability. In deciding liability adjudicators might consider if there is photographic evidence or not; whether it is digitally time stamped; if there was a work instruction or not; whether or not there are invoices from subcontractors; if there is evidence that subcontractors have been paid or not; whether or not there are supporting day sheets to corroborate time and materials, etc. Having a tiered decision-making structure can facilitate handling and manipulation of large amounts of information quickly.

Adjudicators can develop and use templates for collation and presentation of information and sub-decisions. Adjudicators can approximate, filter and prioritise information. For example, in one multi-million-pound dispute where I was adjudicator there were over 20 lever arch files of evidence, covering 16 disputed categories of work, involving about 600 work packages. In the timescale available, it would have been very difficult to assess all of those in any adequate way, and to do so would incur disproportionate and unnecessary expense. I had hoped that Pareto's Law could apply to the value of the work packages. Pareto, an Italian economist and sociologist, proposed the inexact rule identifying the phenomenon that often 80% of the effects come from 20% of the causes. For example, 80% of sales come from 20% of customers or 20% of customers make 80% of complaints. I asked the referring party to collate and present the 600 work packages in each of the 16 disputed categories, which when sorted in order from the highest value to the lowest value revealed that an analysis of about 180 work packages would account for over 80% of the sums in dispute. When asked for submissions on the methodology I proposed, the parties agreed that I should consider the much smaller sample, and to pro rata the financial outcome within each category.

19 Code of Practice Governing the Conduct of Adjudications (21 July 2017).

Adjudicators and indeed experts can use sampling of their own volition, however the sample must be sufficiently representative to enable reliance to be placed on it; and bias must be avoided in the sample selection process.[20]

Adjudicators cannot decide to omit information simply because they do not have time to consider it. However, they can skim and speed-read to make progress. Indexing information as you pass over it takes some time initially, but is often an astute investment. Some adjudicators use "Post-it" notes to mark different information, but the use of these can quickly become unmanageable. In parallel with reading, consider keeping a shadow word processing document that will later allow for keyword searches.

Germans are adequately fluent in German, engineers are adequately fluent in engineering, and lawyers are adequately fluent in law. It is easier to deal with large amounts of information if the adjudicator has the right background and skills. If you have a dispute that requires forensic legal analysis, ask for a lawyer to make the decision; if you have a dispute that requires complex engineering analysis, ask for an engineer to make the decision.

17.5 Communicating complex information

Experts who provide information to people who make important decisions have a responsibility to present it clearly, concisely, in an explanatory fashion and without bias. Normally with presenting complex often-technical information, the goal is to create easily absorbable material for a specific audience. However, in the adjudication context good communication can allow you, the parties, to present information in a way that will persuade the adjudicator to make the decision that you seek for.

Good listening is a foundation for good communication. For there to be adjudication, a dispute must have been crystallised. There must have been an opportunity for each of the protagonists to consider the position adopted by the other, and to formulate reasoned arguments.[21] Active listening is a process where a conscious decision is taken to listen to, and understand the message of the other party, at first remaining neutral and non-judgemental. Good listeners find the opportunity to ask open questions, to clarify, to reflect and to summarise. Although it is not your goal as a party, if you have listened, even though you might not have agreed with it, you should be able to set out the opposing party's position for them. If you have listened, you will have documented your opponent's position in detail, your response and the reasons that you disagree. With that in mind, the potential for an adjudication ambush will be lessened.

Adjudication is a battle for influence, and about the struggle for supremacy of positions. In terms of overarching approach, *Influence: The Psychology of*

20 Amey LG Limited v Cumbria County Council [2016] EWHC 2856 (TCC).

21 Edmund Nuttall Limited v RG Carter Limited [2002] BLR 312.

Persuasion[22] sets out principles for influencing others. According to the concept of reciprocity, the adjudicator is likely to respond favourably to a party who behaves professionally and treats him with respect. According to the concept of commitment or consistency, when faced with similar facts adjudicators will decide similar outcomes. This is why in a series of adjudications between parties, one party will request the nomination of a particular adjudicator, whereas the other party will strongly resist such nomination. Adjudicators must not be biased towards the arguments of one party until they have been presented with all the information. However, according to the concept of social proof, people they see as similar can influence adjudicators, especially when uncertain. This might be in the form of a decision of another adjudicator in a linked dispute, an extract from an authority textbook, the opinion of an expert that the adjudicator respects, or an objective standard. According to the concept of authority, adjudicators are more likely to be persuaded by information presented in an authoritative and professional way. Experts can influence adjudicators because they offer a view which is likely to be correct; especially if it is a view that the adjudicator does not have to think too carefully about. Although it may be uncomfortable to acknowledge, in adjudication the approach and conduct of the parties can subliminally affect how the adjudicator feels about, and responds to, a party.

17.6 *Twelve Angry Men* – and the "TKI"

Twelve Angry Men (1957) is the iconic adaptation of Reginald Rose's teleplay, directed by Sidney Lumet. In the film, a jury of 12 men deliberates the guilt or acquittal of a defendant in a murder trial based on reasonable doubt. The defendant, if found guilty, would receive a mandatory death penalty. The jury explores consensus-building techniques and the problems encountered in the process of achieving agreement from a conflict situation. The dialogue is often confrontational, emotional, tense and laden with prejudice. The jurors with different psychologies use different conflict management styles with varying degrees of success.

Twelve Angry Men portrays the story of a dissenting juror (an architect) who subtly and with unassuming confidence manages to demonstrate to the other jurors that the case is not as obviously clear as it seemed in the courtroom. Whereas "the facts are supposed to determine the case", as the story unfolds, argument, logic and time expose the weaknesses in seemingly incontrovertible evidence. The architect uses active listening, he asks open questions to dissipate challenges, he acknowledges his fallibility and he builds valuable support.

22 Cialdini, Robert B., *Influence: The Psychology of Persuasion*, Harper Business (1 February 2007) ISBN 0-688-12816-5.

According to K. W. Thomas and R. H. Kilmann, there are five different modes for responding to conflict situations which can be articulated on two dimensions: assertiveness and cooperativeness. This psychometric instrument is called the "TKI" which stands for "Thomas-Kilmann Instrument" (Figure 17.6).

Accommodating is unassertive and cooperative – the complete opposite of competing. When accommodating, the individual neglects his own concerns to satisfy the concerns of the other person; there is an element of self-sacrifice in this mode. Accommodating might take the form of selfless generosity or charity, obeying another person's order when you would prefer not to, or yielding to another's point of view.

Avoiding is unassertive and uncooperative – the person neither pursues his own concerns nor those of the other individual. Thus he does not deal with the conflict. Avoiding might take the form of diplomatically sidestepping an issue, postponing an issue until a better time, or simply withdrawing from a threatening situation.

Collaborating is both assertive and cooperative – the complete opposite of avoiding. Collaborating involves an attempt to work with others to find some solution that fully satisfies their concerns. It means digging into an issue to pinpoint the underlying needs and wants of the two individuals. Collaborating between two persons might take the form of exploring a disagreement to learn from each other's insights or trying to find a creative solution to an interpersonal problem.

Competing is assertive and uncooperative – an individual pursues his own concerns at the other person's expense. This is a power-oriented mode in which you use whatever power seems appropriate to win your own position – your ability to argue, your rank, or economic sanctions. Competing means "standing up for your rights", defending a position which you believe is correct, or simply trying to win.

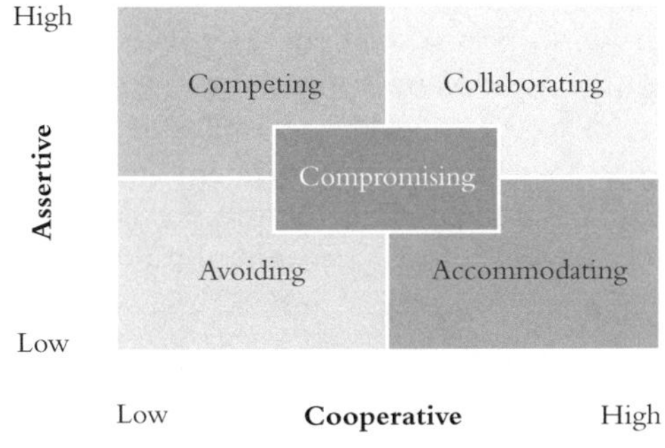

Figure 17.6 Thomas-Kilmann Conflict Mode Instrument (TKI).

> **Compromising** is moderate in both assertiveness and cooperativeness. The objective is to find some expedient, mutually acceptable solution that partially satisfies both parties. It falls intermediate between competing and accommodating. Compromising gives up more than competing but less than accommodating. Likewise, it addresses an issue more directly than avoiding, but does not explore it in as much depth as collaborating. In some situations, compromising might mean splitting the difference between the two positions, exchanging concessions, or seeking a quick middle-ground solution.[23]

In *Twelve Angry Men* the architect while the only person voting not guilty at the outset makes the question of guilt or innocence not about what he or others believe; his approach is to emphasise the importance of collaboration and better understanding through communication: "What do we do now?", "I guess that we talk"; "What do you want?", "I just want to talk"; "We stay here and talk it out!"

The TKI can help construction professionals and lawyers assess their intuitive behaviour in conflict situations. Being aware of your own primary or fallback TKI default pattern can help you become more effective in helping to resolve disputes. The TKI has weaknesses, but it's a good starting point for self-reflection.

17.7 Advocacy in adjudication

Adjudication has been described as a quick and easy, "non-judicial" procedure which does not involve expensive lawyer time.[24] However, the reality of much adjudication is that lawyers play an important role. The understandable reason parties do not want to avail themselves of a lawyer's services is reluctance to commit to expenditure. Feeling a lack of control and not knowing what the fees will finally be exacerbates any problem with the level of professional fees incurred. Lawyers giving a budget for undertaking work helps clients deal with and manage their reluctance to instruct.

If as a party you have a project of reasonable size, and you need more money (or time) for work, you should consider if this relates to just one particular package or if it is a problem across the board. If your problem is deep and wide, having access to the right lawyer can be invaluable to you in determining what your options are, and what strategy and tactics are best to use. The right lawyer will be able to move seamlessly from advisor to advocate, and will be experienced in delivering results using consensual dispute resolution methods such as mediation, as well as delivering results using non-

23 Thomas, Kenneth W. & Kilmann, Ralph H., *Thomas-Kilmann Conflict Mode Instrument*, Xicom (1997) ASIN: B000VSKYIY.

24 Hansard HL, 22 April 1996, Volume No. 571, Column 973.

compulsory dispute resolution methods such as adjudication. The right lawyer will have business acumen and judgement that you respect, will help disassociate your decision-making from the emotion of a dispute, and will be a trusted part of your team.

Adjudication can be used tactically early in the project to address disputes over particular work packages, and in that sense, it can provide short-term relief, mitigating against contractors funding construction without payment. Sometimes, to avoid expense parties will ask a lawyer to give a "legal opinion" on a question that they try to cast in the narrowest terms. This is not using a lawyer strategically. Notwithstanding the importance of the lawyer's strategic input, a party should remain in control, and not delegate it.

Advocacy is the art of speaking on behalf of or in support of another. Strategic advice is important, but advocacy and persuasion are a large part of lawyers' adjudication work. "As a moment's reflection will remind you, virtually all tasks undertaken by lawyers on behalf of clients are for the most part, themselves a form of advocacy – the employment of the technique of persuasion."[25]

Choosing the best person (or team) to support you well in adjudication depends on the nature of the dispute, and your strengths and weaknesses. Essentially, your advocacy options are to undertake the work yourself, employ a construction professional to assist, employ a lawyer to assist, or employ a team. Many construction professionals, as well as lawyers, specialise in adjudication and claims support. Recognising that clients request surveyors to act as advocates, the RICS published guidance and a practice statement setting out the core duties expected.[26] Parties should beware of employing construction professionals who are *only* qualified by experience; this can lead to the dangerous situation where the advice given is incomplete, remiss or just plain wrong. Construction professionals qualified by study and assessment will know about adjudication, and they will be aware of areas where they lack expertise. Courts have demonstrated ire with adjudication advocates who fall short of the mark, waste time trying to trip up the adjudicator in relation to matters of bias or jurisdiction, make quasi-legal points, and fail to take the adjudicator's cue in respect of invitations to make submissions.[27]

The advocate should understand the dispute, and should look at the dispute from the perspective of the adjudicator. This of course begs the question of what the adjudicator's professional background will be. In the UK, the ANB in the contract can determine that; if it is the ICE, it is likely the adjudicator will be an engineer; if it is the RICS, it is likely the adjudicator will be a surveyor; if it is TeCSA, it is likely the adjudicator will be a lawyer. It is

25 Napley, Sir David, *The Technique of Persuasion*, Sweet & Maxwell; 4th revised edition (August 1991) ISBN-10: 042143340X, ISBN-13: 978-0421433403.

26 RICS Surveyors acting as advocates (2nd edition guidance note 1 January 2009).

27 Michael John Construction Limited v Richard Henry Golledge [2006] EWHC 71 (TCC).

unclear what policy and procedure the Chairperson of the Minister's panel in Ireland will work under to appoint adjudicators.

The advocate should connect with the adjudicator. My experience is that the adjudicator will meet the parties and their representatives in less than one out of ten adjudications. Therefore, in most adjudication, advocacy is confined to writing. In written communication, the starting point and the end are telling the adjudicator clearly what you want them to decide. Make clear what relief you want, including apportionment of the adjudicator's fees and expenses, interest, tax, and time for payment of any sums decided. Allow the adjudicator to decide other such sums, as he deems correct. Confidently assert only what is essential. Distil the critical issues into a simple story supported by evidence. Present the facts, and underpin the issues with applicable law. Present your positive case briefly, directly and simply in an easy to follow logical structure. Consider if you should abandon weak claims. Anticipate opposition arguments and plan how to deal with them.

In many disputes, without exercising control the advocate will be involved with collating statements of witnesses of fact and expert witness statements. Witness statements should be concise, directly relevant and targeted to the adjudicator's decision. Witness statements can be poorly prepared, internally inconsistent and at odds with the contemporary documentary evidence. This undermines the goal of advocacy, which is persuasion. The TCC provides helpful guidance for the preparation of witness statements.[28] When collating witness statements to complement the referral, invest in the preparation of witness statements to deal with arguments you envisage the responding party will use in the response.

17.8 The writing process

In 2002, there was a significant dilapidations dispute concerning a Grade I listed property, which resulted in a trial in the Technology and Construction Court in London. As part of pupillage to become a Chartered Arbitrator, His Honour Judge Peter Bowsher QC gave me the wonderful opportunity of writing an arbitration award that he would review. The parties granted me access to the trial bundle, and based on the arguments presented I prepared a preliminary issues award. After reviewing my award Judge Bowsher wrote,

> Taking the Award as a whole, I thought it quite excellent. One can read it from beginning to end and understand a logical argument without looking at any other document. It makes sense as a stand-alone document. Many arbitrators find it very difficult to achieve that.

28 *Technology and Construction Court Guide*, 2nd edition issued 3 October 2005, third revision with effect from 3 March 2014, Section 12.

Although an adjudicator's decision is likely to have multiple audiences, in almost every dispute there will be an authority figure who decides whether their company will comply with the decision, to use the decision as the basis from which to negotiate a settlement, or to reject the decision in its entirety. In many disputes, the authority figure has not been actively involved with the project where the dispute arose, or indeed with the adjudication. Usually other people involved with the adjudication and the project will influence the authority figure. Through their decision, the adjudicator can bypass these people and make a direct connection to the authority figure, providing them with a stand-alone document containing all the information necessary.

The stand-alone document does not have to be unnecessarily long. It should only recount evidence, quotes and submissions in précis. In *Digi-Tech (Australia) Limited v Brand* [2004] Honourable Justices Sheller, Ipp and McColl said "prolixity is an enemy of comprehensibility and, indeed, cogency ... a succinct analysis of the issues and their sequential determination, involving a clear and ordered statement of the facts found is the preferred aim". They also posited that setting out "in great detail the submissions advanced by the parties and the evidence of many of the witnesses ... tends to obscure the essential issues that have to be decided and makes the reasoning process difficult to follow".[29]

Even with commitment and experience, starting to write a claim document for adjudication, or writing the decision itself can be difficult. The adjudicator should start writing the adjudicator's decision as soon as possible. Whereas decisions are bespoke, experienced adjudicators will have decision templates from previous disputes, which they use to set out a basic document structure.

In one paradigm, the writing process is considered as four stages called "Madman, Architect, Carpenter, and Judge".[30] Each stage of the process requires a different approach and a different personality. It is important to keep the roles separate and to allow each personality to perform without interruption from the others:

- The Madman is creative and enthusiastic and generates ideas. In adjudication, each of the parties can be the Madman.
- The Architect reviews what the Madman has created and presented, and uses it to develop the outline of the adjudicator's decision. The role of the Architect is to select the key ideas from the Madman and to arrange them in a pattern that will form discrete questions to be addressed and document the cogent arguments.

29 Digi-Tech (Australia) Limited v Brand [2004] NSWCA 58 (61 IPR 184).

30 Flowers, Betty S. (1979). Madman, Architect, Carpenter, Judge: Roles and the Writing Process, *Proceedings of the Conference of College Teachers of English*, 44: 7–10.

- The Carpenter bonds the key ideas logically. The role of the Carpenter is to write and re-write each sentence so that the adjudicator's decision has clarity, and flows elegantly from beginning to end telling a story.
- The Judge is a critical authority who will ratify that the tone of the adjudicator's decision is appropriate, that the decision reads well, and that the decision demonstrates balance and fairness. The Judge is responsible for capitalisation, consistent terminology, fine detail, grammar, punctuation and spelling.

As adjudicator, one of the most important writing skills is to be able to summarise accurately presenting in your own words the full sense of the information made available, but in a condensed form. This should show the parties that you have listened to what they have said and understood it in full. Aim to use simple direct sentences that are free of ambiguity. Remove all words that are not performing a real function.

In adjudication, the parties may express themselves in an emotive language. In the narrative of a decision, the adjudicator should use descriptive sentences objectively without forming or reinforcing conclusions and emotion. For example, write "the Responding Party says that the Architect did not certify the agreed final value, because the work was not in accordance with the Contract" rather than "the Responding Party says that it would have been irresponsible for the Architect to certify the agreed final value, because the work was not in accordance with the Contract". Consider what positive message you want the reader to get. Avoid double negatives as they can be confusing. For example, write "the evidence is refutable" rather than "the evidence is certainly not irrefutable" or "the investigation report results are conclusive" rather than "the investigation report results are not inconclusive".

Two of the tools that I have found to be invaluable in decision writing are the use of writing prompts and a scratch file. A writing prompt is a simply a topic that requires development. At the highest level, this might be a specific party submission. Often there will be ideas and information swirling around in your head and to avoid losing these, or worrying that they might later become important, save jottings into a scratch file which can subsequently be searched using keywords. Adjudicators should not use the internet or other sources of information to inform their decision. However, although it may at first sound heretical, if you are struggling for the vocabulary to express what you want to write, an internet keyword search might help with that, but then only use authoritative sources. Aim to complete your decision several days before you must send it to the parties; almost always you will return to your decision to enhance and refine it.

It is important to be aware of your writing weaknesses and to find a way to compensate for those. In his feedback on the dilapidations preliminary issues award Judge Bowsher also wrote

> Principal/Principles: Your spelling of those two words is quite haphazard. I realise that one modern view is that spelling does not matter but those two words have quite different meanings and if you do not spell them correctly, the authority of your Award will be injured.

Adjudication is a confidential process, but because of this feedback, I always have a decision proofread before sending it to the parties. Reading a decision aloud will also improve what you write, as it forces you to slow down, and you will discover mistakes that your eye would otherwise miss. If you find it easier to read over your final work using a paper-based document, it is best to use one.

17.9 Format and structure of the decision

It is fundamental that an adjudication decision must be able to withstand judicial scrutiny; a decision which is unenforceable is of no value to the parties.[31] Notwithstanding that, it is clear that the construction industry regards adjudicators' decisions not simply as a station before arbitration or litigation, but as having considerable weight and impact that goes beyond the legal requirement that the decision has to be observed on an interim basis.[32] The parties may use the decision to negotiate an agreement. If there is no arbitration or litigation following the adjudication, the adjudicator's decision will be final and binding.

Not all adjudication (security of payment) law requires that the adjudicator's decision must be in writing, however it would be very unusual if it were otherwise, and in Ireland it is mandatory. Unless the parties dispense with the requirement for reasons, the adjudicator's decision must contain reasons.[33] Apart from the requirement to be in writing and reasoned, the adjudication law[34] does not give guidance as to the form and contents of an adjudicator's decision. However, because that which is applicable in arbitration is basically applicable to adjudication,[35] and because of my training to become a Chartered Arbitrator, I have used the statutory requirements for a valid arbitration award, as set out in the UK Arbitration Act 1996, to inform the way that I set out adjudication decisions.

Adjudicators should craft their decisions so that they are:

- Capable of performance. There should be no linkage to the actions of others.

31 P C Harrington Contractors Limited v Systech International Limited [2012] EWCA Civ 1371.

32 Balfour Beatty Construction Limited v The London Borough of Lambeth [2002] (TCC) BLR 288.

33 Code of Code of Practice Governing the Conduct of Adjudications (21 July 2017).

34 Construction Contracts Act 2013.

35 Deko Scotland Limited v Edinburgh Royal Joint Venture and Others [2003] Scot CS 113.

- Certain. It should be clear and unambiguous. It must identify the parties correctly by their legal name.
- Complete. It must consider all the issues required and decide them.
- Consistent. There should be no internal contradictions.
- Directive. It should be written so as to compel action. For example, where the decision requires the payment of money it should stipulate the time in which that should be done.
- Final. The issues must be decided by the adjudicator, and not someone else. For example, to avoid any doubt the adjudicator should complete even simple mathematical calculations.
- Mandatory. The decision should be expressed in terms of imperatives and not expectations. That is, "the responding party shall pay" not "it is appropriate for the responding party to pay".

The layout of the adjudicator's decision will vary from dispute to dispute but could be structured as follows:

- Cover page – refer to the applicable law, name the parties, identify that the document is an adjudicator's decision, summarise the dispute in a few words to provide a working title, and be dated.
- Index – this lists the different sections setting out the structure of the decision and any appendices.
- Introduction – this should outline the project works; state where the dispute has arisen; identify the contract type (standard form or bespoke); provide the names and addresses of the parties and contract administrator, etc.; refer to the adjudication law or the contract including an agreement to adjudicate; list any special rules the parties have agreed; describe the adjudicator appointment process; refer to preliminary directions on procedural matters; defend against allegations of breaches of natural justice, by clearly showing that each party had the opportunity to set out its case and was treated reasonably; provide the basis for decision (documents only or with meetings); and if the adjudicator has been assisted by the professionalism of the parties' representatives, identify them and record that fact.
- Jurisdiction – where the jurisdiction of the adjudicator to make a decision has been challenged, there should be a section which describes that and how the adjudicator has dealt with it.
- Submissions and facts – take time to clearly and succinctly set out the parties' submissions, and what they want the adjudicator to decide. This will provide a checklist to help ensure that the decision is complete. The facts are those parts of the story which are not controversial.
- Issues – capture and identify the issues, which are those factual and legal matters on which the parties have not agreed.
- Findings – findings of fact, findings of law, and reasons for holding these findings. I often include findings into issues under the sub-heading "adjudicator's decision and reasoning".

- Costs – as empowered, decide who will pay the adjudicator's fees and expenses.
- Interest – consistent with your jurisdiction decide if and what interest is to be paid.
- Tax – if deciding VAT is controversial, in the UK I limit my decision to that the responding party shall pay the referring party value-added tax (VAT) as applicable under the Value Added Tax Regulations 1995 or any amendment or re-enactment thereof.
- Operative decision – summarising all the decisions on the substantive issues, interest, costs of the adjudication. This is the dispositive part of the decision, on the basis of what has been done.
- Signature and date – also identifying the place the decision is made.
- Appendix – if there were jurisdictional challenges that I considered, I would include any associated documents and letters into an appendix.

Although undesirable, the adjudicator may find that they are pressured for time when it comes to completing a decision. However, they should develop and use a quality checklist. My checklist includes:

- Have I answered all the questions?
- Have I collated all the answers into the operative decision?
- Have I correctly numbered the final index?
- Have I introduced case law or used any contract clauses not referred to by the parties?
- Are the factors I have relied upon "fairly in play" or "in the arena"?[36]
- Have I checked the mathematics?
- Has the decision been proofread subject to confidentiality?
- Have I acknowledged the hard work and professionalism of the parties' representatives?

As set out above, I advocate producing a decision where the reader can understand a logical argument without looking at any other document. This can be a somewhat contentious approach as it makes the decision longer than it might otherwise be, and can leave you open to criticism for lack of being concise.

17.10 Adjudicators should make a decision based on the evidence available

The adjudicator is required to reach a decision on the issues in dispute. The adjudicator's decision is the conclusion reached after an evaluation of facts and their effect, the contract and the applicable law. A fact is something that exists or has

36 Grindrod Shipping Pte Limited v Hyundai Merchant Marine Co. Limited [2018] EWHC 1284 (Comm).

happened. Each party should provide a concise statement of the facts on which it relies. Not all facts will be important or material. The material facts are the facts that a party must prove to be successful with its claim. The parties might agree on the facts but disagree as to how the facts should be applied or interpreted. Facts essential to the decision should become part of a connected, logical chain of reasoning. Typically, a dispute will comprise individual claims each requiring separate analyses, and the adjudicator will receive conflicting treatise in respect of entitlement and quantum. A structured referral or response will include a summary of the claim or rebuttal; a description of the facts that support the claim or rebuttal; an analysis that connects the effect of the facts to the contract and the law leading to the conclusion of entitlement or otherwise; a calculation of quantum; and evidence to support the above. In most cases, the dispute will condense to entitlement to money and compensable time.

Sometimes the adjudicator will have to decide on primary issues. For example, in an adjudication the referring party and the responding party had entered into a contract for the renovation of apartments in Central London. The contract terms were the terms of the JCT Minor Works Contract.

The referring party said that responding party repudiated the contract by changing the locks to the building and preventing it from accessing the property to complete the works. The referring party said that it sent a letter accepting the responding party's repudiatory breach of contract.

The responding party said that it was the referring party who repudiated the contract. It said that it did so because of the poor quality sub-standard workmanship, incomplete works, defective works, and the delay to the work undertaken by the referring party being sufficiently severe.

As which party repudiated the contract would affect my decision in respect of other disputed matters, it became a primary issue to be decided. In deciding a primary issue or individual claims for time and money, a method used by competent adjudicators is to resolve what facts are relevant, determine their effect, consider if there are contractual provisions or judicial authority supporting entitlement (or otherwise) and decide.

The decision-making stages are detailed in Table 17.10.

Table 17.10 The adjudicator's decision-making stages

Referring Party's Contentions	*Responding Party's Contentions*
Facts	Facts
Alleged fact: the responding party changed the locks. Supporting evidence: letter from the responding party's lawyer confirming that the locks were changed, and which outlined the responding party's position in respect of the referring party's performance. Detracting evidence: none.	Alleged fact: the referring party's work was defective and severely delayed. Supporting evidence: photographs, a schedule of defects, and remedial costs as estimated by the responding party's replacement contractor. Detracting evidence: the architect had recently certified a progress payment; the architect had not issued a notice specifying a failure to proceed diligently.
Effect	Effect
The referring party could not gain access to the property to complete the works.	The responding party could not rely on the referring party to complete the works, and had to find an alternative way to achieve that.
Contract	Contract
No relevant provisions.	Contract Clause 4.3 "Progress payments and retention" and Contract Clause 6.4 "Default by the contractor".
Law	Law
Parties' submission that: *London Borough of Merton v Stanley Hugh Leach* [1988][37] is authority for, if it is necessary for the employer to cooperate with the contractor, a term will be implied (in the absence of express terms) requiring the employer to do all that is necessary to allow the contractor to complete the work.	Parties' submission that: *Sutcliffe v Chippendale & Edmondson* [1971][38] is authority for, when the quality of work was deteriorating and the number of defects was multiplying, and that the contractor had neither the ability, competence nor the will to complete the work in accordance with the contract, the employer is justified in ordering the contractor off site and in determining the contract.
Reasons	Reasons
	The architect had recently certified a progress payment. When exercising a decision-making function such as authorising

(*Continued*)

37 London Borough of Merton v Stanley Hugh Leach Limited [1986] 32 BLR 51.

38 Sutcliffe v Chippendale & Edmondson [1971] 18 BLR 149.

Table 17.10 (Cont.)

Referring Party's Contentions	*Responding Party's Contentions*
	payments the architect should always be impartial. In terms of authorising payments *Sutcliffe v Chippendale & Edmondson* confirms that "An architect was required in issuing interim certificates to exclude work, which was not properly executed, from the value of work for which he recommended his employer to make payment on account, and if work was defective and unacceptable as it then stood, it had to be classified as work not properly executed until the defect had been remedied." The architect's contemporaneous assessment of the quality of the works undertaken, and progress was persuasive evidence.
Decision	
I decide that the referring party's work was not defective and severely delayed. I decide that changing the locks is evidence that the responding party intended not to be bound by the contract and repudiated it.	

Deciding extension of time

In analysing entitlement to an extension of time and to prolongation costs there are some essential steps for the adjudicator to take. Depending on the circumstances, the order of the steps can vary. The first question to ask is "does the referring party have a contractual entitlement to claim for an extension of time?", and the second question is "did the referring party comply with the contract extension of time notice provisions?" The third step is to consider if there was a "relevant event" which delayed the works, and the fourth step is to assess if responsibility rests with the responding party or if the "relevant event" is neutral. The fifth step is based on the facts to decide what works were affected by the "relevant event". The sixth step is to decide if the referring party's delay analysis demonstrates the

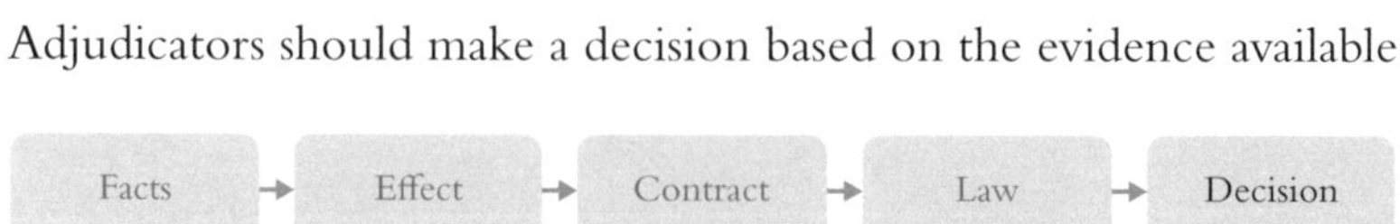

Figure 17.10 Adjudicators should make a decision based on the evidence available.

delay effect of the "relevant event" on the programme's critical path as the referring party states. If so, the seventh and final step is to calculate the quantum of the referring party's compensation for prolongation costs which are properly associated with any extension of time.

At the early stages of a project when the parties are developing their working relationship, for a range of reasons contractors are often disinclined to advance claims against employers by issuing notices of delay. Although this is very understandable, it does not relieve a contractor of its contractual obligations, and where the contractor holds back issuing requisite notices of delay it can often bar the contractor for advancing a claim for loss and expense associated with delay. In adjudication I have decided the referring party would have been entitled to an extension of time, and payment of substantial loss and expense costs *but for* the lack of requisite notices of delay.

17.11 Structured decision-making

A decision is a position arrived at after consideration. The role of the adjudicator is to make the best decision possible, under law, without emotion or subjectivity, and in a transparent way. Complexity is often inherent in decision-making. Adjudicators might address complexity intuitively in an informal manner using common sense, or use formal models to capture it.

Decision analysis (structured decision-making) is "a formalization of common sense for decision problems which are too complex for informal use of common sense".[39] In making decisions in adjudication, marginalise common sense or intuition. Whereas "common sense" is a quality seemingly revered, it is subjective. A losing party is unlikely to agree that the adjudicator's "common sense" perspective was either "common" or "sense". In adjudication, it is good discipline to formalise "common sense" into "reasoned sense", which is using sound judgement through analysis and structured decision-making.

A structured approach to decision-making in adjudication has six steps:

1) Decompose the problem. The parties will identify the constituent parts of a dispute, albeit sometimes poorly. However, a constituent part can represent a complex problem. Decomposing the problem is simply breaking it into smaller parts. Remember if the answer to the dispute were apparent or easy, the parties would have found it themselves.
2) Define the constituent questions to address. If you have not clearly and precisely stated or understood the questions, then the decision will almost certainly be wrong. Try-out expressing the questions in different ways to

39 Keeney, Ralph L. (1982). Decision Analysis: An Overview, *Operations Research*, 30(5) (September–October): 803–838.

clarify meaning and scope. In disputes, questions usually require consideration from multiple viewpoints.

3) Decide what criteria to use. The criteria will be determined and evaluated according to the question. For example, if the parties disagree as to the value of a "thing", criterion might be objective standards such as pricing guide data, or what something similar costs. Each objective standard can have positive features and negative features to commend it or denigrate it. Not all criteria carry the same weight. It can help to list the criteria in a simple table. Criteria mitigate subjective assessment and views.
4) Collate the information required. Information will comprise alleged facts, law and evidence. Responsibility for providing the information lies with the parties. However, what the parties provide is often laden with language-based ambiguity, inadequate, incomplete and obscure. The adjudicator can take the initiative by requesting information, and is empowered to draw inferences if the parties fail to comply with his request.
5) Identify the alternative answers. The answers in adjudication might be almost certain, possible or highly unlikely. The answers in adjudication can simply be yes or no. However, there will be times where there are several plausible alternatives. The alternatives can present themselves as positions on a continuum.
6) Make the decision. If the adjudicator maps the supporting information to the criteria, usually the answer will reveal itself allowing the adjudicator to make his decision. The adjudicator should document the decision and support it with reasons. This will usually be in the form of a narrative.

A structured approach to decision-making constrains "common sense" and provides a framework for improvement and learning. A non-structured approach to decision-making can be opaque and conceal value judgements.

17.12 Reasons

The Act provides that "unless the parties agree otherwise in writing, the decision shall include the reasons for the decision". In short, the reasons are why we make our decision.

The Merriam-Webster dictionary defines reason as a

> statement offered in explanation or justification, a rational ground or motive, a sufficient ground of explanation or of logical defense; especially something (as a principle or law) that supports a conclusion or explains a fact, the thing that makes some fact intelligible.

The importance of having reasons was underlined by the UK Court of Appeal when it said "The essential requirement is that the terms of the judgment should enable the parties and any appellate tribunal readily to analyse the reasoning that was essential to the Judge's decision".[40]

However, in adjudication, reasons do not have to show that the adjudicator's conclusion is right or wrong. Decisions will withstand challenge even if they are wrong in fact or law. The courts will resist any temptation to ask – was the adjudicator's reasoning correct? To allow this would undermine the purpose of adjudication, which is to provide a swift and binding, albeit provisional, resolution to a construction dispute.[41]

Reasons must be adequate; that is, sufficient to show that the adjudicator has dealt with the disputed issues referred.[42] If the adjudicator does not deal with all the issues and show that he has done so by giving reasons, and this causes substantial prejudice to one of the parties, the court may decline to enforce the decision.[43]

Reasons can be brief. All that is necessary is that based on the evidence adjudicators set out what did or did not happen, and succinctly explain why in the light of what happened, they have reached their decision, and what that decision is.[44]

Reasons must be intelligent. The adjudicator fails to give "reasons" when purported "reasons" are so incoherent that it is impossible for the reasonable reader to make sense of them.[45]

In practice, in most jurisdictions, adjudicators will be required to give reasons. From a party's point of view, it is sensible to request reasons, particularly if the dispute is complex or multi-faceted. Asking for reasons can be a way of ensuring or guaranteeing that the adjudicator does their job properly, but also for the adjudicator to demonstrate they have applied rigour and to validate time spent.

Reasons make the decision easier to understand and make the decision easier to explain to others who may not have had any part to play in the adjudication itself, but are interested in the outcome.

Sometimes in respect of deciding a particular issue, it can be very difficult to find good reasons or even any reason at all. However, adjudicators can tackle decision-making in different ways: head-on providing reasons through conscious thought (black box), and subliminally through subconscious thought (glass box). The black box and glass box approach to solving problems recognises that the human mind works with conscious thought and subconscious thought. Even when our conscious mind is not working, the subconscious mind is problem-solving, and that is why we have moments of inspiration when the answer leaps out at us.[46]

40 English v Emery Reimbold & Strick Limited [2002] EWCA Civ 605.

41 Pihl UK Limited v Ramboll UK Limited [2012] CSOH 139.

42 Thermal Energy Construction Limited v AE & E Lentjes UK Limited [2009] EWHC 408 (TCC).

43 Carillion Construction Limited v Devonport Royal Dockyard Limited [2005] EWCA Civ 1358.

44 Bremer Handelsgesellschaft v Westzucker (No. 2) [1981] 2 Lloyd's Report 130.

45 Gillies Ramsay Diamond v PJW Enterprises Limited [2003] P629/01.

46 Christopher, John & John, Jones, *Design Methods: Seeds of Human Futures*, Wiley & Sons, English (1 January 1970) ISBN-10: 0471447900; ISBN-13: 978-0471447900.

For me what works well is not to complete the decision, but to write as much of it as soon as I practically can, and to leave it. Often when I am not thinking about the decision new things, new ideas, or new or improved reasons will leap into my conscious mind. The decision becomes earlier to conclude and more compelling. Working early on the decision is good practice, notwithstanding that the parties might suffer additional costs for the adjudicator's time, if they compromise the dispute during the adjudication.

Reasons are important because:

1. Committing them to writing supports me to be open-minded and compels me to make sure that all the dots in my analysing and processing are "joined up" correctly. There have been occasions when I have finished writing reasons and arrived at a different conclusion to the one in my mind before I started to type.
2. They are important because they give you the opportunity to demonstrate to the parties that you have listened to them, you have understood their arguments, and taken them into account. If the parties ask themselves the questions "has the adjudicator considered every possible argument, has the decision-maker really understood our position?"… the answers should be yes. I do not repeat the parties' arguments verbatim, but rephrase them succinctly.
3. Consistent with maintaining the integrity of the decision, reasons should so far as possible protect the individuals involved in the formation of the dispute and adjudication, and help those participating in the adjudication to sell the decision internally.

17.13 Mistakes

An unfortunate characteristic of adjudication is that adjudicators will make mistakes, and their decisions are wrong in fact or law.

> Sometimes those mistakes will be glaringly obvious and disastrous in their consequences for the losing party. The victims of mistakes will usually be able to recoup their losses by subsequent arbitration or litigation, and possibly even by a subsequent adjudication. Sometimes, they will not be able to do so, where, for example, there is intervening insolvency, either of the victim or of the fortunate beneficiary of the mistake.[47]

Repeatedly, the courts have reinforced the principle that as long as the adjudicator has and exercises jurisdiction properly, and addresses the question within his terms of reference, answering the relevant factual or legal questions incorrectly, will not prevent the decision from being enforced.

47 Bouygues UK Limited v Dahl-Jenson UK Limited [1999] EWHC Technology 182.

The Act provides that the adjudicator "may correct his or her decision so as to remove a clerical or typographical error arising by accident or omission but may not reconsider or re-open any aspect of the decision". This enshrines the "slip rule", empowering the adjudicator to deal with patent errors such as the arithmetical miscalculations, failing to give credit for sums they decided were paid, or the wrong transposition of names.

The "slip rule" does not enable an adjudicator who has had second thoughts and intentions to change his decision. However, consequential corrections to costs and interest to avoid internal inconsistency following a "gateway error" should be possible.[48] The time for revising a decision by way of the "slip rule" will be what is reasonable in all the circumstances. It will be an exceptional and rare case where the adjudicator can revise his decision more than a few days after making it. The purpose of adjudication is to facilitate cash flow. If an adjudicator were able to revise his decision 21 or 28 days later, it would slow down and interfere with the speedy enforcement of adjudicators' decisions.[49]

My terms for acting as adjudicator include:

> The Adjudicator may on his own initiative, or at the request of either party, correct a decision to remove any obvious clerical mistake, error or ambiguity provided that the initiative is taken, or the request is made within five days. Such correction shall not affect the validity of the decision or its enforceability.

However, where an adjudicator has refused to correct a mistake that was a significant error, and the question did not involve a substantial dispute of fact, the courts have intervened to make a final decision on a question decided by the adjudicator.[50]

If the adjudicator considers an issue and decides it, in a subsequent adjudication they cannot reconsider that issue, alter their view and come to a different decision. An adjudication decision is binding. You cannot adjudicate a decided matter again in any later adjudication.[51]

17.14 Managing costs and time

Conceptually you can split the problem of managing adjudication costs and time into three phases. The first phase is the period before the adjudication, the second phase is the adjudication, and the third phase is the period after the adjudication.

48 Axis M&E UK Limited v Multiplex Construction Europe Limited [2019] EWHC 169 (TCC).

49 YCMS Limited v Grabiner & Another [2009] EWHC 127 (TCC).

50 Geoffrey Osborne Limited v Atkins Rail Limited [2009] EWHC 2425 (TCC).

51 Vertase F.L.I. Limited v Squibb Group Limited [2012] EWHC 3194 (TCC).

Phase one – before the adjudication

Accountability and costs management are "hand in glove". The person in charge of managing the dispute should be accountable for the costs incurred in doing so. If there is no budget and you do not have visibility of dispute costs and their accrual, they are difficult if not impossible to manage. There is an extremely close relationship between accountability, relevant information (actual and budget) and overall dispute costs.

In the period before the adjudication, the parties alone are responsible for costs incurred. Costs include the expense of company personnel, external consultants, experts' fees and legal costs. The best way to minimise costs is to negotiate an agreement with the other party quickly. Initiate a formal "dispute resolution" negotiation to be conducted on a "without prejudice and subject to contract" basis. This helps to ensure that any admissions or concessions tabled during the dispute resolution negotiation cannot be used later in any proceedings. In addition, the terms agreed only become binding when incorporated into a signed document.[52] If you conclude a settlement agreement, consider if it estops either party later referring the dispute to adjudication, or if the settlement agreement itself can be the subject of adjudication.

Having undertaken a formal review of risk, parties should carefully consider what percentage of their irrecoverable costs they should contribute to settling the dispute before the detailed preparation for adjudication begins. If the only real winners financially are the adjudicator, claims consultants, experts and lawyers, you have every entitlement to feel aggrieved and disappointed. It takes a mature and wise perspective to avoid this unhappy circumstance.

The Mediation Act 2017, which came into effect on 1 January 2018, encourages parties to resolve their difficulties through mediation, with the aim of reducing legal costs and resolving disputes in a more efficient manner. One of the objectives of the Act was to promote mediation, which had long before been recognised by the Irish Courts as an effective means of dispute resolution. Pursuant to the Act, parties are required to consider mediation in advance of issuing proceedings and to confirm to the court that they have considered it. Where the dispute is of sufficient size or importance, mediation should be considered. The cost of settling a dispute using mediation should be considerably less than the expense of settling a dispute using adjudication, arbitration or litigation. If the other party refuses to attend settlement discussions or unreasonably refuses to participate in mediation, that can have adverse costs consequences for them.[53] It is necessary to carefully consider whether the mediator should be a construction and engineering professional or a lawyer. The decision in this regard will depend on the nature of the

52 Oceanbulk Shipping & Trading SA (Respondents) v TMT Asia Limited and others (Appellants) [2010] UKSC 44.

53 Halsey v Milton Keynes General NHS Trust [2004] EWCA Civ 576.

dispute between the parties. The recently published CIC model mediation agreement and procedure reflects that where the dispute revolves around issues of fact, the best mediation approach is evaluative carried out by an experienced construction and engineering professional.[54]

If the formal negotiation/mediation is not successful, even at this early stage before any adjudication, if you believe that the other party has some entitlement, consider making a formal written settlement proposal in the form of a Calderbank offer. Whereas in adjudication the parties are responsible for their own legal and other costs incurred in connection with the adjudication,[55] ultimately a dispute could end up at arbitration or litigation where typically costs "follow the event". This means that costs are awarded to a successful party, which is one who achieves an award against another. Having issued a competitive Calderbank offer early on in the dispute might prove useful much later on.

A Calderbank offer is effectively a written offer made on a "without prejudice save as to costs" basis, to satisfy the whole or part of your opponent's claim. The offer should be a genuine compromise, carefully drafted in clear terms capable of being accepted. State that it is your intention, subsequent to any adjudication and where required, to bring the offer to the arbitrator's or judge's attention, in accordance with the principles expressed in *Calderbank v Calderbank*.[56] If the offer is not accepted, and your opponent fails to "beat the offer" at arbitration or litigation, the offer can then be produced to the arbitrator or judge in support of an application to fix your opponent with the costs of the proceedings. This is on the basis that the proceedings were unnecessary to begin with where an offer in excess of the sum awarded was previously made.

Therefore, it is essential that your costs record system accurately, clearly and easily distinguishes between costs incurred during the various stages, namely negotiation, mediation, adjudication and any subsequent arbitration or litigation.

Phase two – the adjudication

Sometimes the parties will spend considerably more on costs and fees than can be recovered in adjudication.[57] However, as far as he can exercise authority, the adjudicator must use reasonable endeavours to conduct the adjudication at the lowest cost and in the shortest time.[58] Parties can control costs by having work carried out by a capable person. Preparing for adjudication yourself should save costs, but in appropriate circumstances, employing a specialist(s) is cost-efficient. Do not commence adjudication until you are

54 http://cic.org.uk/admin/resources/cic-model-mediation-agreement-and-procedure-first-edition-11th-june-2019.pdf.

55 Construction Contracts Act 2013.

56 Calderbank v Calderbank [1975] 3 All ER 333.

57 Primus Build Limited v Pompey Centre Limited (& Slidesilver Limited) [2009] EWHC 1487 (TCC).

58 Construction Contracts Act 2013, Code of Practice Governing the Conduct of Adjudications.

ready, getting the wrong answer can be a costly mistake. Only include documents upon which you rely. Use a coherent system for referring to and identifying such documents. Do not duplicate referral documents in the response or any reply. Bespoke contracts can have unpredictable consequences; use of a standard form of contract can save costs. Get an adjudicator who will understand your dispute quickly, and who confirms he has time available to deal with it. The complexity of a dispute may warrant it, but increasing the time for a decision to more than 28 days is likely to increase the associated costs. Comply with the adjudicator's directions. Avoid argument by making spurious challenges to jurisdiction. Minimise the issues to be decided. Limit the length and number of submissions, expert reports and witness statements. Request that the adjudicator decide the dispute on a documents-only basis. Consider asking the adjudicator to decide just cost principles in a payment dispute, which the parties can then apply to the disputed matter and other similar matters.

Phase three – after the adjudication

What have you learned from the adjudicator's decision? Do not consider the outcome at a summary level, drill down into the detail of the decision. Does the detail change or inform your opinion as to the other party's entitlement? If so, should you revise the terms of your initial Calderbank settlement proposal? Go back to basics; be dispassionate and realistic! What are your chances of defeating the adjudicator's decision in enforcement proceedings? As both parties' costs of an unsuccessful challenge will be yours, is refusal to pay a worthwhile risk to take?

The success of adjudication will be determined by the Irish Courts. Adjudication will be successful if the courts put into place an effective, efficient, fast procedure for dealing with enforcement of an adjudicator's decision.

In subsequent arbitration or litigation, the dispute is considered "*ab initio*". In light of the adjudicator's decision, if there are facets of the overall dispute that are without merit, abandon them quickly. In respect of entitlement to recover arbitration or litigation costs, the costs "follow the event" principle is the starting point and plays a significant role; however, courts can depart from it.[59] The UK Civil Procedure Rules "encourage the court to consider making orders, which reflect a more detailed analysis of success and failure and, in particular, to make costs orders by reference to certain issues or by way of percentages".[60]

It has been suggested in Ireland that where a trial is clearly increased in length as a result of the successful party raising some issues upon which it failed, then the costs awarded to the successful party should reflect a refusal of costs attributable to the elongated hearing and payment of the unsuccessful party's costs for the portion of the hearing spent on the unsuccessful issues. This will be assessed on a case-by-case basis.

59 A.E.I. Rediffusion Music Limited v Photographic Performance Limited [1999] 1 WLR 1507.
60 Ian McGlinn v Waltham Contractors Limited & Others (No. 5) [2007] EWHC 698 (TCC).

17.15 Managing unequal power

In Ireland, the adjudicator must observe the principles of procedural fairness, act impartially, independently and without bias.[61]

Adjudication was conceived to be a quick and easy, "non-judicial" procedure, which "does not involve expensive lawyer time".[62] Believing that adjudication is a simple process enabling disputes to be resolved inexpensively and quickly, some parties come to adjudication without professional help. Considering that their dispute is straightforward they decide it is not necessary to incur the cost of obtaining professional assistance from claims consultants, lawyers or other specialists. However, adjudication is a serious process and mistakes can be very costly. Sometimes parties do seek assistance, but misguidedly instruct a "third party" who knows relatively little about adjudication. The "third party" understanding of adjudication first principles can seem almost non-existent; they will copy large swathes of articles from the internet and offer them nearly unaltered as argument to try to persuade the adjudicator. Sometimes parties will write to the adjudicator (genuinely or disingenuously) saying that as they have never been involved with adjudication before, please would the adjudicator help with guidance on the procedures and responsibilities.

There can be a power imbalance where the parties are in dispute over relatively significant sums of money, but they do not have the ability to put forward a cogent case establishing or defending such claim. A power imbalance might exist because of a party's conduct, or because of a lack of external or specialist help. The parties' state of mind may differ in that one party may be averse to, and uncomfortable with any third-party intervention. One party may have better information and had longer to prepare. The imbalance could be intellectual. One party in adjudication may be a repeat player (serial offender) with greater experience and knowledge. Being financially weaker can translate to needing a swift outcome, and payment of sums due.

What can or should the adjudicator do, if anything, in such circumstances? Most adjudicators, same as other highly skilled professionals, enjoy a life of privilege, and a realistic starting point is to be cognisant that the world is not fair. As adjudicators introspectively recognise, there is a power imbalance when some people have authority and others do not. A premise of the legal system is that people freely make choices and bear responsibility for the consequences. There are many factors, conscious and unconscious, which lead parties to their decisions, and which create disputes.[63] The adjudicator should be mindful that they are not responsible for the consequences of the parties' decisions.

61 Code of Practice Governing the Conduct of Adjudications (21 July 2017).

62 Hansard Housing Grants, Construction and Regeneration Bill [H.L.] HL Deb, 22 April 1996, Volume No. 571, Column 973.

63 Greenfield, Kent, *The Myth of Choice: Personal Responsibility in a World of Limits*, reprint edition, Yale University Press (13 November 2012) ISBN-10: 9780300169867; ISBN-13: 978-0300169867.

Most adjudicators will want to observe the principles of procedural fairness, not acting as an advocate for one of the parties, or making good any deficiencies in their submissions. As adjudication curtails a party's power to self-determine, most adjudicators will also want to act in an even-handed fair manner, without showing favouritism, and in conformity with the law. If remaining neutral leaves a seemingly unfair distribution of power intact, so be it. It is not the role of the adjudicator to protect the weak. However, whereas adjudication does not finally determine the parties' rights, neither should the adjudicator hide behind this as if it were a shield against unfairness.

Maintaining procedural fairness, the adjudicator is empowered to invite written submissions and evidence from both parties, and request any reasonable supporting or supplementing documents. The adjudicator has more pervasive power in that they may take the initiative in ascertaining the facts and matters necessary to determine the dispute. Irrespective of the adjudicator's sense of unequal power, and unfairness that might flow from that, my conduct has been to exercise inquisitorial powers with the lightest of touches.

Where there was a deficiency of evidence, I have asked the recalcitrant party to supply copies of original agreements, invoices, timesheets, etc. by a particular date. I accompany this with a request that the other party makes any further submissions on the disclosure by a later date. I have referred the parties to case law, for example

> In addition, before 5.00 p.m. Monday 4th June 2018, I ask both parties to make any submissions that they wish to make on the relevance of [Insert case law reference] to this adjudication, and in particular, paragraphs 38 to 40, where [Insert judge's name] said …

In terms of responding to requests for guidance on the procedures and responsibilities in adjudication, as a professional courtesy I will refer both parties to the CIC Users' Guide to Adjudication: Ireland which I attach in my correspondence to them.[64] This guide provides a general introduction to adjudication in the context of construction contracts, and in particular, the right to adjudication provided by the Act.

Faced with a situation which the adjudicator considers unconscionably unfair, as an ultimate sanction, the adjudicator can resign.[65] In such a situation, the referring party is faced with commencing the adjudication over again.

17.16 Experts in adjudication

The Civil Procedure Rules of England and Wales define an expert witness as "a person who has been instructed to give or prepare expert evidence for the

64 CIC Users' Guide to Adjudication: Ireland (26 July 2017).
65 Construction Contracts Act 2013.

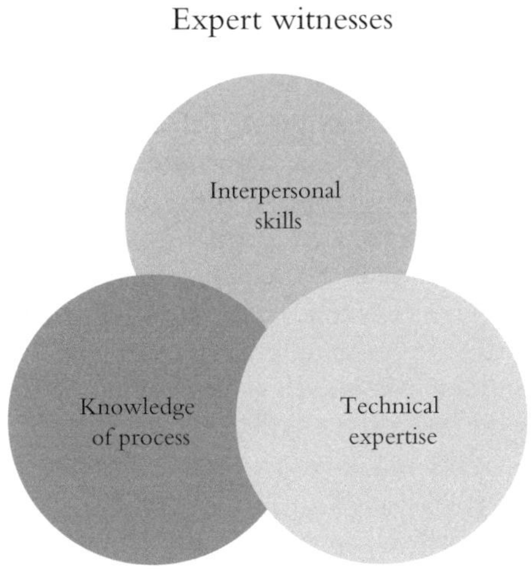

Figure 17.16 Expert witness attributes.

purposes of proceedings". Expert witnesses have professional obligations to the party instructing them. They must comply with their professional code of conduct, and have a duty to exercise reasonable skill and care. Their overriding duty is to assist the court on matters within their expertise.

Expert witnesses should possess and keep a high standard of up-to-date technical knowledge and practical experience in their professional field. Other important attributes are that they should have good interpersonal skills, such as being able to negotiate, write reports and withstand critical challenges. They should also have a good understanding of the dispute resolution process they are involved in, and their role in that.

The Code of Practice of the Academy of Experts sets out minimum standards of practice, which include the expert acting with impartiality, independence, integrity and objectivity. Practice Direction PD 1 of 2015 in Northern Ireland introduced a requirement for all experts giving evidence in Northern Ireland to undertake appropriate training and to seek accreditation as an expert witness, and this is increasingly being evaluated as part of assessing experts' competency to offer expert testimony.

In the landmark *Ikarian Reefer* case[66] Mr Justice Creswell lists the duties and responsibilities of expert witnesses in civil cases, subsequently reinforced by the Civil Justice Council's "Protocol for the Instruction of Experts to Give Evidence in Civil Claims" (June 2005). For example, the expert should

66 The Ikarian Reefer – National Justice Compania Naviera SA v Prudential Assurance Company Limited [1993] 2 Lloyd's Rep 68 (Commercial Court); [1995] 1 Lloyd's Rep 455 (CA).

confine his evidence to matters within his expertise, and make it clear when a question falls outside his expertise. The evidence should identify the assumptions or facts used, and consider facts that could detract from the opinion expressed. The evidence should be supported with sufficient reliable data, and identify where such data is not available. The expert's report should include documents referred to. The expert should never assume the role of advocate promoting the point of view of the party instructing them.[67]

Incredible although it might seem, some "experts" will use intemperate language. In one adjudication an "expert" instructed by the responding party stated "I find it absurd … I find that contention preposterous … this is a fabrication". The "expert" then attempted to usurp the adjudicator's authority by telling the adjudicator what his decision should be. Such conduct or colourful language does not assist the adjudicator, and it can leave the impression that the evidence is not from an independent "expert", but metaphorically speaking is from a "hired assassin".

The TCC Guide says that the quality and reliability of expert evidence will depend upon (a) the experience and the technical or scientific qualifications of the expert and (b) the accuracy of the factual material used by the expert for his assessment.[68]

Good experts should be authoritative, coherent, convincing, decent, helpful, independent, impartial, knowledgeable, pragmatic, serious, straightforward and thoughtful. They should have analysed all the issues in question. They should not be academic, argumentative, combative, partisan and unsure of themselves under examination.[69] Good experts will be reasonable, and be prepared to make appropriate concessions at the experts' "without prejudice meetings".

Notwithstanding such clear guidance, the reality is that clients and lawyers reappoint experts who would give evidence that supports the client's position. However, clients and claims support professionals need to be aware that where they appoint the same expert repeatedly an adverse inference might be drawn that the expert was not a properly independent witness.[70]

To provide good expert opinion, a construction professional should have relevant experience. If the construction professional embraces the role of expert, it is not necessary to have experience of acting as expert. Indeed some of the most experienced experts can be cynical and unconvincing; they are polished but not practical. In addition, as systems and technology move forward and fragment, it can become impossible to find knowledgeable professionals with prior experience of acting as expert. My experience is that if

67 Anglo Group plc v Winther Brown & Co. Limited and BML (Office Computers) Limited [2000] EWHC Technology 127.

68 *Technology and Construction Court Guide*, 2nd edition, second revision, October 2010.

69 The Board of Trustees of National Museums and Galleries on Merseyside v AEW Architects and Designers Limited & PIHL UK Limited & Another [2013] EWHC 2403 (TCC).

70 The Bank of Ireland UK Plc v Watts Group Plc [2017] EWHC 1667 (TCC).

an expert cannot describe technical subject matter in a way such that lay people can easily comprehend, then the expert may not understand the subject matter himself.

However, novice experts should tread carefully, they must not allow themselves to be used, or act as the mouthpiece of the party instructing them. If they allow themselves to be hoodwinked, or prepare their expert report to try to plug gaps in a party's witness evidence, they run the risk of having their evidence adjudged "entirely worthless" and disregarded "in full".[71]

The role of expert in civil litigation is clear. However, there is no clarity from the courts or case law as to the role of the expert witness in adjudication,[72] and the adjudicator often faces "expert witness" reports, which contradict each other on matters which seem straightforward to him. Lord Woolf said that most of the problems with expert evidence arise because "the expert is initially recruited as part of the team which investigates and advances a party's contentions and then has to change roles and seek to provide the independent expert evidence which the court is entitled to expect".[73] The contemporaneous nature of construction claims preparation and adjudication can lead to claims documents simply being reassigned as an "expert report".

With this in mind, adjudicators should develop their own techniques for evaluating expert evidence. They should ensure that the expert report states the instructions the expert has received and that the expert report includes a signed declaration of the expert's independence, verified by an adequate statement of truth.[74] In assessing credibility, the adjudicator should consider if an "expert report" refers to documents that are detrimental to the instructing party. If an adjudicator needs further information that he considers vital, he must take the initiative and ask for it directly.[75] In the tight timescale for adjudication a meeting with the parties and experts is rare, and there is opportunity for "experts" to act as advocates hiding behind the process. Faced with conflicting written evidence, ideally, the adjudicator could direct that the party's experts concurrently give oral evidence – "hot-tubbing". The adjudicator can question both parties' experts, and either party can do the same with the other's expert. In order to make this happen efficiently, the adjudicator needs to proactively manage the process, and at the beginning set a date in the timetable for a meeting with the parties and any experts, to be confirmed or vacated as required.

71 Van Oord UK Limited and SICIM Roadbridge Limited v Allseas UK Limited [2015] EWHC 3074 (TCC).

72 Horne, Robert & Mullen, John, *The Expert Witness in Construction*, Wiley-Blackwell (October 2013) ISBN: 978-0-470-65593-1.

73 Lord Woolf, "Access to Justice": Interim Report to the Lord Chancellor on the Civil Justice System in England and Wales (June 1995) Chapter 23; and Final Report to the Lord Chancellor on the Civil Justice System in England and Wales (July 1996) Chapter 13. See also White Book §35.0.2.

74 Civil Justice Council's Guidance for the instruction of experts in civil claims 2014, Paragraph 53.

75 Volker Stevin Limited v Holystone Contracts Limited [2010] EWHC 2344 (TCC).

The adjudicator can decide which expert's evidence to accept completely or in part. In resolving a conflict in expert evidence, the adjudicator is entitled to consider particular evidence in the context of all of the evidence that is before him. In addition, the adjudicator may prefer the evidence of one expert on one matter, and the evidence of the other expert on another matter. If one party provides expert evidence which has not been challenged, but which is not inherently implausible, the adjudicator would need to have a good reason to reject it; for example, because it does not fit with other facts which the adjudicator has decided.

What is the adjudicator to do when faced with two equally experienced, impressive, professional expert witnesses, but who have arrived at different conclusions because of a different approach in methodology and principle?

One approach is for the adjudicator to work his way through the specific issues collating positive and negative features of the experts' positions and in the light of other evidence, find a way to a conclusion, without resorting to the burden of proof.[76]

A Hong Kong case outlined three possible approaches:

> Firstly, the [adjudicator] should endeavour to decide on one figure or the other, based on the evidence. Secondly, if the evidence does not justify a simple choice of one or the other, then an interim figure may be selected, again, based on the evidence. Thirdly, and as a last resort, where neither expert has persuaded the [adjudicator] of the merits of their approach to the exclusion of the other, split the difference. Hopefully, this third option should only apply where the difference is relatively small.[77]

The UK Court of Appeal does not support splitting the difference:

> I do not wish to be misunderstood as commending any unprincipled splitting of differences; such is a practice which furious advocates sometimes suspect to be in hidden operation, which encourages submission of unreasonable figures and which brings the law into disrepute.[78]

It is not clear whether this approach will be followed by the Irish Court albeit that the decision would be of persuasive authority. The writer is aware of one High Court matter where a judge appeared to have endorsed a Circuit Court decision where when confronted with two different valuations, a Circuit Court judge effectively split the difference between the

76 Stephens v Cannon [2005] EWCA Civ 222.

77 Chan Shun Kei, trading as Chan Shun Kei Construction Works v Hong Kong Construction (Hong Kong) Limited [2014] HKEC, 7 April 2014.

78 Stephens v Cannon [2005] EWCA Civ 222.

prices arrived at by the valuers.[79] It would seem, however, that for policy reasons, the court might prefer not to generally endorse this approach.

What if in adjudication one party presents an expert report that the other party has not seen? Whereas the time available to a responding party to deal with an expert report revealed for the first time in the referral is limited, the time available to a referring party to deal with an expert report revealed for the first time in the response is even more limited. Because time is limited, this can raise an allegation of abuse of process, or "ambush". On more than one occasion, I have seen an astute responding party offer previously undisclosed expert evidence as a fair and objective basis for rejecting the referring party's contentions. Irrespective of any vehement objections, providing the original claim is factually robust, and is consistent with the notice of intention and the referral, there seems no procedural reason why statements of witness of fact and/or expert reports be disclosed prior to adjudication.[80]

In conclusion, a few words of warning to instructing parties or lawyers. Where an adjudicator does not have confidence that a purported expert witness has discharged the difficult role of providing evidence in a way that is impartial or that does not have a bias in favour of the instructing party, you run the risk of the adjudicator deciding that an expert report does not assist the adjudicator, and having the reports contents dismissed in their entirety.

Whereas in a multi-million pound adjudication the responding party's lawyers asserted that 'their' expert's report was clear and logical, and relied entirely on its contents to establish that the referring party's extension of time claim must fail, I decided the expert report did not assist me. In the adjudicator's decision I explained

> Unfortunately, the adjudicator does not have confidence that in this adjudication the Expert has discharged the difficult role of providing Expert Witness evidence in a way that is impartial or that does not have a bias in favour of the Responding Party ... When the Expert wrote his Expert Report he either did not have access to all relevant materials, or selectively declined to consider relevant emails in his report ... The Expert's evidence is of little value to the adjudicator.

17.17 How to become an adjudicator and appointment

I summarise the advantages of being arbitrator as first being in control of your own calendar. The arbitrator and the parties can usually agree a timetable that meets everyone's needs without difficulty. Second, the arbitrator's work is both challenging and interesting, if the answer to the parties' dispute were apparent or easy, they would have found it for themselves.

79 O'Gorman & Co Limited v JES Holding Limited [2005] 11 ICLMD 124.
80 Edmund Nuttall Limited v R G Carter Limited [2002] EWHC 400 (TCC).

Third, the arbitrator is well paid for his expertise and time. Whereas the role of the adjudicator is equally challenging, interesting and his remuneration is good, the adjudicator does not enjoy the same discretion in agreeing a timetable with the parties. The adjudicator can only accept an appointment if they are available to decide a dispute within a short period, usually 28 days.

Experienced practitioners will be asked for advice on what path might be taken to become an adjudicator, and to provide other support information. I suggest that initially the candidate participates in some entry-level courses to get a better feel for the ADR space, and the people within it. Does that experience confirm your anticipated interest in ADR generally, in adjudication, arbitration or mediation specifically? This should lead to the candidate investigating and documenting the education and training requirements, identifying alternative routes to achieve the requirements, the associated cost, and the time commitment necessary. One request for advice included, "my basic question is un-blushingly financial, if I'm to do this, I need to understand the potential earnings :-)". My answer was that despite appearances, very few people make a full-time living from being an adjudicator. The cost to become an adjudicator is high, the road is hard, and there is no guarantee that there will be a monetary payback on your investment. So if your motivation is "un-blushingly financial", embarking on the journey to become an adjudicator is a leap of faith. Many people do not reach the destination. My reasons for learning about ADR including adjudication were acquisition of knowledge and skills, and self-improvement.

In Ireland, in accordance with the Act, adjudicators are appointed in two ways. The first is that the parties agree to appoint an adjudicator they choose, and the second is that the Chair of the Minister's panel appoints the adjudicator. Although in Ireland there are other organisations that have established adjudication panels, the Chair of the Minister's panel will make the vast majority of adjudicator appointments. Therefore, the key to acting as adjudicator in Ireland is to be accepted as a member of the Minister's panel.

To be a member of a credible recognised panel of adjudicators, having requisite experience and expertise in construction dispute resolution is essential. Candidates should expect to demonstrate an understanding, to degree level, of the laws of contract, tort and evidence. They should know adjudication law, and adjudication practice and procedure, which I split into knowledge, skills and process. Candidates may have vast knowledge, but be unsuitable to be adjudicator because they do not have the personal characteristics to allow them to control and manage adjudication dilemmas in real time. This can be tested through a rigorous face-to-face interview, with a panel of experienced adjudicators who will challenge you. Candidates should have demonstrated their ability to write an adjudication decision that will withstand judicial scrutiny. An anonymised decision can be the platform for the face-to-face interview.

Barristers and solicitors will have covered much of the above through their development and education. To be suitable as an adjudicator, construction professionals have to invest considerable time and effort to reach the required level of good-standing. In addition to the wide range of university-based construction law courses, there are tailored construction adjudication courses. For example, the College of Estate Management offers a Postgraduate Diploma in Adjudication; the Royal Institute of Chartered Surveyors offers a Diploma in Adjudication in the Construction Industry; and the Chartered Institute of Arbitrators has a Fellowship Pathways programme in Construction Adjudication. Completing one of these courses does not guarantee acceptance onto a panel of adjudicators; further assessment and mentoring may be required. Being a member of a panel of adjudicators does not ensure that appointments as adjudicator will follow.

To be successful as an adjudicator, set achievement goals; taking a long-term view, ten years is a reasonable period. Analyse what other people are doing. Use basic business analysis and planning tools such as SWOT (strengths, weaknesses, opportunities and threats). Visualise different kinds of futures, and develop scenarios around those. To get work you have to promote yourself in the field of adjudication and within your own profession. If you have unique experience of an area of construction, can you build a reputation as an expert? Some people get closer to adjudication by advising and representing parties in adjudication. Networking is critical to the success of any type of professional services business; begin building your network of contacts immediately, both offline and online using platforms such as LinkedIn. You can influence people by having meaningful endorsements and recommendations. Attend seminars and workshops. Write articles for construction industry publications. Develop, improve and publish your curriculum vitae on an ongoing basis. A construction professional can write to construction industry companies and law firms, asking them to keep your details on record. Keep an up-to-date written record of any interesting or relevant experiences, they can be forgotten easily. Are there industry-linked or relevant volunteering opportunities available? You can promote yourself by "doing a good job", and consistently demonstrating your competence and experience. Find ways to allow people to judge the quality of your services. A professionally produced document provides tangible evidence subliminally reinforcing the quality of your work. Use your personality to your commercial advantage. Be professional and behave with integrity.

17.18 CPD and quality standards

For professionals, continuing professional development ("CPD") is a career-long process. If a professional body requests you to provide a CPD record and you are taking CPD seriously, then the required information will be readily to hand. CPD takes different forms, achieved through participating in professional activity, work-based learning, formal education, self-directed learning, and in other ways such as public service or voluntary work. CPD helps maintain user confidence in the adjudication system, helping ensure that practitioners preserve and enhance knowledge and skills to deliver a professional service.

The Royal Institute of British Architects ("RIBA") is an exemplar for adjudication panel CPD oversight. The RIBA requires its adjudication panel members to be able to demonstrate 24 hours of relevant annual CPD allocated to three activity categories. Category one is "Practical adjudication experience": that is, acting as adjudicator in a construction dispute, or representing a party to an adjudication in a construction dispute. Category two is "Other appropriate CPD": attending lectures or courses on adjudication or relevant aspects of construction law, or attending adjudication workshops or surgeries. Category three is "Other appropriate CPD": reading articles, papers, books, law reports, etc. on adjudication or relevant aspects of construction law, or lecturing on adjudication including preparation, writing articles, papers, books, etc. on adjudication, or serving on committees concerned with adjudication.

For a range of reasons, the performance of an adjudicator can deteriorate, and bad habits and practice can start to permeate their work. Seniority is no guarantee of good performance, the further a practitioner moves away from basic textbook guidance, the more the risk of bad practice could increase. The deterioration may be simply self-induced, taking place over many years. Sometimes the deterioration is because of a life-changing personal event such as bereavement or long-term illness; it might also be related to mental and physical health. In most commercial and professional environments adequate oversight and supervision exists so that the implementation of any necessary corrective measures takes place. For example, the Institution of Chemicals Engineers provides that inclusion on any of its ADR Lists is for a period of five years save for

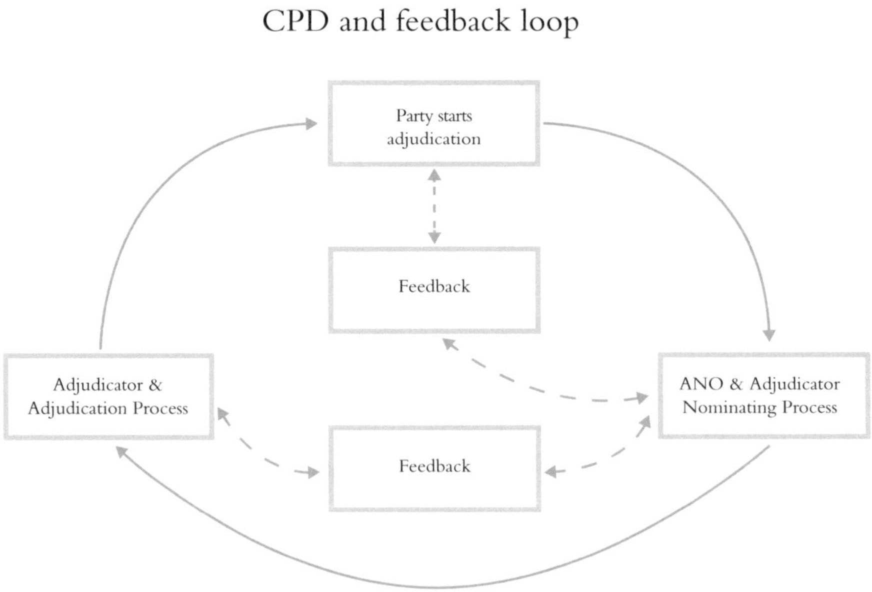

Figure 17.18 Adjudication feedback loop.

any candidate who is aged 70 or over at the time of appointment (including reappointment), in which case inclusion is for a period of two years.

However, a particular difficulty in adjudication is that the main work output, the adjudicator's decision is usually confidential as between the adjudicator and the parties, and therefore rarely subject to external monitoring and scrutiny. Usually, the adjudicator receives little or no feedback, which if given can help improve accountability and effectiveness.

Quality feedback can help protect the adjudicator as well as the parties. The feedback loop shown in Figure 17.18 allows information to be gathered from the parties in respect of the performance of the adjudicator, and the performance of the organisation responsible for the adjudicator nominating process. In addition, the adjudicator could provide information about adjudication in the public interest. For example, what the nature of the dispute was, which party was successful, fees charged, who paid the fees, etc. In Ireland the Chairperson of the Minister's panel has put in place arrangements to seek details of adjudication cases and anonymised information from adjudicators which will be used for the purpose of compiling statistical information relevant to adjudications conducted in accordance with the Act.

Recognising that adjudications are generally not subject to public scrutiny, the RIBA invites the parties to complete a short post-adjudication questionnaire. This helps the RIBA to monitor that the adjudication has met the reasonable expectations of the parties, and to identify any serious shortcomings on the part of the adjudicator. The questionnaire covers competence, costs, fairness, procedure and time. It also asks if both parties accepted the adjudication decision as final in determining the dispute.

Some ANBs also receive a copy of the adjudicator's decision. The Malaysian Construction Industry Payment and Adjudication Act 2012 provides that the adjudicator shall serve a copy of the adjudication decision on the parties and the Director of the AIAC. Whereas the AIAC serves as the administrative authority responsible for any functions as may be required for the efficient conduct of adjudication, it cannot and does not influence the content of adjudication decisions. However, the AIAC is in regular contact with its adjudicators, it monitors all AIAC adjudications, and it reviews the quality of all adjudication decisions.

In 2018, the RIBA introduced a re-assessment process on a rolling three-year basis to review the competence and standards of the adjudicators on the RIBA panel. The RIBA intends to ensure that standards are maintained and that its panel is made up of active and engaged adjudicators.

In pursuit of quality standards, in 2010 the Institution of Civil Engineers introduced a five-year rolling assessment of its panel of adjudicators. When selected, panel members are required to submit a recent adjudicator's decision rendered anonymous. Three experienced peer practitioners assess the decision using defined criteria under the headings: content, presentation and overall quality. Following the review, the adjudicator has an oral examination. Adjudicators either pass performing to a satisfactory standard, or if they fall short of the required standard, are assigned development objectives and reassessed.

While such oversight is to be welcomed, on its own it is not sufficient. Except in situations where the adjudicator's decision collides with a competent court during enforcement proceedings, its quality is seldom tested considering the evidence presented by the parties. Therefore, whereas an adjudicator's decision may be internally consistent and logical, well-reasoned and written, it might still fall far short of what is required. It is the adjudicator's responsibility to be vigilant as to best practice, and to ensure that they have the skills and knowledge to ensure that.

17.19 Adjudicators' fees and being paid

The success of adjudication depends on factors including the quality of the adjudicators who decide disputes. Becoming a construction adjudicator requires a significant investment in education and training, which is expensive and time consuming. Skilled adjudicators continue such investments so that they are always kept abreast with developments and updates in the industry. The parties in adjudication should expect to pay an hourly rate commensurate with employing someone qualified to a senior level in their primary profession, and who has additional expertise and skills. Whereas there is some consistency as to the hourly charge of lawyer adjudicators, there is a wide diversity in the hourly charge rate of construction profession adjudicators. Some of whom take the view that as they operate at the same level and perform the same work as lawyer adjudicators, as well as utilising their sector-specific knowledge and expertise, the hourly charges should be similar or higher. Other construction profession adjudicators charge an hourly rate consistent with the lower rates for providing architectural, engineering or quantity surveying services.

At the outset, the adjudicator should present the parties with their proposed terms and conditions of appointment, including the basis for his fees, costs and expenses.[81] Using an hourly rate for time actually spent to ascertain the adjudicator's final fee entitlement is sensible and reasonable. However, it is possible that an inexperienced adjudicator could use time less productively than an experienced adjudicator could, resulting in spending hours in an extravagant and objectively unwarranted fashion.[82]

Some adjudicators request via their terms and conditions an advance for the due payment of anticipated fees and expenses. Other adjudicators, some parties and their representatives frown at this practice. In the UK, support for both approaches is found in ANB procedural rules and standard contract forms. For example, the TeCSA Adjudication Rules provide that the adjudicator may not request and/or require any advance payment of or security for his fees.[83]

81 Code of Practice Governing the Conduct of Adjudications (21 July 2017).

82 London & Amsterdam Properties Limited v Waterman Partnership Limited [2003] EWHC 3059 (TCC).

83 The Technology and Construction Solicitors Association Adjudication Rules 2018, Version 3.2.2.

However, the NEC3 Adjudicator's Contract and the NEC4 Dispute Resolution Service Contract provides that each time a dispute is referred, the referring party makes an advance payment of the amount stated in the contract data.[84] In Malaysia, the standard terms for acting as adjudicator are the Asia International Arbitration Centre (AIAC) Adjudication Rules & Procedure, in accordance with which the parties shall contribute and deposit with the AIAC the adjudicator's fees and expenses in full as estimated by the adjudicator, and the administrative fee payable to the AIAC.[85]

Acting as adjudicator is a commercial activity and my standard schedule of conditions request an advance, offering a reasonable hourly rate, commensurate with that charged by lawyers. My experience is completely mixed in that in about 50% of disputes both parties accept my conditions, in about 25% of disputes either the referring party or responding party will accept my conditions, and in about 25% of disputes neither party accepts. Sometimes a party will say that it does not agree to pay an appointment fee and/or any advance payment of fees and expenses. Sometimes the parties simply do not give the adjudicator any response to the schedule of conditions they offer. Irrespective of the parties' response, or lack of response, I do not pursue the parties for their agreement of the schedule of conditions, and I continue to act as adjudicator.

Compared to other jurisdictions, being paid for acting as adjudicator in Northern Ireland can be particularly problematic. Some parties and more regrettably their professional advisors seem to consider adjudication as akin to the National Health Service in that it should be free at the point of use. For example, in a "pay less notice" dispute I decided the responding party would pay the referring party the full amount asked for, £46,386.49 (exclusive of VAT), plus interest. I computed my fees and expenses in the amount £2,775.00. In my decision I decided "The Adjudicator will invoice 'Referring Party' £2,775.00. 'Referring Party' will recover the £2,775.00 through this Decision. 'Referring Party' is invited to pay the fees promptly." Whereas I assume the referring party and its advisors were adamant that according to the decision the responding party should pay it £46,386.49 (exclusive of VAT), plus interest; the referring party refused to pay the adjudicator's fees, requiring the adjudicator to waste time in debt recovery action. After one adjudication, a party argued that because of Northern Ireland's unique location far from London people earned less money, and it should enjoy a discount on the adjudicator's normal fee rate.

When parties refuse to pay the adjudicator's fees their most usual arguments are that the hourly rate in unreasonable, and that the adjudicator

84 NEC3 Adjudicator's Contract, April 2013 and NEC4 Dispute Resolution Service Contract, June 2017.

85 Construction Industry Payment & Adjudication Act 2012 & Asia International Arbitration Centre (AIAC) Adjudication Rules & Procedure 2014.

expended an unreasonable amount of time. If a party wishes to object to the adjudicator's hourly rate, it should do so at the beginning of the adjudication and not at the end. This allows the adjudicator the opportunity to adjust the hourly rate if they choose to do so. It is for the objecting party to demonstrate that the adjudicator's fees are unreasonable. Evidence showing that the hourly rate is clearly and significantly more than the general level of fees charged by other similarly experienced adjudicators will support any objection. To be able to demonstrate that the time used was reasonable, the adjudicator should keep accurate timesheets, and allocation of time spent on tasks. In my decisions I always indicate separately the amount of time spent both on the substantive decision, and the amount of time spent on jurisdictional challenges, and other conduct or procedural matters. This enables the parties to hold their adjudication advisors to account for their mien during the adjudication.

Absent agreement on the adjudicator's hourly rate the Code of Practice provides that:

> The Adjudicator's fees, costs and expenses shall be reasonable in amount having regard to the amount in dispute, the complexity of the dispute, the time spent by the Adjudicator and other relevant circumstances.

Helpful guidance as to what is reasonable is as follows:[86]

- The court's approach in assessing what is reasonable should be a robust one, with a considerable "margin of appreciation".
- Amongst other things, a factor in determining what is reasonable is the seniority and experience of the adjudicator.
- There is no basis to interfere, even if it could be shown that a different adjudicator, especially an adjudicator with different qualifications, may have charged less or even significantly less.

To maintain confidence and good order in the adjudication system, adjudicators should be mindful that their fees and expenses should be proportionate to the disputed amount. Provided that the parties' conduct has not wasted my time, where required, I will usually abate my fees to between 10% and 15% of the amount in dispute. This is ordinarily relevant where the sum disputed is relatively small, say less than €50,000. The approach of discounting fees has enjoyed court support.[87]

However, my debt recovery contretemps described in Section 16.17 ("Bullying and intimidatory tactics toward adjudicators") has changed the way I go about any abatement. In that dispute I computed my fees to be

86 Fenice Investments Inc v Jerram Falkus Construction Limited [2011] EWHC 1678 (TCC).

87 Barrie Green v (1) GW Integrated Building Services Limited and (2) G&M Floorlayers (Derby) Limited [2001] LE014261 Leicester CC.

£6,600.00 and abated them to £4,600.00 (each party to pay 50%). I decided "With that in mind and in anticipation of prompt payment I abate my fees to £4,600.00". Now I will decide "With that in mind and *on the condition of prompt payment* I abate my fees to £4,600.00". In that way, where debt recovery action is required the adjudicator can apply to the court for the higher sum. To maintain their professional good-standing and to help eliminate undesirable conduct, I advocate adjudicators adopt a "zero tolerance" approach where parties unreasonably refuse to pay their fees and expenses.

Adjudicators should be mindful as to whether they have been properly appointed so as to enjoy threshold jurisdiction. If you did not have jurisdiction, you were never adjudicator, and whereas you can decide not to continue, you cannot resign. In such circumstances, you have no entitlement to be paid fees under the Act or in accordance with the Code. Notwithstanding that, under the doctrine of "*quantum meruit*" you should be entitled to payment of a reasonable sum of money for the services rendered.

Unlike an arbitrator, an adjudicator is not entitled to a lien on his decision pending payment of his fees. The adjudicator is required to reach a decision within 28 days beginning with the day on which the referral is made, and the potential to create an open-ended extension of time between the date of the adjudicator's decision and the decision being transmitted to the parties would be contrary to the whole principle of adjudication as described in the Act.[88]

The parties are "jointly and severally" liable for the adjudicator's fees. As the general rule under English Law is that costs follow the event, when I decide which party will pay my fees and expenses, I apply that norm. For example, if the referring party establishes its entitlement to a significant sum of money in accordance with my adjudication decision, it means that the referring party has largely won in the dispute. Because of that, I decide that the responding party is responsible for the adjudicator's fees and expenses. Even if I decide that one party shall pay all the adjudicator's fees and expenses, I might invoice each of them for 50% of my fees and expenses, and add that amount to the sum that the losing party has to pay the winning party. As I usually decide that all sums due shall be paid within seven days of the adjudicator's decision, even if the losing party does not want to pay the adjudicator promptly, so as to be seen to be complying with the adjudicator's decision the winning party has the incentive to so do. I have also decided that the winning party should pay all of the adjudicator's fees and expenses, and recover the sum through the decision.

In the UK, the Court of Appeal has decided that if the adjudicator produces an unenforceable decision, they are entitled to their fees.[89] In due course, the Irish Courts will decide what the "bargained-for performance" between the adjudicator and the parties is. Although enforcement hearings generally involve

88 Cubitt Building & Interiors Limited v Fleetglade Limited [2006] EWHC 3413 (TCC).

89 PC Harrington Contractors Limited v Systech International Limited [2012] EWCA Civ 1371.

only the referring party and the responding party (usually body corporate), the outcome affects the adjudicator (an individual), his reputation, and his entitlement to be paid. The Irish Constitution mandates that all citizens in Ireland shall be held equal before the law. I consider that the situation where the adjudicator can be substantially affected by, but is not involved in, enforcement hearings breaches the principle of equality before the law. Such a situation offers the real possibility of substantial unfairness, in that the adjudicator without any control has to rely on the competence and submissions of the "adjudication winning party". Such competence is far from guaranteed.

As adjudicator I have decided that a losing party should pay my fees, and where they have failed to do so, I have commenced proceedings against both parties because they remained jointly and severally liable for the same. In such circumstances, where a winning party honourably elects to settle the adjudicator's invoice, it will be able to recover its legal costs in defending and settling the proceedings brought by the adjudicator from the losing party.[90]

17.20 Checklists and golden rules

Avoiding adjudication

- Have a commercially aware organisational culture, embracing ADR practice and training.
- Conduct yourself in an ethical way, be willing to compromise, fair and reasonable.
- Use mediation, or agree to use a mutually respected independent person to offer a non-binding recommendation.
- If compromise is illusive, move confidently to adjudication, before arbitration or litigation.
- Senior management are responsible for disputes.
- Use standard forms of contract.
- Put effort into developing team working and relationships, both on and off site.
- Develop and maintain a comprehensive documents management system.
- Make payment applications as provided for in the contract.
- Get your notices right, and operate notice time clauses in an apt manner.

Before adjudication

- Make sure you have correctly and completely identified the corporate entity you have contracted with, especially when you encounter trading names.
- Prefer to use claims support firms and lawyers with real adjudication experience.

90 BM Services Inc. Limited v Greyline Builders Limited [2018] EWHC 3884 (TCC).

- Ensure that there are clear contracts between parties and any advisors.
- Beware of accepting at face value the opinions of claims consultants.
- Evaluate risk via a methodical investigation.
- Use formal dispute negotiation and mediation as early as possible.
- Clearly define the dispute and identify the questions you want answered.
- Prepare pleadings and present evidence well.
- Use diagrams to explain and reinforce concepts.
- Include appropriate date-and-time-stamped photographs, they can be highly persuasive and evidentially powerful.
- Undertake an adjudication audit to objectively assess how strong your case is.
- Do not overreach.
- Appoint somebody who is accountable for managing the cost of the dispute.
- Both parties should decide wisely what percentage of their irrecoverable costs should be contributed to the "disputes pot" to help mitigate against the situation where the only real winners financially are the adjudicator, claims consultants, experts and lawyers.

Adjudication – parties

- When you make an application for the appointment of an adjudicator, provide comprehensive contact information for both parties and their representatives (if any).
- Get the right decision maker.
- Be realistic about the timetable.
- Comply with directions.
- Do not try to speak with the decision maker, or have any ex-parte communication.
- Have a professional manner, be polite and respectful.
- Do not be needlessly argumentative or obtuse with the adjudicator, it will not assist you.
- Do not waste money and time with spurious jurisdictional challenges.
- Use active judgement and sense to evaluate your position, and if necessary use colourful language to compel the adjudicator's attention.
- Track carefully through the response to ensure that you have not included information which damages or undermines your position.
- Avoid bombarding the adjudicator with ad-hoc submissions.
- Warn your family that during the adjudication you will be suffering from stress, and your demeanour may be changeable.

Adjudication – adjudicator conduct

- Act with confidence and courage.
- Consider duties – conflict of interest, sufficient time, etc.

- Beware of unconscious bias.
- Confirm appointment, jurisdiction, and terms and conditions.
- Take control of the process – set a timetable and procedure, think clearly, be firm but polite.
- Comply with any notice requirements, for example in respect of conflict of interest, receipt of the referral.
- Address challenges to jurisdiction clearly and as early as possible.
- Read submissions quickly and in full.
- Prepare properly for any meetings.
- Deal with all matters referred.
- Make sure the decision is clear, complete and issued within the time allowed.
- Tell a complete contained story in the decision, focused on the parties' decision makers.
- Finish your decision one or two days before you need to send it to the parties.
- When writing early drafts of your decision, tag the locations of evidence, if you want to refer back to it, it can be difficult to locate later.
- Consistent with the principles of natural justice, consider sending your decision to the parties early, it might assist in minimising expense.
- Correct any mathematical or typographical mistakes quickly.
- Act fairly.
- Develop mechanisms to manage the stress of decision-making.

Adjudication – checklist for decision

- Check that the index is page numbered correctly.
- Have you answered *all* the questions you must address?
- Have you collated *all* the answers into the operative decision?
- Is the chronology clear and does it demonstrate clearly that the process has been fair?
- Have you introduced case law not referred to by the parties, does it have a decisive effect on the decision?
- Have you used any contract clauses not referred to by the parties, does it have a decisive effect on the decision?
- Have you considered all documents/submissions?
- Check the mathematics.
- Proofread for clarity, grammar and spelling.
- Acknowledge the hard work and professionalism of the parties' representatives, thank them.

After adjudication – parties

- Do not write to the adjudicator asking for clarifications, elaborations or additional reasons.

- Do not present the adjudicator with documents repleading your case, or trying to make good deficiencies.
- Do not resist the adjudicator's decision by participating in unsuccessful enforcement proceedings.
- Undertake a post-decision workshop.
- Implement changes to work practices.
- Put effort into rebuilding damaged relationships.
- Learn for the future.

Appendix A

Number 34 *of* 2013

CONSTRUCTION CONTRACTS ACT 2013

ARRANGEMENT OF SECTIONS

Acts Referred to

Bankruptcy Act 1988	1988, No. 27
Building Control Act 2007	2007, No. 21
Companies Act 1963	1963, No. 33
Companies (Amendment) Act 1990	1990, No. 27
Organisation of Working Time Act 1997	1997, No. 20
State Authorities (Public Private Partnership Arrangements) Act 2002	2002, No. 1

Number 34 *of* 2013

CONSTRUCTION CONTRACTS ACT 2013

AN ACT TO REGULATE PAYMENTS UNDER CONSTRUCTION CONTRACTS AND TO PROVIDE FOR RELATED MATTERS.

[29*th July*, 2013]

BE IT ENACTED BY THE OIREACHTAS AS FOLLOWS:

1.—(1) In this Act— Interpretation.

"construction contract" means (subject to *subsection (2)* and *section 2)* an agreement (whether or not in writing) between an executing party and another party, where the executing party is engaged for any one or more of the following activities:

(a) carrying out construction operations by the executing party;

(b) arranging for the carrying out of construction operations by one or more other persons, whether under subcontract to the executing party or otherwise;

(c) providing the executing party's own labour, or the labour of others, for the carrying out of construction operations;

"construction operations" means, subject to *subsections (3)* and *(4)*, any activity associated with construction, including operations of any one or more of the following descriptions:

(a) construction, alteration, repair, maintenance, extension, demolition or dismantling of buildings, or structures forming, or to form, part of the land (whether permanent or not);

(b) construction, alteration, repair, maintenance, extension, demolition or dismantling of works forming, or to form, part of the land, including (without prejudice to the foregoing) walls, roadworks, power-lines, telecommunications apparatus, aircraft runways, docks and harbours, railways, inland waterways, pipe-lines, reservoirs, water-mains, wells, sewers, industrial plant and installations for purposes of land drainage, coast protection or defence;

(c) installation in any building or structure of fittings forming part of the land, including (without prejudice to the foregoing) systems of heating, lighting, air-conditioning, thermal insulation, ventilation, power supply, drainage, sanitation, water supply or fire protection, or security or communications systems;

(d) external or internal cleaning of buildings and structures, so far as carried out in the course of their construction, alteration, repair, extension or restoration;

(e) operations which form an integral part of, or are preparatory to, or are for rendering complete, such operations as are previously described in this subsection, including site clearance, earth-moving, excavation, tunnelling and boring, laying of foundations, erection, maintenance or dismantling of scaffolding, site restoration, landscaping and the provision of roadways and other access works and traffic management;

(f) painting or decorating the internal or external surfaces of any building or structure;

(g) making, installing or repairing sculptures, murals and other artistic works that are attached to real property;

"executing party", in relation to a construction contract, means—

(a) where the parties to the construction contract are a contractor and the person for whom the contractor is doing work under the contract, the contractor, or

(b) where the parties to the construction contract are a contractor and a subcontractor or are 2 subcontractors, the subcontractor or whichever of the subcontractors agrees to execute work under the contract;

"main contract" means a construction contract such as is referred to in *paragraph (a)* of the definition of "executing party";

"Minister" means the Minister for Public Expenditure and Reform;

"other party", in relation to a construction contract, means the party to the construction contract who is not the executing party;

"payment claim" means a claim to be paid an amount under a construction contract;

"payment claim date", in relation to a construction contract, means the date when a payment claim in relation to an amount due under the construction contract is required to be made;

"payment claim notice" has the meaning assigned to it by *section 4*;

"payment dispute" has the meaning assigned to it by *section 6*;

"subcontract" means a construction contract such as is referred to in *paragraph (b)* of the definition of "executing party";

"subcontractor" means a person to whom the execution of work under a construction contract is subcontracted by the contractor or another subcontractor;

"work", in relation to a construction contract, means any act done in furtherance of the construction contract under the terms of the construction contract.

(2) In this Act references to a construction contract include an agreement, in relation to construction operations, to do work or provide services ancillary to the construction contract such as—

(a) architectural, design, archaeological or surveying work,

(b) engineering or project management services, or

(c) advice on building, engineering, interior or exterior decoration or on the laying-out of landscape.

(3) Subject to *subsection (4)* references in this Act to construction operations do not include the manufacture or delivery to a construction site of—

(a) building or engineering components or equipment,

(b) materials, plant or machinery, or

(c) components for systems of heating, lighting, air-conditioning, ventilation, power supply, drainage, sanitation, water supply or fire protection, or for security or communications systems.

(4) In this Act references to construction operations do include a case where the things referred to in *subsection (3)* are supplied under a contract which also provides for their installation.

2. —(1) A contract is not a construction contract— Construction contracts: exceptions, etc.

(a) if the value of the contract is not more than €10,000, or

(b) if—

(i) the contract relates only to a dwelling, and

(ii) the dwelling has a floor area not greater than 200 square metres, and

(iii) one of the parties to the contract is a person who occupies, or intends to occupy, the dwelling as his or her residence.

(2) A contract of employment (within the meaning of the Organisation of Working Time Act 1997) is not a construction contract.

(3) A contract between a State authority and its partner in a public private partnership arrangement, as those terms are defined in the State Authorities (Public Private Partnership Arrangements) Act 2002, is not a construction contract.

(4) Where a contract contains provisions in relation to activities other than those referred to in the definition of a construction contract and *section 1(2)*, it is a construction contract only so far as it relates to those activities.

(5) This Act applies to a construction contract whether or not—

(a) the law of the State is otherwise the applicable law in relation to the construction contract, or

(b) the parties to the construction contract purport to limit or exclude its application.

Payments under construction contracts.

3.—(1) A construction contract shall provide for—

(a) the amount of each interim payment to be made under the construction contract, and

(b) the amount of the final payment to be made under the construction contract,

or for an adequate mechanism for determining those amounts.

(2) A construction contract shall provide for—

(a) the payment claim date, or an adequate mechanism for determining the payment claim date, for each amount due under the construction contract, and

(b) the period between the payment claim date for each such amount and the date on which the amount is so due.

(3) The *Schedule* shall apply to a main contract if and to the extent that it does not make provision for the matters specified in *subsections (1)* and *(2)*.

(4) The *Schedule* shall apply to a subcontract except to the extent that it makes provision which is more favourable to the executing party than that which would otherwise be made by the *Schedule*.

(5) Except after the occurrence of the circumstances specified in *subsection (6)*, a provision in a construction contract is ineffective to the extent that it provides that payment of an amount due under the construction contract, or the timing of such a payment, is conditional on the making of a payment by a person who is not a party to the construction contract.

(6) The circumstances referred to in *subsection (5)* are:

(a) where the other person is a company other than an unregistered company—

(i) the commencement of its winding up pursuant to section 251 of the Companies Act 1963 where no declaration of solvency has been made under section 256 of the Companies Act 1963,

(ii) the presentation of a petition to wind it up pursuant to section 213 of the Companies Act 1963,

(iii) the appointment of a receiver in respect of any of its property or assets, or

(iv) the presentation of a petition for the appointment of an examiner under the Companies (Amendment) Act 1990 in relation to it;

(b) where the other person is an unregistered company, the commencement of its winding up pursuant to section 345 of the Companies Act 1963;

(c) where the other person is an individual or partnership, the making of an application for adjudication under the Bankruptcy Act 1988 in relation to it;

(d) the making of a winding up or similar order by a court in relation to the other person;

(e) the occurrence of any event corresponding to those specified in this subsection under the law of any state to which Council Regulation (EC) No. 1346/2000 of 29 May 2000[1] on insolvency proceedings applies.

4.—(1) This section applies where, not later than 5 days after the payment claim date, an executing party to a construction contract delivers a payment claim notice relating to a payment claim to the other party or another person specified under the construction contract. Payment claim notices.

(2) A payment claim notice is a notice specifying—

(a) the amount claimed (even if the amount is zero),

(b) the period, stage of work or activity to which the payment claim relates,

(c) the subject matter of the payment claim, and

(d) the basis of the calculation of the amount claimed.

(3) If the other party or specified person referred to in *subsection (1)* contests that the amount is due and payable, then the other party or specified person—

(a) shall deliver a response to the payment claim notice to the executing party, not later than 21 days after the payment claim date, specifying—

(i) the amount proposed to be paid,

(ii) the reason or reasons for the difference between the amount in the payment claim notice and the amount referred to in *subparagraph (i)*, and

(iii) the basis on which the amount referred to in *subparagraph (i)* is calculated,

and

(b) if the matter has not been settled by the day on which the amount is due, shall pay the amount referred to in *paragraph (a)* to the executing party not later than on that day.

(4) Where a reason for the different amount in the response is attributable to a claim for loss or damage arising from an alleged breach of any contractual or other obligation of the executing party (under the construction contract or otherwise), or any other claim

[1]OJ No. L160, 30 June 2000, p.1.

that the other person alleges against the executing party, the response shall also specify—

(a) when the loss was incurred or the damage occurred, or how the other claim arose,

(b) the particulars of the loss, damage or claim, and

(c) the portion of the difference that is attributable to each such particular.

(5) The rights and obligations conferred or imposed by this section are additional to any conferred or imposed by the terms of the construction contract.

Right to suspend work for non-payment.

5.—(1) Where any amount due under a construction contract is not paid in full by the day on which the amount is due, the executing party may suspend work under the construction contract by giving notice in writing under *subsection (2)*.

(2) Notice under this subsection shall specify the grounds on which it is intended to suspend work and shall be delivered to the other party—

(a) not earlier than the day after the day on which the amount concerned is due, and

(b) at least 7 days before the proposed suspension is to begin.

(3) Work may not be suspended under *subsection (1)*—

(a) after payment by the other party of the amount due, or

(b) after notice has been served by a party to the construction contract under *section 6(2)* in relation to a dispute relating to payment of the amount concerned.

(4) Where work is suspended under *subsection (1)* and the ability of the executing party to complete work within a contractual time limit is affected by the suspension of work, the period of suspension shall be disregarded for the purpose of computing the contractual time limit unless the suspension of work is unjustified in the circumstances.

(5) Where work is suspended under *subsection (1)* and the ability of a subcontractor to complete work within a contractual time limit is affected by the suspension of work, the period of suspension shall be disregarded for the purpose of computing the contractual time limit.

(6) A period of suspension of work under *subsection (1)* shall also be disregarded for the purpose of computing the time taken to complete the work under another construction contract where—

(a) the construction contract the work under which is suspended is a subcontract,

(b) the other construction contract is also a subcontract and the other party to that other subcontract is the same as the other party to the subcontract the work under which is suspended, and

(c) the ability of the executing party under that other subcontract to complete work within a contractual time limit is affected by the suspension of work.

(7) This section is without prejudice to the right of the other party to the construction contract under which work is suspended to claim for compensation or damages for any loss due to a suspension of work that is unjustified in the circumstances.

Right to refer payment disputes to adjudication.

6.—(1) A party to a construction contract has the right to refer for adjudication in accordance with this section any dispute relating to payment arising under the construction contract (in this Act referred to as a "payment dispute").

(2) The party may exercise the right by serving on the other person who is party to the construction contract at any time notice of intention to refer the payment dispute for adjudication.

(3) The parties may, within 5 days beginning with the day on which notice under *subsection (2)* is served, agree to appoint an adjudicator of their own choice or from the panel appointed by the Minister under *section 8*.

(4) Failing agreement between the parties under *subsection (3)*, the adjudicator shall be appointed by the chair of the panel selected by the Minister under *section 8*.

(5) The party by whom the notice under *subsection (2)* was served—

(a) shall refer the payment dispute to the adjudicator within 7 days beginning with the day on which the appointment is made, and

(b) shall at the same time provide a copy of the referral and all accompanying documents to the person who is party to the construction contract.

(6) The adjudicator shall reach a decision within 28 days beginning with the day on which the referral is made or such longer period as is agreed by the parties after the payment dispute has been referred.

(7) The adjudicator may extend the period of 28 days by up to 14 days, with the consent of the party by whom the payment dispute was referred.

(8) The adjudicator shall act impartially in the conduct of the adjudication and shall comply with the code of practice published by the Minister under *section 9*, whether or not the adjudicator is a person who is a member of the panel selected by the Minister under *section 8*.

(9) The adjudicator may take the initiative in ascertaining the facts and the law in relation to the payment dispute and may deal at the same time with several payment disputes arising under the same construction contract or related construction contracts.

(10) The decision of the adjudicator shall be binding until the payment dispute is finally settled by the parties or a different decision is reached on the reference of the payment dispute to arbitration or

in proceedings initiated in a court in relation to the adjudicator's decision.

(11) The decision of the adjudicator, if binding, shall be enforceable either by action or, by leave of the High Court, in the same manner as a judgment or order of that Court with the same effect and, where leave is given, judgment may be entered in the terms of the decision.

(12) The decision of the adjudicator, if binding, shall, unless otherwise agreed by the parties, be treated as binding on them for all purposes and may accordingly be relied on by any of them, by way of defence, set-off or otherwise, in any legal proceedings.

(13) The adjudicator may correct his or her decision so as to remove a clerical or typographical error arising by accident or omission but may not reconsider or re-open any aspect of the decision.

(14) The adjudicator is not liable for anything done or omitted in the discharge or purported discharge of his or her functions as adjudicator unless the act or omission is in bad faith, and any employee or agent of the adjudicator is similarly protected from liability.

(15) Each party shall bear his or her own legal and other costs incurred in connection with the adjudication.

(16) The parties shall pay the amount of the fees, costs and expenses of the adjudicator in accordance with the decision of the adjudicator.

(17) An adjudicator may resign at any time on giving notice in writing to the parties to the dispute and the parties shall be jointly and severally liable for the payment of the reasonable fees, costs and expenses incurred by the adjudicator up to the date of resignation.

(18) The parties to a dispute may at any time agree to revoke the appointment of the adjudicator and the parties shall be jointly and severally liable for the payment of the reasonable fees, costs and expenses incurred by the adjudicator up to the date of the revocation.

Right to suspend work for failure to comply with adjudicator's decision.

7.—(1) Where any amount due pursuant to the decision of the adjudicator is not paid in full before the end of the period of 7 days beginning with that on which the decision is made, the executing party may suspend work under the construction contract by giving notice in writing under *subsection (2)*.

(2) Notice under this subsection shall specify the grounds on which it is intended to suspend work and shall be delivered to the other party not later than 7 days before the proposed suspension is to begin.

(3) Work may not be suspended under *subsection (1)*—

(a) after payment by the other party of the amount due, or

(b) after the decision of the adjudicator is referred to arbitration or proceedings are otherwise initiated in relation to the decision.

(4) Where work is suspended under *subsection (1)* and the ability of the executing party or a subcontractor to complete work within a contractual time limit is affected by the suspension of work, the period of suspension shall be disregarded for the purpose of computing the contractual time limit.

(5) A period of suspension of work under *subsection (1)* shall also be disregarded for the purpose of computing the time taken to complete the work under another construction contract where—

(a) the construction contract, the work under which is suspended, is a subcontract,

(b) the other construction contract is also a subcontract and the other party to that other subcontract is the same as the other party to the subcontract the work under which is suspended, and

(c) the ability of the executing party under that other subcontract to complete work within a contractual time limit is affected by the suspension of work.

Selection of panel of adjudicators.

8.—(1) The Minister shall from time to time select persons to be members of a panel (in this section referred to as the "panel") to act as adjudicators in relation to payment disputes and shall select one of those persons to chair the panel.

(2) Persons selected under *subsection (1)* shall be members of the panel for a period of 5 years commencing on the date of selection and shall be eligible for reselection at the end of the period of 5 years.

(3) The Minister may, for good and sufficient reason, remove a member of the panel.

(4) A member of the panel may at any time resign by giving notice in writing to the Minister.

(5) In selecting persons to be members of the panel, the Minister shall have regard to their experience and expertise in dispute resolution procedures under construction contracts; and a person may not be selected to be a member of the panel unless the person is a person of any of the descriptions specified in *subsection (6)*.

(6) The descriptions of persons referred to in *subsection (5)* are as follows:

(a) a registered professional as defined in section 2 of the Building Control Act 2007;

(b) a chartered member of the Institution of Engineers of Ireland;

(c) a barrister;

(d) a solicitor;

(e) a fellow of the Chartered Institute of Arbitrators;

(f) a person with a qualification equivalent to any of those specified in *paragraphs (a)* to *(e)* duly obtained in any other Member State of the European Union.

Code of practice for adjudication.

9.—The Minister may prepare and publish a code of practice governing the conduct of adjudications under *section 6*.

Delivery of notices, etc.

10.—(1) The parties to a construction contract may agree on the manner by which notices under this Act shall be delivered.

(2) If or to the extent that there is no such agreement, a notice may be delivered by post or by any other effective means.

(3) Where under this Act a notice is required to be delivered not later than a specified number of days after a particular date and the last of those days is a day which is a Saturday or Sunday or a public holiday (within the meaning of the Organisation of Working Time Act 1997), the notice shall be taken to be validly delivered if delivered on the next day which is not such a day.

Expenses.

11.—The expenses incurred by the Minister in the administration of this Act shall be paid out of moneys provided by the Oireachtas.

Short title and commencement.

12.—(1) This Act may be cited as the Construction Contracts Act 2013.

(2) This Act applies in relation to construction contracts entered into after such day as the Minister may by order appoint.

SCHEDULE

Section 3.

Provisions to Apply to Matters Regarding Payments

1. The payment claim dates under a construction contract shall (subject to *paragraph* 2) be as follows:

(a) 30 days after the commencement date of the construction contract;

(b) 30 days after the date referred to in *clause (a)* and every 30 days thereafter up to the date of substantial completion;

(c) 30 days after the date of final completion.

2. Where a construction contract provides, or the parties to a construction contract otherwise agree, that the duration of the work under the construction contract is or is estimated to be less than 45 consecutive days, the payment claim date shall be 14 days following completion of the work under the construction contract.

3. The date on which payment is due in relation to an amount claimed under a construction contract shall be no later than 30 days after the payment claim date.

4. The amount of an interim payment under a construction contract shall (subject to *paragraph 5*) be the difference between—

(a) the aggregate of the gross value (determined in accordance with the construction contract) of the work done under the construction contract at the payment claim date concerned together with any additional amounts in the interim payment under the construction contract, less any deductions from payment provided for by the construction contract, and

(b) the aggregate amount of interim payments that have already been made at that payment claim date.

5. The aggregate of payments made under a construction contract shall not exceed—

(a) the amount provided for in the construction contract as originally concluded, and

(b) amounts provided for by any amendments to that contract agreed between the parties.

Appendix B

Construction Contracts Act, 2013

Code of Practice Governing the Conduct of Adjudications, 25th July 2016

An Roinn Post, Fiontar agus Nuálaíochta
Department of Jobs, Enterprise and Innovation

Table of Contents **Page**

Code of Practice Governing the Conduct of Adjudications under Section 6 of the Construction Contracts Act 2013

This Code of Practice is made pursuant to section 9 of the Construction Contracts Act, 2013 and should be read in conjunction with that Act.

Definitions

1. A reference in this Code of Practice to:

   a) "the Act" means the Construction Contracts Act, 2013;
   b) "Adjudicator" means an Adjudicator who is appointed to a payment dispute in accordance with section 6 of the Act;
   c) "Chairperson" means the Chairperson of the Construction Contracts Adjudication Panel who is appointed by the Minister under section 8(1) of the Act;
   d) "Construction Contracts Adjudication Service" means the section of the Department of Jobs, Enterprise and Innovation responsible for, inter alia, processing applications to the Chairperson under section 6(4) of the Act and contact details are available at www.djei.ie;
   e) "Minister" means the Minister or Minister of State with responsibility for the Construction Contracts Act, 2013;
   f) "Notice of Intention" means the notice of intention to refer a payment dispute for adjudication, referred to in section 6(2) of the Act;
   g) "Panel" means the panel of Adjudicators referred to in section 8(1) of the Act, the members of which are appointed by the Minister; and
   h) "Payment dispute" has the meaning assigned to it by section 6 of the Act.

General

2. The procedures set out in this Code of Practice shall apply to each individual payment dispute arising under the Act. In accordance with section 6(9) of the Act, an Adjudicator may deal at the same time with several payment disputes arising under the same construction contract or related construction contracts.

3. No liability whatsoever shall extend to the Minister, Chairperson or to the Department of Jobs, Enterprise and Innovation in respect of this Code of Practice or for any loss that arises from the operation of this Code of Practice. The Minister reserves the right to make changes to this Code of Practice.

Preliminary

4. A party to the construction contract (known as "the Referring Party") commences adjudication pursuant to section 6(2) of the Act by serving a written Notice of Intention on the other party or parties to the construction contract (known as the "Responding Party/Parties") under which an individual payment dispute arises.

5. A Notice of Intention shall include:

   (i) the name, address and contact details of each party to the construction contract;
   (ii) relevant details of the payment dispute to include the amount in dispute (even if the amount is zero), the nature of the payment dispute, and the site address;
   (iii) a copy of the relevant payment claim notice, and any response to that payment claim notice as provided for in section 4 of the Act; and
   (iv) relevant details to identify the construction contract and any supporting information that may assist an Adjudicator in understanding the nature of the payment dispute. Where a written construction contract exists, this must be attached.

Prospective Adjudicator responsibilities to the parties to a payment dispute

6. A prospective Adjudicator should only accept an appointment to a payment dispute under the Act if he/she:

   (i) is able to give the adjudication the time and attention which the parties to the payment dispute are reasonably entitled to expect;
   (ii) believes that he/she is competent to determine the issues in dispute; and
   (iii) is satisfied that no conflict of interest exists between him/her and the parties subject to paragraphs 11 and 20 of this Code of Practice.

7. A prospective Adjudicator shall not contact any party to a payment dispute under the Act in order to solicit appointment as an Adjudicator to that dispute.

The Appointment of an Adjudicator – by agreement of the parties

8. The parties to the construction contract may, within five days beginning with the day on which Notice of Intention is served, agree to appoint an Adjudicator of their own choice and he/she may be a person referred to in the construction contract to perform that role, a person from the Panel or he/she may be another suitably qualified person.

9. A person who is requested to accept an appointment as Adjudicator following an agreement by the parties to the construction contract in accordance with section 6(3) of the Act shall, within two days of such a request and prior to accepting the appointment, write to the parties to ask them to disclose any information indicating any potential conflict of interest that may arise from the person's appointment as Adjudicator. He/she shall draw the attention of the parties to the provisions of paragraph 32 of this Code of Practice. The prospective Adjudicator shall, at the same time provide the parties with his/her proposed terms and conditions of appointment, including the basis for his/her fees, costs and expenses.

10. Each party shall within three days of the communication from the prospective Adjudicator decide if the appointment of the prospective Adjudicator is to proceed and inform the prospective Adjudicator in writing of their decision.

11. If a potential conflict of interest is disclosed by any party, the prospective Adjudicator may subject to the consent of all the parties, and on satisfying any professional and/or ethical concerns he/she may have, accept the appointment.

12. If the appointment of the prospective Adjudicator is to proceed, the prospective Adjudicator shall write to each party to accept the appointment and the date of the letter of acceptance sent to the parties shall be deemed to be the date on which the appointment of the Adjudicator is made for the purposes of section 6(5)(a) of the Act. Such acceptance, anonymised in terms of the details of the parties to the dispute, shall be notified by the Adjudicator to the Construction Contracts Adjudication Service of the Department of Jobs, Enterprise and Innovation for the purpose of compiling statistical information relating to the Act.

The Appointment of an Adjudicator – by the Chairperson

13. Failing agreement by the parties to select an Adjudicator in accordance with section 6(3) of the Act, a party to the construction contract may apply to the Chairperson to seek the appointment of an Adjudicator from the Panel in accordance with section 6(4) of the Act. Relevant contact details are available on the website of the Department of Jobs, Enterprise and Innovation at www.djei.ie.

14. If an application is to be made under section 6(4) of the Act to the Chairperson, it shall be made not earlier than five days from and including the day on which the Notice of Intention was served.

15. An application to the Chairperson to appoint an Adjudicator from the Panel to a payment dispute shall be in writing and submitted to the Chairperson in accordance with the application procedures set out by the Construction Contracts Adjudication Service of the Department of Jobs Enterprise and Innovation from time to time. Such application, shall be copied by the applicant to the other party/parties to the payment dispute at the same time and shall include:

    (i) the name, address and contact details of each party to the construction contract;
    (ii) relevant details of the payment dispute to include the amount in dispute (even if the amount is zero), the nature of the payment dispute, and the site address;
    (iii) a copy of the Notice of Intention including any accompanying documents attached to that Notice;
    (iv) the date as to when the Notice of Intention was served on the Responding Party/Parties and how this was done; and
    (v) relevant details to identify the construction contract and any supporting information that may assist an Adjudicator in understanding the nature of the payment dispute. Where a written construction contract exists, this must be attached.

16. The Chairperson and/or the Construction Contracts Adjudication Service of the Department of Jobs, Enterprise and Innovation may seek further information or clarification(s) from the applicant relevant to the nature of the dispute and such information or clarification(s) should be provided promptly by the applicant and copied to the other party/parties to the payment dispute at the same time. No additional or other supporting information should be submitted by the applicant without a specific request for such information from the above-mentioned in this paragraph.

17. The Chairperson shall, following receipt of a completed application from a party to the construction contract made in accordance with paragraph 15 of this Code of Practice and subject to paragraph 16 of this Code of Practice, appoint an Adjudicator from the Panel.

18. The appointment of an Adjudicator from the Panel shall be made by the Chairperson and notified in writing by the Construction Contracts Adjudication Service of the Department of Jobs, Enterprise and Innovation to the parties, normally within seven days after the receipt of the application to the Chairperson, subject to paragraph 16 of this Code of Practice. The date of the letter from the Construction Contracts Adjudication Service to the parties shall be deemed to be the date on which the appointment of the Adjudicator is made for the purposes of section 6(5) (a) of the Act.

19. An Adjudicator appointed by the Chairperson to a payment dispute shall, within two days of such appointment, request of the parties in writing to disclose any information indicating any potential conflict of interest that may arise from the person's appointment as Adjudicator. He/she shall draw the attention of the parties to the provisions of paragraph 32 of this Code of Practice. The Adjudicator shall at the same time provide the parties with his/her terms and conditions of appointment, including the basis for his/her fees, costs and expenses.

20. If the information disclosed indicates a potential conflict of interest, the Adjudicator may only proceed with the adjudication where he/she is satisfied that the disclosures are frivolous or vexatious; that no professional or ethical concerns arise; and that no actual conflict of interest exists.

Referral of a payment dispute to an Adjudicator

21. Following the appointment of an Adjudicator, the Referring Party shall in accordance with section 6(5) of the Act refer the payment dispute to the Adjudicator in writing within seven days of the Adjudicator's appointment, and the Referring Party shall provide a copy of all such documentation to the Responding Party/Parties at the same time.

22. The referral of the payment dispute to the Adjudicator shall include:

(i) the name, address and contact details of each party to the construction contract;
(ii) relevant details of the payment dispute to include the amount in dispute (even if the amount is zero), the nature of the payment dispute, and the site address;
(iii) a copy of the Notice of Intention including any accompanying documents attached to that Notice;
(iv) the date when the Notice of Intention was served on the Responding Party/Parties and how this was done;
(v) the contentions on which the Referring Party intends to rely upon to support their case; and
(vi) relevant details to identify the construction contract and any supporting information that may assist an Adjudicator in understanding the nature of the payment dispute. Where a written construction contract exists, this must be attached.

Adjudication of a payment dispute – Procedures and Decision

23. The Adjudicator in any payment dispute under the Act shall be impartial, independent and only adjudicate where satisfied that no actual conflict of interest exists. He/she shall observe the principles of procedural fairness, which shall include giving each party a reasonable opportunity to put their case and to respond to the other party's case.

24. For the purposes of the adjudication proceedings, the Adjudicator may:

(i) request any reasonable supporting or supplementing documents pertaining to the payment dispute detailed in the Notice of Intention and/or in the referral of the payment dispute to the Adjudicator;
(ii) take the initiative in ascertaining the facts and matters required for a decision and make use of his/her own specialist knowledge, if it is appropriate to do so. If the Adjudicator uses any such specialist knowledge he/ she shall disclose this to the parties as appropriate;

(iii) appoint experts, assessors or legal advisers, provided that the parties have been notified of their identity and their terms of reference;
(iv) make site visits and inspections or carry out tests, subject to prior notification to the parties and obtaining any necessary consent from a third party or parties;
(v) invite written submissions/representations and evidence from the parties, if appropriate;
(vi) meet jointly with the parties and their representatives, if any, to enable further investigation;
(vii) hold a teleconference with the parties, with the consent of the parties; and
(viii) hold an oral hearing, where appropriate.

25. The Adjudicator shall upon receipt of the referral of a payment dispute from the Referring Party, inform the parties in writing of the date on which it was received by the Adjudicator. This date of receipt shall be regarded as the date on which the referral of the payment dispute to the Adjudicator has been made for the purposes of section 6(6) of the Act.

26. The Adjudicator shall, at the same time, also inform the parties of the procedures that he/she intends to apply during the adjudication process. This shall include directions as to the timetable for the adjudication and any deadlines to be adhered to by the parties and/or limits as to the length of written documents. The Adjudicator shall draw the attention of the parties to the provisions of paragraph 32 of this Code of Practice. The Adjudicator may revise his/her guidance to the parties on the above mentioned matters in circumstances where he/she considers it necessary to do so and he/she shall inform the parties of any such change as appropriate.

27. The Adjudicator shall ensure that the procedure adopted is commensurate with the nature and value of the payment dispute and he/she shall be mindful of whether or not an oral hearing is required having regard to matters such as to whether or not there is a conflict of fact or other relevant matter that requires such a hearing.

28. The Adjudicator shall use reasonable endeavours to process the payment dispute between the parties in the shortest time and at the lowest cost. He/she shall promptly notify the parties of any matter that will slow down or increase the cost of making a determination.

29. The parties may agree to revoke the appointment of the Adjudicator in accordance with section 6(18) of the Act and shall be jointly and severally liable for the payment of the reasonable fees, costs and expenses incurred by the Adjudicator up to the date of the revocation.

30. In accordance with section 6(17) of the Act, the Adjudicator may resign for reasonable cause at any time on giving notice in writing to the parties to the payment dispute. Such resignation, anonymised in terms of the details of the parties to the dispute, shall be notified by the Adjudicator to the Construction Contracts Adjudication Service of the Department of Jobs, Enterprise and Innovation for the purpose of compiling statistical information relating to the Act.

31. Upon such resignation the adjudication is at an end subject to the payment by the parties of the reasonable fees, costs and expenses incurred by the Adjudicator up to the date of resignation. The parties shall be jointly and severally liable for the payment of the reasonable fees, costs and expenses incurred by the Adjudicator up to the date of resignation.

32. If a party to the adjudication, without showing sufficient cause, fails to:

    (i) attend a meeting; or
    (ii) comply with any directions of the Adjudicator made in accordance with paragraph 26 of this Code of Practice; or
    (iii) disclose any information indicating a potential conflict of interest as required to do in accordance with paragraphs 9 and 19 of this Code of Practice; or
    (iv) produce any document or written statement requested by the Adjudicator;

    the Adjudicator may:

    a) continue the adjudication in the absence of a party;
    b) continue the adjudication without the document or written statement requested;
    c) draw such inferences from that failure to comply as circumstances may, in the Adjudicator's opinion, be justified;
    d) make a decision on the basis of the material properly provided; and
    e) make a decision apportioning the fees, costs and expenses of the Adjudicator, as appropriate.

33. The Adjudicator shall, in accordance with the Act, reach a decision within 28 days beginning with the day on which the referral is made or such longer period as is agreed by the parties after the payment dispute has been referred. The Adjudicator may extend the period of 28 days by up to 14 days, with the consent of the Referring Party.

34. The decision of the Adjudicator on a payment dispute shall be in writing and it shall be signed and dated by the Adjudicator. Unless the parties agree otherwise in writing, the decision shall include the reasons for the decision.

35. The Adjudicator's decision shall allocate such fees, costs and expenses of the Adjudicator as he/she has authority to allocate under section 6(16) of the Act and under the provisions of this Code of Practice.

36. The Adjudicator's fees, costs and expenses shall be reasonable in amount having regard to the amount in dispute, the complexity of the dispute, the time spent by the Adjudicator and other relevant circumstances.

37. Any document or information supplied for and/or disclosed in the course of the adjudication shall be kept confidential by the Adjudicator. He/she will only disclose such document or information if required to do so by law, or pursuant to an order of a court, or with the consent of all the parties to the payment dispute.

38. The parties are responsible for their own legal and other costs incurred in connection with the adjudication in accordance with section 6(15) of the Act.

Reporting on the Conduct of Adjudication Cases

39. The Chairperson may seek or, put in place arrangements to seek, details of adjudication cases from Adjudicators and which shall not include the names of the parties to a payment dispute. An Adjudicator, regardless of whether appointed to a payment dispute under section 6(3) or 6(4) of the Act, shall provide such anonymised information to the Construction Contracts Adjudication Service of the Department of Jobs, Enterprise and Innovation on each adjudication case within 21 days of the completion of the case. This will be used for the purpose of compiling statistical information relevant to adjudications conducted in accordance with the Act.

I, PAT BREEN, Minister of State at the Department of Jobs, Enterprise and Innovation, in exercise of the powers conferred on me by section 9 of the Construction Contracts Act, 2013 (No. 34 of 2013) and the Construction Contracts (Transfer of Departmental Administration and Ministerial Functions) Order 2014 (S.I. No. 476 of 2014) (amended by the Construction Contracts (Transfer of Departmental Administration and Ministerial Functions) Order 2015 (S.I. No. 173 of 2015)) and the Jobs, Enterprise and Innovation (Delegation of Ministerial Functions) Order 2016 (S.I. No. 333 of 2016) make this Code of Practice.

The Construction Contracts Act 2013 Code of Practice Governing the Conduct of Adjudications made on the 5th day of July 2016 is revoked.

PAT BREEN,
Minister of State at the Department of Jobs, Enterprise and Innovation.

Date: 25th July 2016

The images on the cover of this document are of drawings by Gabriel Hayes who was commissioned in 1941 to design and complete a range of carved stonework for the facade of the Department of Industry and Commerce building at Kildare Street.

Department of Jobs Enterprise and Innovation

Construction Contracts Adjudication Service
Earlsfort Centre, Lower Hatch Street
Dublin 2, D02 PW01

Tel: **+353 1 631 2121**
LoCall: **1890 220 222**

Email: info@djei.ie
www.djei.ie

An Roinn Post, Fiontar agus Nuálaíochta
Department of Jobs, Enterprise and Innovation

References

Books

Books	Paragraph Number Part 1	Section Number Part 2
Abrahamson, Max, *Engineering Law and the I.C.E. Contract*, 4th edition, Routledge (1979) ISBN 0 419 16080 9		17.3
Aeschylus, *The Oresteia*, translated by Hugh Lloyd-Jones (1 December 1993) ISBN 13: 9780520083288		17.1
Apollodorus: The Library, Volume II: Book 3.10-16/Epitome (Loeb Classical Library, No. 122), translated by J. G. Frazer (31 January 1921) ISBN-10: 0674991362.		15.1
Apple, James G. and Deyling, Robert P., *A Primer on the Civil-Law System*, Federal Judicial Center (1995) ASIN: B00010XEHK		15.9
Avruch, Kevin, *Culture & Conflict Resolution*, United States Institute of Peace Press (12 January 1998) ISBN-10: 1878379828; ISBN-13: 978-1878379825		17.2
Cialdini, Robert B., *Influence: The Psychology of Persuasion*, Harper Business (1 February 2007) ISBN 0-688-12816-5		17.5
Confucian Analects – The Great Learning of the Doctrine of the Mean: Chinese Classics, new edition, translated by James Legge, Dover Publications Inc. (28 March 2003) ISBN-10: 0486227464; ISBN-13: 978-0486227467		16.1
Coulson, Peter, *Coulson on Construction Adjudication*, 3rd edition, Oxford University Press (2011) ISBN-13: 978-0199597222; ISBN-10: 0199597227		16.8, 16.20
Christopher, John and John, Jones, *Design Methods: Seeds of Human Futures*, Wiley & Sons, English (1 January 1970) ISBN-10: 0471447900; ISBN-13: 978-0471447900		17.12
Gross, Bertram, *The Managing of Organisations*, 1st printing edition, Free Press/Collier (1964) ASIN: B000IQ72CK		17.4
Flowers, Betty S. (1979). Madman, Architect, Carpenter, Judge: Roles and the Writing Process, *Proceedings of the Conference of College Teachers of English*, 44: 7–10		17.8
Horne, Robert and Mullen, John, *The Expert Witness in Construction*, Wiley-Blackwell (October 2013) ISBN: 978-0-470-65593-1		17.16

Macaulay, Thomas Babington, *The Miscellaneous Writings and Speeches of Lord Macaulay*, Contributions to the Edinburgh Review, Volume II History (May 1828) 15.8

Murray, Alexander S., *Manual of Mythology*, David McKay (1895) ASIN: B000M0JF6I 17.1

Fellows, Richard and Liu, Anita, *Trust and Money: 20 Years of (No) Progress?* (June 2011) ISBN: 9789052693958 15.4

Fisher, Roger and Ury, William, with Patton, Bruce, *Getting to Yes: Negotiating Agreement Without Giving In*, 3rd edition, Penguin Books (3 May 2011) ISBN-10: 0143118757; ISBN-13: 978-0143118756 15.3

Frank, Jerome, *Courts on Trial*, Princeton, NJ: Princeton University Press (1950), pp. 80–102. 15.8

Greenfield, Kent, *The Myth of Choice: Personal Responsibility in a World of Limits*, reprint edition, Yale University Press (13 November 2012) ISBN-10: 9780300169867; ISBN-13: 978-0300169867 17.15

Napley, Sir David, *The Technique of Persuasion*, Sweet & Maxwell; 4th revised edition (August 1991) ISBN-10: 042143340X, ISBN-13: 978-0421433403 17.7

O'Day, Rosemary, *The Professions in Early Modern England, 1450–1800* (Themes in British Social History), 1st edition, Longman (21 November 2000) ISBN-10: 0582292654; ISBN-13: 978-0582292659 15.7

Ovid, *Metamorphoses* (43 BC – 17 or 18 AD), translated by Sir Samuel Garth, John Dryden, et al. London (1717) 15.1

Thomas, Kenneth W. and Kilmann, Ralph H., *Thomas-Kilmann Conflict Mode Instrument*, Xicom (1997) ASIN: B000VSKYIY 17.6

Institutional rules, guidelines and procedures

CEDR Rules for Construction Adjudication Incorporating Provisions for Mediation 2016 16.16

CIC Model Adjudication Procedure Revision 5, October 2011 15.19

CIC Users' Guide to Adjudication: Ireland, 26 July 2017 16.14, 17.15

Asia International Arbitration Centre (AIAC) Adjudication Rules & Procedure, 2014 17.19

Guidance for Adjudicators, Construction Umbrella Bodies' Adjudication Task Group, July 2002 16.17

IBA Guidelines on Conflicts of Interest, 23 October 2014, ISBN: 978-0-948711-36-7 15.14

Institution of Civil Engineers Adjudication Procedure, 30 April 2012, ISBN 978-0-7277-5779-1 15.13

Institution of Mechanical Engineers Code of Conduct, 9 April 2008 17.2

NEC3 and NEC4 Engineering and Construction Contracts 17.3

NEC3 Adjudicator's Contract, April 2013 and NEC4 & Dispute Resolution Service Contract, June 2017 — 17.19

RIBA Standard Form of Agreement for the Appointment of an Architect (SFA/99) Updated April 2004: Article 9.2.1. Adjudication: England and Wales — 16.9

RICS Surveyors acting as advocates (2nd edition guidance note, 1 January 2009) — 17.7

Rules for adjudication. For use with the JCT Building contract for a home owner/occupier and JCT Consultancy agreement for a home owner/occupier, JCT 2011 — 16.9

Society of Construction Law Delay and Disruption Protocol (2nd edition, February 2017) ISBN 978-0-9543831-2-1 — 17.3

Solicitors Regulatory Authority Code of Conduct, 2011 — 17.2

Technology and Construction Solicitors Association Adjudication Rules, 2018, Version 3.2.2 — 17.19, 15.17,

TeCSA Adjudication Rules 2018, Version 3.2.2 — 15.19, 15.20, 16.9

Court guidance

Civil Justice Council's Guidance for the instruction of experts in civil claims, 2014, Paragraph 53 — 17.16

Technology and Construction Court Guide, 2nd edition issued 3 October 2005, third revision with effect from 3 March 2014, Section 12 — 17.7

Technology and Construction Court Guide, 2nd edition, second revision, October 2010 — 17.16

Articles, reports and others

Abrahamson, M. (1973). Contractual Risks in Tunneling: How They Should Be Shared, *Tunnels & Tunneling* (November): 587–598 — 15.5

Ackoff, R. L. (1989). From Data to Wisdom, *Journal of Applied Systems Analysis*, 15: 3–9 — 17.4

Ajibade Ayodeji Aibinu, Avoiding and Mitigating Delay and Disruption Claims Conflict: Role of Pre-contract Negotiation Journal of Legal Affairs and Dispute Resolution in Engineering and Construction, ASCE, February 2009 — 17.3

Beijing Statement of Principles of the Independence of the Judiciary, 6th Biennial Conferences of Chief Justices of Asia and the Pacific, Beijing, 19 August 1995 — 15.14

Block, Randolph, (1980). Stump v Sparkman and the History of Judicial Immunity, *Duke Law Journal*, November 1980, No. 5: 879 — 15.19

Downey, Gary Lee et al. (2006). The Globally Competent Engineer: Working Effectively with People Who Define — 17.2

Problems Differently, *Journal of Engineering Education*, April: 107–122

Fox, George, Harvey, Thomas Edmund and Norman, Penney (Editors), *The Short Journals and Itinerary Journals of George Fox: In Commemoration of the Tercentenary of His Birth (1624–1924)* (Cambridge Library Collection – Religion), Cambridge University Press (31 October 2010) ISBN-10: 1108015328; ISBN-13: 978-1108015325	15.7
Final Report All-Party Inquiry into Late Payments in Small and Medium-Sized Enterprises Debbie Abrahams MP, July 2013	15.4
Gast, Dr Henrik (2010). Trust in Leadership, Some Observations on the Process of Trust Building in the German Bundestag and the German Cabinet, *Irish Journal of Public Policy*, 4(1): ISSN: 2009-1117	15.5
Hansard HC Debate, 7 May 1996, Volume No. 277, Column 51	15.6
Hansard HL, 22 April 1996, Volume No. 571, Column 973	17.7
Hansard HL, 22 April 1996, Volume No. 571, Column 989	15.6
Hansard Housing Grants, Construction and Regeneration Bill, HC Deb, 8 July 1996, Volume No. 281, Columns 78–88	15.19
Hansard Housing Grants, Construction and Regeneration Bill [H.L.], HL Deb, 22 April 1996, Volume No. 571, Column 973	17.15
Hansard Housing Grants, Construction and Regeneration Bill, HL Deb, 22 April 1996, Volume No. 571, Columns 975–977	15.19
House of Commons Business and Enterprise Committee Construction matters, Ninth Report of Session 2007–08, Volume I Report, together with formal minutes, printed 8 July 2008	15.4
House of Commons Business. Carillion Second Joint report from the Business, Energy and Industrial Strategy and Work and Pensions Committees of Session 2017–19, HC 769, published 16 May 2018	15.4
Howarth, David, *Law as Engineering: Thinking About What Lawyers Do*, Edward Elgar Publishing Limited (28 Feb 2014) ISBN-10: 178254013X, ISBN-13: 978-1782540137	17.2
Improving Adjudication in the Construction Industry. A Consultation Document, January 2003, ISBN 0 7559 0653 5	16.9
Keeney, Ralph L. (1982). Decision Analysis: An Overview, *Operations Research*, 30(5) (September–October): 803–838	17.11
Latham, Sir Michael, Constructing the Team, Joint Review of Procurement and Contractual Arrangements in the United Kingdom Construction Industry, July 1994	15.2, 15.4
Lenzen, Faris, Interest and Islamic Banking, *International Construction Law Review*, Vol. 20, Jan 2003	15.18
Meldrum, Christopher, Civil Law Contracts	15.9
Nouel, Gide Loyrette, World Bank Toolkit (2006) – Approaches to Private Participation in Water Services, presentation to IFC on Some Differences between Civil Law and Common Law in a "nutshell", 2007	15.9

Rana, Rashda, Is Adjudication Killing Arbitration?, 75 Arbitration No. 2, pp. 223–230, May 2009 15.7
Redmond, John, *Adjudication in Construction Contracts*, Blackwell Science Limited (1 Nov 2001), English ASIN: B00L6Z4H1Q 15.20
Summary of the Regulatory Impact Analysis on the Construction Contracts Bill 2010, published by the Department of Public Expenditure and Reform, September 2011 15.2
Trushell, I., Milligan J. L. & Cattanach L., Glasgow Caledonian University Adjudication Reporting Centre Report 12 – October 2012 (Research analysis of the progress of adjudication based on returned questionnaires from adjudicator nominating bodies (ANBs) and from a sample of adjudicators) ISBN-9781905866632 17.4
Lord Woolf, "Access to Justice": Interim Report to the Lord Chancellor on the Civil Justice System in England and Wales (June 1995), Chapter 23; and Final Report to the Lord Chancellor on the Civil Justice System in England and Wales (July 1996), Chapter 13. See also White Book §35.0.2 17.16
Zack, J., Jr. (1993). "Claimsmanship": Current Perspective, *Journal of Construction Engineering and Management*, 119(3): 480–497. ISSN (print): 0733–9364 15.4
Zaheer, Akbar, McEvily, Bill and Perrone, Vincenzo (1998). Does Trust Matter? Exploring the Effects of the Effects of Interorganizational and Interpersonal Trust on Performance, *Organization Science*, 9(2): 141–159 15.4

Websites

http://author.acc.com/legalresources/quickcounsel/clc.cfm
www.bailii.org 15.9
http://cic.org.uk/admin/resources/cic-model-mediation-agreement-and-procedure-first-edition-11th-june-2019.pdf 15.10 15.3
http://cic.org.uk/admin/resources/cic-lvd-mapconsultation-draft-1.pdf 15.17,
http://cic.org.uk/admin/resources/cic-model-mediation-agreement-and-procedure-first-edition-11th-june-2019.pdf 16.9 17.14
www.engineersireland.ie/services/dispute-resolution/documentation.aspx 15.3
http://ppp.worldbank.org/public-private-partnership/legislation-regulation/framework-assessment/legal-systems/common-vs-civil-law 15.9
www.singaporeconvention.org
https://treaties.un.org/Pages/ViewDetails.aspx?src=TREATY&mtdsg_no=XXII-1&chapter=22&clang=_en 15.3 15.3
www.ucc.ie/academic/law/irlii/index.php
15.10

Index

Note: Page numbers referring to a table are in *italics*; page numbers that indicate a figure are **bold**.